STEINER TREE PROBLEMS IN COMPUTER COMMUNICATION NETWORKS

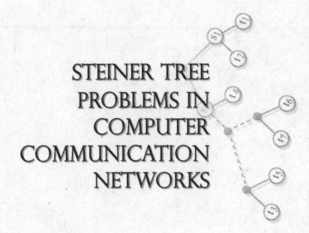

STEINER TREE PROBLEMS IN COMPUTER COMMUNICATION NETWORKS

Dingzhu Du

University of Texas at Dallas, USA

Xiaodong Hu

Chinese Academy of Sciences, China

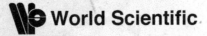

NEW JERSEY · LONDON · SINGAPORE · BEIJING · SHANGHAI · HONG KONG · TAIPEI · CHENNAI

Published by

World Scientific Publishing Co. Pte. Ltd.

5 Toh Tuck Link, Singapore 596224

USA office: 27 Warren Street, Suite 401-402, Hackensack, NJ 07601

UK office: 57 Shelton Street, Covent Garden, London WC2H 9HE

British Library Cataloguing-in-Publication Data
A catalogue record for this book is available from the British Library.

ISBN-13 978-981-279-144-3
ISBN-10 981-279-144-2

Printed in Singapore.

Preface

The classical Steiner tree problem is defined as follows: Given a set of points in a metric space, find the shortest network interconnecting all given points. Such a network is called a Steiner tree on the given set.

The Steiner tree problem, which is one of most well known combinatorial optimization problems, has a long research history [33; 183; 250]. It can be considered as a generalization of Fermat's problem. Three hundred years ago, Fermat (1601-1665) proposed a problem of finding a point to minimize the total distance from this point to three given points in the Euclidean plane; Clearly, its solution yields the Steiner minimum tree on the three points. The general form of Steiner tree problem was proposed by Gauss (1777-1855) [69]. But somehow, Courant and Robbins [70] referred to it as the Steiner tree problem. The popularity of their famous book was responsible for bringing the Steiner tree problem to people's attention. Two important papers in the 1960's further laid a solid groundwork for further study. Melzak [206] first gave a finite algorithm for the Euclidean Steiner trees. Gilbert and Pollak [114] made an excellent survey of the problem where they raised many new topics including Steiner ratio problem and extended the Euclidean Steiner tree problem to other metric spaces. Since then, lots of research papers have been contributed to the Steiner tree problem. One may refer to the book by Hwang et al. [144] for a comprehensive survey on related works before 1992.

The study of the Steiner tree problem received much more attentions in 1990s since many important open problems, including Gilbert-Pollak conjecture on the Euclidean Steiner ratio, the existence of better approximation algorithms, and the existence of polynomial time approximation schemes, have all been solved, respectively. Those achievements have greatly influenced not only the general theory of designs and analysis of approximation algorithms for combinatorial optimizations but also the discovery and study on many new important applications, including VLSI designs, optical and wireless communication networks. Those applications usually require some modifications on the classical Steiner tree problem and hence demand new techniques for solving them. As a result, studying various variations of Steiner tree problems forms a very hot topic in the past twenty years.

In this book we will discuss the above mentioned significant achievements in the study of the classical Steiner tree problems and some of their applications in computer communication networks. The book can be divided into three parts:

(1) Fundamentals of classical Steiner tree problem;
(2) Variations of classical Steiner tree problem;
(3) Steiner tree based problems.

Each part consists of three or four chapters, each of them focuses on either one approach or one problem for the study or the application of Steiner tree problems. Roughly speaking, problems addressed in the first few chapters are more closely related to classical Steiner tree problem than that addressed in the last few chapters.

In the first part, we will study some powerful approaches introduced in the study of the classical Steiner tree problems and discuss some open problems that may be solved by using those approaches. It includes the first three chapters. In Chapter 1, we will study the minimax approach for determining the Steiner ratio in the Euclidean plane. In Chapter 2, we will study some techniques for designing approximation algorithms for Steiner tree problem with performance ratios better than the inverse of Steiner ratio. In Chapter 3, we will study two more techniques that turn out to be very useful for designing good approximation algorithms not only for the Steiner tree problem but also many other geometric optimization problems.

In the second part, we will discuss three variations of classical Steiner tree problem, all of them have objectives different from that of the classical Steiner tree problem. In Chapter 4, we will study the Steiner tree problem for minimum cost of grade of service. In Chapter 5, we will study the Steiner tree problem for minimal number of Steiner points. In Chapter 6, we will study the Steiner tree problem for minimal longest edge of trees with at most k Steiner points.

In the third part, we will study four optimization problems arising in the design and applications of all-optical fiber networks and wireless sensor networks. They all use Steiner trees (or networks) as basic models, but two of them aim at computing a set of Steiner trees, instead of computing one optimal Steiner tree, for different objectives. In Chapter 7, we will study the Steiner k-tree problem that asks for a set of Steiner trees of minimum total cost each containing at most k terminal points in given terminal-set. In Chapter 8, we will study two Steiner tree coloring problems. One asks for a maximal number of sets of Steiner trees each spanning a given terminal-set such that they could be properly colored with given colors; The other asks for a set of Steiner trees each spanning a given terminal-set such that Steiner trees associated with all terminal-sets could be properly colored with minimal number of colors. In Chapter 9, we will study Steiner tree scheduling problem that asks for the Steiner tree such that data transmission over the tree could be finished in minimal time. In Chapter 10, the last chapter, we will study multi-connected Steiner network problem for minimal length.

Almost every chapter includes the motivation of study, related works, problem formulation, complexity analysis, algorithm design and performance analysis, and related problems. So all chapters are basically self-contained, readers may read any of them while skipping the rest.

Since a book like this can not be expected to cover and discuss all progresses achieved in the past twenty years in the study and applications of Steiner tree problems, we list in the appendix some important extensions and variations of classical Steiner tree problem which are not addressed in Chapters 1-10. Of course, the list of the problems is far from complete and corresponding results may not be updated. Readers could refer to some other monographs or books for further discussions in the diversity of Steiner tree problems. For example, *The Steiner Tree Problem* by Hwang et al. (1992) [144], *Minimal Networks: The Steiner Problem and Its Generalizations* by Ivanov and Tuzhilin (1994) [149], *Steiner Minimal Trees* by Cieslik (1998) [68], *Advances in Steiner Trees* edited by Du et al. (2000) [87], *Steiner Trees in Industry* edited by Cheng and Du (2001) [55], *The Steiner Ratio* by Cieslik (2001) [69], and *The Steiner Tree Problem* by Prömel and Steger (2002) [228].

Acknowledgment

This book has benefited a lot from our joint work with many friends and former students (in Department of Computer Science and Engineering, University of Minnesota, and Institute of Applied Mathematics, Chinese Academy of Sciences). In particular, we would like to thank Donghui

Chen, Xujin Chen, Xiuzhen Cheng, Al Borchers, Xiufeng Du, Qing Feng, Biao Gao, Ronald L. Graham, Jun Gu, D. Frank Hsu, Frank Hwang, Xiaohua Jia, Yoji Kajitani, Joon Mo Kim, Guohui Lin, Zicheng Liu, Bing Lu, Hung Quang Ngo, Weiping Shang, Eugene B. Shragowitz, Tianping Shuai, Pengjun Wan, Lusheng Wang, Guoliang Xue, Muhong Zhang, Jianming Zhu, and Yanjun Zhang.

Dingzhu Du's work is supported in part by NSF of United States under Grant No. CCF0627233 and CCF0728851 and NNSF of China under Grant No. 60573021, respectively. Xiaodong Hu's work is supported in part by NNSF of China under Grant No. 70221001, 10531070 and 10771209.

It was a pleasure to work with Yixin Tan, Huey Ling Gow and Rebecca Fu, all from World Scientific Publishing Co Pte Ltd, on editorial matters. Their supports and help are especially impressive.

In the end, we thank our family members, old and young, whose love, patience, and encouragement have made this book possible. We affectionately dedicate this book to them.

Dingzhu Du

September 2007   *Xiaodong Hu*

Contents

Chapter 1

Minimax Approach and Steiner Ratio

Given a set P of points in a metric space[1], the *classical Steiner tree problem* is to find a shortest network interconnecting the points in P, that is, the total length of edges in the network is minimal (the length of an edge is the distance between its two endpoints). The optimal solution of this problem has a tree structure, so it is called the *Steiner minimum tree* (SMT) on P and denoted by $T_{smt}(P)$. An SMT may have some points not in P, which are called *Steiner points* while the points in P are called *terminal points* (also called *regular points*) and P is called *terminal set*. The problem is formally defined as follows.

Problem 1.1 *Steiner Tree Problem in Metric Spaces*

Instance A set $P = \{t_1, t_2, \cdots, t_n\}$ of terminals in a metric space M.
Solution A Steiner tree T for P.
Objective Minimizing the total length of the edges in T, $l(T) \equiv \sum_{e \in T} l(e)$.

A closely related problem is *minimum spanning tree problem*. A *spanning tree* on P is a tree interconnecting all points in P under the restriction that all edges have endpoints in P. In other words, no Steiner point is allowed to use. The *minimum spanning tree* (MST) is the shortest spanning tree, denoted by $T_{mst}(P)$. The problem is formally defined as follows.

Problem 1.2 *Minimum Spanning Tree Problem in Metric Spaces*

Instance A set $P = \{t_1, t_2, \cdots, t_n\}$ of terminals in a metric space M.
Solution A spanning tree T for P.
Objective Minimizing the total length of the edges in T, $l(T) \equiv \sum_{e \in T} l(e)$.

---

[1]A *metric space* M is a set S with a metric that is a nonnegative, symmetric real function $f(,)$ satisfying the *triangle inequality*; That is, for every three points x, y, z in S, $f(x, y) = 0$ if and only if $x = y$, $f(x, y) = f(y, x)$, and $f(x, y) + f(y, z) \geq f(x, z)$.

Fig.1.1 shows a simple example of the above two problems. Terminal set P consists of three points which form a regular triangle of unit side. (a) is an MST of length 2. (b) is an SMT of length $\sqrt{3}$. Note that there is a Steiner point at which three edges meet at each two forming an angle of 120°. (In general, the Steiner point can be determined in a geometrical way as shown in (c).)

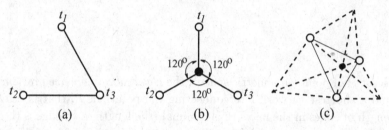

Fig. 1.1 (a) An MST, (b) the SMT, and (c) a geometrical method to determine the Steiner point.

While the Steiner tree problem is intractable (i.e., NP-hard [109]), the minimum spanning tree problem is polynomial-solvable (by using the well-known greedy algorithms due to Kruskal [182] and Prim [226]). Thus, we can use an MST to approximate the SMT. The *Steiner ratio* in a metric space M is the largest lower bound for the ratio between lengths of an SMT and an MST for the same set of points in the metric space. It can be more formally defined as follows.

$$\rho_M \equiv \inf \left\{ \frac{l\big(T_{smt}(P)\big)}{l\big(T_{mst}(P)\big)} \,\middle|\, P \text{ in } M \right\}, \qquad (1.1)$$

where $l(T_{smt}(P))$ and $l(T_{mst}(P))$ denote the length of $T_{smt}(P)$ and $T_{mst}(P)$ under metric space M, respectively.

Steiner ratio can be considered as a measure of performance for an MST as a polynomial time approximation of the SMT. An algorithm A is called a *polynomial time α-approximation algorithm* for a minimization problem if given any instance I of the problem, it finds a solution for I with cost $c(I) \leq \alpha\, c_{opt}(I)$, where $c_{opt}(I)$ is the cost of an optimal solution for I, and the running time of algorithm A is bounded by a polynomial in the input size of I. In addition, a problem is called α-*approximable* (or *approximable within ratio* α) if there is an α-approximation algorithm for the problem.

Clearly, $\alpha \geq 1$ and the smaller this ratio α, the better approximation algorithm A. It immediately follows from the definition (1.1), the minimum spanning tree algorithm [182; 226] is a polynomial time $(1/\rho_M)$-approximation algorithm for Steiner tree problem.

Determining the Steiner ratio in each metric space is a traditional problem on the study of Steiner tree problem ([83; 69; 67]). In 1976, Hwang [142] determined that the Steiner ratio in the rectilinear plane[2] is 2/3.

Surprisingly, however, it took 22 years for determining the Steiner ratio in the Euclidean plane. In 1968, Gilbert and Pollak [114] conjectured that the Steiner ratio in the Euclidean plane is $\sqrt{3}/2$. (Fig.1.1 gives an example with $\rho_M = \sqrt{3}/2$.) Through continuous efforts made by Graham and Hwang [119], Pollak [225], Chung and Hwang [64], Du and Hwang [80], Du et al. [85], Chung and Graham [63], Friedel and Widmayer [102], Booth [38], and Rubinstein and Thomas [240; 241], the conjecture was finally proved by Du and Hwang [81] in 1990. Their proof is based on a new minimax theorem about minimizing the maximum value of several concave functions over a simplex. Moreover, its significance stems also from the potential applications of the new approach included in the proof.

In this chapter we will first prove some minimax theorems and then describe how they were used to solve Gilbert and Pollak's long standing conjecture on Steiner ratio in the Euclidean plane [75]. At the end we will discuss some open problems related to Steiner ratios.

1.1 Minimax Approach

Minimax approach is one of the most important techniques for solving optimization problems [72]. There are two fundamental ideas used in studying *minimax problems*. The first one is the search for a *basis*. That is, for the problem

$$\min_{x \in X} \max_{y \in Y} f(x, y), \qquad (1.2)$$

determine first a finite subset B of X such that

$$\min_{x \in X} \max_{y \in Y} f(x, y) = \min_{x \in B} \max_{y \in Y} f(x, y)$$

and then search for an optimal solution x^* to problem (1.2) from B in finitely many steps. The second is the determination of a *saddle point*. That

---

[2]Given two points p_1 and p_2 in a plane with coordinates (x_1, y_1) and (x_2, y_2), the *rectilinear distance* (also known as *Manhattan distance*) between pq is defined as $l(pq) \equiv |x_1 - x_2| + |y_1 - y_2|$.

is, a point (x^*, y^*) on the set $X \times Y$ that satisfies the following condition,

$$f(x^*, y) \leq f(x^*, y^*) \leq f(x, y^*), \quad \text{for any } x \in X \text{ and } y \in Y. \qquad (1.3)$$

By the definition, we deduce that for any saddle point (x^*, y^*),

$$\min_{x \in X} \max_{y \in Y} f(x, y) = f(x^*, y^*) = \max_{y \in Y} \max_{x \in X} f(x, y). \qquad (1.4)$$

These two ideas have resulted in two important mathematical branches. P. L. Chebyshev (1821-1894) is probably the first person who made an important contribution to the first idea. He discover the theory of best approximation. J. V. Neumann (1903-1957) is the person who made a fundamental contribution to the second idea. He initiated the game theory. Since then, many efforts have been made to find various sufficient conditions for a point being a saddle point. This involves a great deal of mathematics including the fixed point theory.

While there are a huge amount of materials about minimax approach in the literature, we study only a small part in this section. In particular, we restrict ourselves only to some recent developments on the first idea, which leads to the settlement of a long standing conjecture about Steiner ratio in the Euclidean plane.

1.1.1   *Chebyshev Theorem*

The original problem considered by Chebyshev is as follows: Given a list of values of some real function, $y_i = f(x_i)$, $i = 0, 1, \cdots, m$, find a polynomial $p(\cdot)$ of degree at most $n < m$ which provides the *best approximation* at these m points, that is, polynomial $p(\cdot)$ minimizes

$$\max_{i=0,1,\cdots,m} \left| y_i - p(x_i) \right|. \qquad (1.5)$$

Chebyshev gave a beautiful result about the solution of this approximation problem.

First, consider $m = n + 1$. In this case, there exists a unique polynomial of the best approximation. Chebyshev proved that a polynomial $p(\cdot)$ is the best approximation if and only if for some h,

$$(-1)^i h + p(x_i) = y_i, \quad \text{for } i = 0, 1, \cdots, n+1.$$

Furthermore, h and p both can be determined explicitly. Such a polynomial $p(\cdot)$ is called a *Chebyshev interpolating polynomial*.

Next, for general m, a subset of $(n+2)$ points is called a *basis*. Each basis σ determines a Chebyshev interpolating polynomial $p_\sigma(\cdot)$ and a value

$$h(\sigma) = \max_{x_i \in \sigma} |y_i - p_\sigma(x_i)|.$$

A basis σ^* is called an *extremal basis* if

$$h(\sigma^*) = \max_\sigma h(\sigma),$$

where σ is taken over all possible bases. Chebyshev proved the following theorem.

Theorem 1.1 *There exists a unique polynomial of best approximation. A polynomial $p(\cdot)$ is the polynomial of best approximation if and only if $p(\cdot)$ is a Chebyshev interpolating polynomial for some extremal basis.*

There are some other ways to characterize the extremal basis (refer to [72]). In fact, Chebyshev also proved that σ^* is an extremal basis if and only if

$$h(\sigma^*) = \max_{i=0,1,\cdots,m} |y_i - p_{\sigma^*}(x_i)|.$$

For each polynomial $p(\cdot)$, define

$$I(p) \equiv \left\{ j \,\middle|\, |y_j - p(x_j)| = \max_{i=0,1,\cdots,m} |y_i - p(x_i)| \right\}.$$

$I(p)$ is called *maximal* if no polynomial $q(\cdot)$ exists such that $I(p) \neq I(q)$ and $I(p) \subset I(q)$. From the second characterization of the extremal basis, we can deduce the following theorem.

Theorem 1.2 *Basis σ^* is extremal if and only if $I(p_{\sigma^*})$ is maximal.*

Chebyshev's theorem can be transformed into a *Linear Programming* (LP) problem as follows:

$$\begin{aligned}
\text{Min} \quad & z \\
\text{subject to} \quad & -z \leq a_0 + a_1 x_i + \cdots + a_n x_i^n - y_i \leq z \\
& i = 0, 1, \cdots, m.
\end{aligned}$$

Note that the above LP problem has $(n+2)$ variables and $2(m+1)$ constraints. For an extremal basis σ^*, $p_{\sigma^*}(\cdot)$ would make $(n+2)$ constraints *active* (i.e., the equality signs would hold true for those constraints). This means that each extremal basis corresponds to a feasible basis of the above LP problem in the following standard form:

$$\text{Min} \quad z$$
$$\text{subject to} \quad u_i - z = a_0 + a_1 x_i + \cdots + a_n x_i^n - y_i = z - v_i$$
$$u_i \geq 0, \, v_i \geq 0;$$
$$i = 0, 1, \cdots, m.$$

On the other hand, LP problems are closely related to minimax problems. In fact, there are several ways to transform a LP problem to a minimax problem. For example, consider the LP problem in *standard form*:

$$\text{Min} \quad z = c^T x$$
$$\text{subject to} \quad Ax = b$$
$$x \geq 0.$$

and its *dual form*:

$$\text{Max} \quad b^T y$$
$$\text{subject to} \quad A^T y \leq c.$$

For any feasible solution x of the original LP problem and any feasible solution y of the dual LP problem, $c^T x \geq b^T y$. The equality sign holds only if the two feasible solutions are actually optimal solutions for the two problems, respectively. This is equivalent to the following minimax problem achieving the minimax value 0.

$$\min_{(x,y)} \max \left\{ c^T x - b^T y, -x, Ax - b, b - Ax, A^T y - c \right\}. \tag{1.6}$$

In the above, we see that two problems (1.5-6) can be formalized as the following minimax problem:

$$\min_{x \in X} \max_{i=1,2,\cdots,m} f_i(x). \tag{1.7}$$

In the next subsection, we will extend Chebyshev's idea to problem (1.7).

1.1.2 *Du-Hwang Theorem*

We consider minimax problem (1.7) with a few general conditions. Assume that X is a polytope in $\Re^n$ and the $f_i(x)$'s are continuous *concave functions* of x. That is, $f_i(\lambda x_1 + (1-\lambda)x_2) \geq \lambda f_i(x_1) + (1-\lambda)f_i(x_2)$ for any $\lambda \in [0,1]$ and $x_1, x_2 \in X$. (Note that $f(\cdot)$ is a *convex function* if "$\leq$" is satisfied; Clearly, $f(\cdot)$ is a convex function if and only if $-f(\cdot)$ is a concave function.) Fig.1.2(a) shows a concave function.

To extend Chebyshev's idea to problem (1.7), we start with the simplest case of $m = n = 1$. As shown in Fig.1.2(a), the minimum value of a concave function $f_1(x)$ on the interval $[a, b]$ is achieved at either a or b (maybe both). For $m = 1$ and general n, it is well-known that the minimum value of $f_1(x)$ is achieved at a vertex of the polytope X. What we are interested in the following discussion is the case $m > 1$. If $m > 1$ and $n = 1$, then as shown in Fig.1.2(b) with $m = 3$, $g(x) = \max\{f_i(x) \mid i = 1, 2, \cdots, m\}$ is a *piecewise concave function*. Thus, the minimum value of $g(x)$ on the interval $[a, b]$ is achieved at an endpoint of a concave piece (one of four dark points in Fig.1.2(b)).

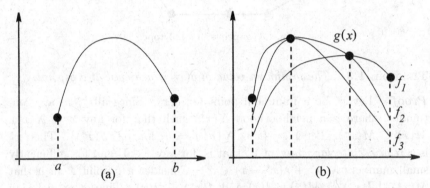

Fig. 1.2   (a) Minimum point of a concave function, and (b) minimum point of a piecewise concave function.

Similarly, for $m > 1$, the polytope X can be divided into small regions in each of which $g(x)$ is concave. These small regions can be defined by $X_i = \{x \in X \mid f_i(x) = g(x)\}$. However, they may not be convex. Thus, where the minimum value of $g(x)$ can be achieved is not easy to be determined. Du and Hwang [81] found that the minimum value of $g(x)$ can still be achieved at an extreme point of some small regions where the extreme point is defined in the following way.

Consider the polytope $X = \{x \mid a_i^T x \geq b_i, i = 1, 2, \cdots, m\}$. Denote $J(x) = \{j \mid a_j^T x = b_j\}$. A point x in X is an *extreme point* if $J(x)$ is maximal, i.e., there does not exist $y \in X$ such that $J(x)$ is a proper subset of $J(y)$. Note that such a definition is different from the traditional one (e.g., refer to [198]): x is an extreme point if $x = \frac{1}{2}y + \frac{1}{2}z$ for $y, z \in X$ implies $x = y = z$. However, they are equivalent for polytopes.

Now, a point x in X is called a *g-point* if $J(x) \cup M(x)$ is maximal where $M(x) = \{i \mid f_i(x) = g(x)\}$, so that $M(x) \cap J(x) = \emptyset$. Fig.1.3 shows some

g-points of function $g(x)$ with $m = 5$ defined in a polytope specified by seven hyperplanes.

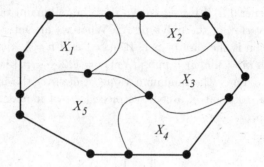

Fig. 1.3 g-points in a polytope.

Theorem 1.3   *The minimum value of $g(x)$ is achieved at a g-point.*

Proof.   Let x^* be a minimum point for $g(x)$. Since all $f_i(x)$ are continuous, there is a neighborhood $N(x^*)$ such that for any $x \in N(x^*)$, $M(x) \subseteq M(x^*)$. Let $Y = \{x \in X \mid a_j^T x = b_j$ for $j \in J(x^*)\}$. Then x^* is a *relative interior point* of Y, that is, for any $x \in Y$ and for sufficiently small numbers λ, $x^* + \lambda(x^* - x) \in Y$. Consider a g-point $\bar{x}$ such that $M(x^*) \cup J(x^*) \subseteq M(\bar{x}) \cup J(\bar{x})$ and $J(x^*) \subseteq J(\bar{x})$. The latter inclusion implies that $\bar{x} \in Y$. To prove the theorem, it suffices to show that $\bar{x}$ is also a minimum point.

Suppose, by contradiction, that $\bar{x}$ is not a minimum point. Choose a positive λ sufficiently small such that $x(\lambda) = x^* + \lambda(x^* - \bar{x}) \in N(x^*) \cap Y$. Thus we have $M(x(\lambda)) \subseteq M(x^*) \subseteq M(\bar{x})$. Consider an index $i \in M(x(\lambda))$. Since x^* is a minimum point of $g(x)$, we obtain $f_i(x^*) < f_i(\bar{x})$ and $f_i(x^*) \leq f_i(x(\lambda))$. Note that

$$x^* = \frac{\lambda}{1+\lambda}\bar{x} + \frac{1}{1+\lambda}x(\lambda).$$

By the concavity of $f_i(x)$, we have

$$f_i(x^*) \geq \frac{\lambda}{1+\lambda}f_i(\bar{x}) + \frac{1}{1+\lambda}f_i(x(\lambda)) > f_i(x^*),$$

which is a contradiction. Hence $\bar{x}$ is a minimum point.   □

In fact, we can prove the following stronger result by relaxing the concavity of $f_i(\cdot)$ for each i. A function $f(\cdot)$ is *pseudo-concave* in a region if for any x and y in the region and for any $\lambda \in [0, 1]$,

$$f(\lambda x + (1 - \lambda)y) \geq \min\{f(x), f(y)\}.$$

Clearly, pseudo concavity is weaker than concavity.

Theorem 1.4 *Let $g(x) = \max\{f_i(x) \mid i \in I\}$ where the $f_i's$ are continuous pseudo-concave functions and I is a finite set of indices. Then the minimum value of $g(x)$ over a polytope is achieved at a g-point.*

Proof. To prove the theorem, we can apply and modify the proof of Theorem 1.3 as follows: Choose a minimum point x^* with maximal $J(x)$ and a point $\bar{x}$ in Y with $M(x^*) \subseteq M(\bar{x})$. By the pseudo-concavity of $f_i(x)$, for $i \in M(x(\lambda))$, $x(\lambda) = x^* + \lambda(\bar{x} - x^*) \in Y \cap N(x^*)$, and for any $\lambda > 0$, we have

$$f_i(x^*) \geq \min\{f_i(\bar{x}), f_i(x(\lambda)) \geq f_i(x^*).$$

It follows that for $x(\lambda) \in Y \cap N(x^*)$, $x(\lambda)$ is a minimum point. Note that all minimum points form a closed set. There exists a maximum value λ^* such that $x(\lambda^*)$ such that $x(\lambda^*)$ is a minimum point. Clearly, $x(\lambda^*)$ cannot be a relative interior point of Y since, otherwise, we can obtain a larger λ such that $x(\lambda)$ is minimum point. Therefore, $J(x^*)$ is a proper subset of $J(x(\lambda^*))$, contradicting the choice of x^*. The proof is then finished. $\square$

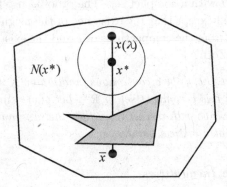

Fig. 1.4 $I(x)$ is defined on a subset of X.

By the definition of g-points, an interior point x in X is a g-point if and only if $M(x)$ is maximal. In general, for any g-point, there exists an extreme subset Y of X such that $M(x)$ is maximal over Y. A point x in X is called a *critical point* if there exists an extreme set Y such that $M(x)$

is maximal over Y. Thus, every g-point is a critical point. However, the inverse is false. As shown in Fig.1.3, the interior boundary of X_2 consists of critical points which are not g-vertices.

A similar result holds true for a more general minimax problem as follows:

$$\min_{x \in X} \max_{i \in I(x)} f_i(x), \tag{1.8}$$

where $I(x)$ is a finite index set varying as x varies. The following theorem is a useful version of Theorem 1.3, whose proof, as shown in Fig.1.4, is similar to the proof of Theorem 1.3.

Theorem 1.5   *Let $g(x) = \max\{f_i(x) \mid i \in I(x)\}$, where f_i's are continuous and pseudo-concave functions in a convex region X and $I(x)$ is a finite index set defined on a compact set X' of P. Denote $M(x) = \{i \in I(x) \mid f_i(x) = g(x)\}$. Suppose that for any $x \in X$, there exists a neighborhood of x such that for any point y in the neighborhood, $M(y) \subseteq M(x)$. If the minimum value of $g(x)$ over X is achieved at at interior point of X', then this minimum value is achieved at a critical point, i.e., a point with maximal $M(x)$ over X'. Moreover, if x is an interior minimum point in X' and $M(x) \subseteq M(y)$ for some $y \in X'$, then y is a minimum point.*

In addition, we can prove a general version of Theorem 1.3 that replaces the finite index set I with a compact set. The proof is also the same as the proof of Theorem 1.3 except that the existence of the neighborhood $N(x^*)$ needs to be derived from the compactness of I and the existence of $\bar{x}$ needs to be derived from Zorn's lemma.

Theorem 1.6   *Let $f(x, y)$ be a continuous function on $X \times I$ where X is a polytope in $\Re^m$ and I is a compact set in $\Re^n$. Let $g(x) = \max\{f(x, y) \mid y \in Y\}$. If $f(x, y)$ is concave with respect to x, then the minimum value of $g(x)$ over X is achieved at a critical point.*

1.1.3   Geometric Inequalities

Du-Hwang Theorem was first used in the proof of a geometric inequality. This inequality was proposed by Debrummer in 1956 and by Oppenhein in 1960 [41]. Since it appeared in *American Mathematics Monthly* as the 4964-th problem in 1961, it obtained several proofs given by Dresel, Breusch, Croft, Zalgaller, and Szekers. Using Du-Hwang Theorem, we can give another proof. Although it is not the simplest one, it is more general. In fact, it is applicable for proving some other similar geometric inequalities.

Theorem 1.7 *Let $a', b',$ and c' be three points on three edges bc, ca, and ab of triangle $\triangle abc$, respectively. Let $l(\triangle abc)$ denote the perimeter of $\triangle abc$. Then $l(\triangle a'b'c') \geq \min\{l(\triangle ab'c'), l(\triangle bc'a'), l(\triangle ca'b')\}$.*

Proof. Let us first fix $a', b',$ and c', and then treat points $a, b,$ and c as variables. Consider the following function

$$f(a, b, c) = \min\{l(\triangle ab'c'), l(\triangle bc'a'), l(\triangle ca'b')\}.$$

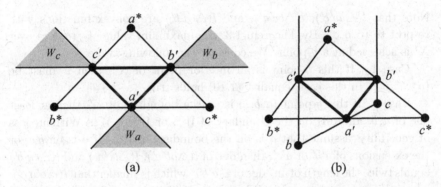

Fig. 1.5 $a, b,$ and c are considered as three variables in $W_a, W_b,$ and W_c, respectively.

As shown in Fig.1.5(a), point a is located in the area W_a bounded by $b'c'$ and extensions of $a'b'$ and $a'c'$. Similarly, b and c are located in the areas of W_b and W_c, respectively. Let

$$X \equiv \left\{ (a, b, c) \in W_a \times W_b \times W_c \;\middle|\; \begin{array}{l} b, a', c \text{ are colinear,} \\ c, b', a \text{ are colinear,} \\ a, c', b \text{ are colinear.} \end{array} \right\}.$$

We want to prove that for any $(a, b, c) \in X$,

$$f(a, b, c) \leq l(\triangle a'b'c'). \tag{1.9}$$

Note that three points (x_1, y_1), (x_2, y_2), and (x_3, y_3) are collinear if and only if

$$\begin{vmatrix} x_1 & y_1 & 1 \\ x_2 & y_2 & 1 \\ x_3 & y_3 & 1 \end{vmatrix} = 0.$$

Thus X is a polyhedran of dimension three, which is an unbounded region. To obtain a polytope, consider $\triangle a^*b^*c^*$ with a', b', and c' as the middle points of corresponding three edges, respectively. Let W_a' be the bounded part obtained from cutting W_a by a line L_a parallel to $b'c'$. If L_a is sufficiently far away from $b'c'$, then a^* is an interior point of W_a', Similarly, we can define polygons W_b' and W_c', respectively. Now define polytope X' in the same way as X by using $W_a' \times W_b' \times W_c'$ to replace $W_a \times W_b \times W_c$. Clearly, to prove inequality (1.9) it suffices to prove that for every X',

$$\max_{(a,b,c)\in X'} f(a,b,c) = l(\triangle a'b'c'). \qquad (1.10)$$

Note that $l(\triangle ab'c')$, $l(\triangle bc'a')$, and $l(\triangle ca'b')$ are convex functions with respect to (a, b, c). By Theorem 1.3, the maximum value of $f(a, b, c)$ over X' is achieved at a g-point. We consider the following two cases.

Case 1. If this g-point is an interior point of X', then it must be (a^*, b^*, c^*). In this case, equality (1.10) holds true.

Case 2. If this g-point (a, b, c) is on the boundary of X', then at least one of a, b, and c is on the boundary of W_a', or W_b' or W_c'. Without loss of generality, assume that a is on the boundary of W_a'. If a is on $b'c'$ or the extensions of $a'b'$ or $a'c'$, then one of $l(\triangle ab'c')$, $l(\triangle bc'a')$ and $l(\triangle ca'b')$ equals twice the length of an edge of $\triangle a'b'c'$ which is smaller than $l(\triangle a'b'c')$. Thus, this a must be on L_a. When L_a is sufficiently far from $b'c'$, bc' and cb' are almost parallel. In the extreme case that bc' and cb' are parallel as shown in Fig.1.5(b), either b is different from b^* and lies in $\triangle b^*a'c'$, or c is different from c^* and lies in $\triangle c^*b'a'$. Thus, either $l(\triangle ba'c') < l(\triangle b^*a'c') = l(\triangle a'b'c')$, or $l(\triangle cb'a') < l(\triangle a'b'c')$. Therefore, in this case equality (1.10) also holds true when L_a moves sufficiently far from $\triangle a^*b^*c^*$. The proof is then finished. $\qquad \square$

Let (x_a, y_a), $(x_{b'}, y_{b'})$, and $(x_{c'}, y_{c'})$ be the coordinates of points a, b' and c', respectively. Denote by $s(\triangle xyz)$ the size of $\triangle xyz$. Then we have the following formula:

$$s(\triangle ab'c') = \frac{1}{2} \begin{vmatrix} x_a & y_a & 1 \\ x_{b'} & y_{b'} & 1 \\ x_{c'} & y_{c'} & 1 \end{vmatrix},$$

which is a linear function with respect to a. Thus, a similar argument yields the following theorem.

Theorem 1.8 *Let a', b', and c' be three points on three edges bc, ca, and ab of triangle $\triangle abc$, respectively. Then $\max\{s(\triangle ab'c'), s(\triangle bc'a'), s(\triangle ca'b')\} \geq s(\triangle a'b'c') \geq \min\{s(\triangle ab'c'), s(\triangle bc'a'), s(\triangle ca'b')\}$.*

Since $s(\triangle ab'c')$ is linear and $l(\triangle ab'c')$ is convex with respect to a, the ratio $l(\triangle ab'c')/s(\triangle ab'c')$ is pseudo-convex with respect to a. Let $r(\triangle xyz)$ denote the radius of the circle inscribed in $\triangle xyz$. Note that $l(\triangle ab'c')/s(\triangle ab'c') = 2/r(\triangle ab'c')$. Therefore, the above argument also yields the following theorem.

Theorem 1.9 *Let a', b', and c' be three points on three edges bc, ca and ab of triangle $\triangle abc$, respectively. Then $\max\{r(\triangle ab'c'), r(\triangle bc'a'), r(\triangle ca'b')\} \geq r(\triangle a'b'c')$.*

1.1.4 *Analysis of Approximation Performance*

Du-Hwang Theorem is the crux of their proof [81] for Gilbert-Pollak conjecture on Steiner ratio in the Euclidean plane [114]. In general, for NP-hard optimization problems, such as Steiner tree problem, their optimal solutions are unlikely to be computable in polynomial time. For such problems, polynomial time approximate solutions are useful. One way to design a polynomial time approximation algorithm is as follows: Put some restriction on feasible solutions so that the optimal solution under this restriction can be computed in polynomial time, and then use this optimal solution for the restricted problem to approximate the optimal solution for the original problem.

To be more explicit, consider the following minimization problem

$$\min_{k \in K} \phi_k(x). \tag{1.11}$$

Suppose that finding the minimum solution of (1.11) is NP-hard. Let $I \subseteq K$ such that

$$\min_{i \in I} \phi_i(x) \tag{1.12}$$

can be computed by a polynomial time algorithm A. Now we use the optimal solution to problem (1.12) as an approximate solution to problem (1.11). By the definition of α-approximation algorithms, the performance ratio of algorithm A is equal to the inverse of the following ratio:

$$\rho = \min_x \frac{\min_{k \in K} \phi_k(x)}{\min_{i \in I} \phi_i(x)}. \tag{1.13}$$

Clearly, the larger this ratio, the better is the approximation by A.

Proving a lower bound for ratio (1.13) can be transformed to a minimax problem. For this purpose, suppose that we want to prove $\rho \geq \rho_0$. Then it suffices to prove that for any x,

$$\min_{k \in K} \phi_k(x) \geq \rho_0 \min_{i \in I} \phi_i(x).$$

This is equivalent to requiring that for any x and $k \in K$,

$$\phi_k(x) - \rho_0 \min_{i \in I} \phi_i(x) \geq 0,$$

that is,

$$\max_{i \in I} \left\{ \phi_k(x) - \rho_0 \, \phi_i(x) \right\} \geq 0.$$

Thus, it suffices to prove that for any $k \in K$,

$$\min_x \max_{i \in I} \left\{ \phi_k(x) - \rho_0 \, \phi_i(x) \right\} \geq 0. \tag{1.14}$$

In the next section we will show how to determine the Steiner ratio in the Euclidean plane using this approach.

1.2 Steiner Ratio in the Euclidean Plane

Du-Hwang's proof of Gilbert-Pollak conjecture consists of four main steps as follows:

(S1) Transforming the Steiner ratio problem to minimax problem (1.8), where $f_i(x) =$ (the length of a Steiner tree) -(the Steiner ratio)·(the length of spanning tree with graph structure i) and x is a vector whose components are edge-lengths of the Steiner tree

(S2) Reducing the minimax problem to the problem of finding the minimax value of the concave functions at critical points through the minimax theorem (Theorem 1.5).

(S3) Transforming each critical point back to a set of terminal points with a special geometric structure;

(S4) Verifying the conjecture on special geometric structures, by proving the nonnegativeness of minimax value of some concave functions.

In the following subsections we will show in details how to implement each of the above four steps.

1.2.1 Equivalent Minimax Problem

The *topology* of a tree is the adjacency relation or the adjacency matrix of the tree. Let $t^*(P)$ denote the minimum tree with topology t on terminal set P. Let $l(t^*(P))$ denote the length of $t^*(P)$. Suppose that all topologies of trees interconnecting P form a set K and all topologies of spanning trees on P form a set I. Clearly, I is a subset of K. Then the Steiner minimum tree problem and the minimum spanning tree problem can be represented respectively as follows:

$$\min_{t \in K} l\big(t^*(P)\big) \text{ and } \min_{s \in I} l\big(s^*(P)\big).$$

Denote by $l_{smt}(P)$ and $l_{mst}(P)$ the lengths of the SMT and the MST on terminal set P, respectively. By equality (1.13) and inequality (1.14), to prove a lower bound ρ_0 for the Steiner ratio, it suffices to prove that

$$\min_{P} \max_{s \in I} \big\{ l(t^*(P)) - \rho_0\, l(s^*(P)) \big\} \geq 0, \text{for any topology } t \in K. \quad (1.15)$$

A tree topology $t \in K$ is called *full* if ever terminal point is a leaf. If a terminal point is not a leaf, then this topology can be decomposed at this point into two or more subtree topologies. In this way, ever topology $t \in K$ can be decomposed into edge-disjoint full topologies $t_1, t_2, \cdots, t_h$, respectively, interconnecting subsets $P_1, P_2, \cdots, P_h$ of P. Note that the union of MSTs for $P_1, P_2, \cdots, P_h$ is a spanning tree for P. Hence, if for each $1 \leq i \leq h$, $l(t_i^*(P_i)) \geq \rho\, l_{mst}(P)$, then $l(t^*(P)) \geq \frac{\sqrt{3}}{2} l_{mst}(P)$. It follows that to prove the lower bound ρ_0 for the Steiner ratio, it suffices to prove that

$$\min_{P} \max_{s \in I} \big\{ l\big(t^*(P)\big) - \rho_0 l\big(s^*(P)\big) \big\} \geq 0, \text{for any full topology } t \in K. \quad (1.16)$$

The following lemma [119] gives three well-known simple (but important) structural properties of SMTs in the Euclidean plane (refer to Fig.1.1), which will be used implicitly in our discussion.

Lemma 1.1 *Each Steiner minimum tree on P in the Euclidean plane satisfies the following conditions,*
(i) *All leaves are terminal points.*
(ii) *Any two edges meet at an angle of at least $120°$.*
(iii) *Each Steiner point is incident to exactly three edges, and any two of them must meet at an angle of $120°$.*

By the above lemma we may only consider those Steiner trees that satisfy the above conditions (i-iii). For the simplicity of presentation, we assume that all Steiner trees satisfy those three conditions. A Steiner tree on P is further called a *full Steiner tree* if every terminal point in P is a leaf. Clearly, every angle in a full Steiner tree equals 120°, and every full Steiner tree has n terminals (as leaves) along with at most $(n-2)$ Steiner points. Thus the full Steiner tree can be determined by lengths of (at most) $(2n-3)$ edges provided the topology of the tree is fixed.

To make the presentation easier, we introduce some notations. Let $t(x)$ denote the full Steiner tree with topology t and edge-length $x = (x_1, x_2, \cdots, x_{2n-3})$, let $P(t; x)$ denote the set of all leaves of tree $t(x)$, and let $s(t; x)$ denote the spanning tree with topology s for the terminal set $P(t; x)$. Now inequality (1.14) can be rewritten as following:

$$\min_{x} \max_{s \in I} \left\{ \sum_{i=1}^{2n-3} x_i - \frac{\sqrt{3}}{2} l\big(s(t; x)\big) \right\} \geq 0, \qquad (1.17)$$

where I is the set of spanning tree topologies for the set of n points. Note that for any $\alpha > 0$, $P(t; \alpha x)$ is similar to $P(t; x)$. Thus, $l(s(t; \alpha x)) = \alpha\, l(s(t; x))$. This means that among all similar point sets, we need to consider only one set. So it suffices to consider x with $x_1 + x_2 + \cdots + x_{2n-3} = 1$. Define the following function

$$f_{t,s}(x) = 1 - \frac{\sqrt{3}}{2} l\big(s(t; x)\big) \text{ and let } X = \Big\{ x \in \Re^{2n-3} \,\Big|\, \sum_{i=1}^{2n-3} x_i = 1, x_i \geq 0 \Big\}.$$

To show inequality (1.17), it suffices to prove that for every full Steiner tree topology t,

$$\min_{x \in X} \max_{s \in I} f_{t,s}(x) \geq 0. \qquad (1.18)$$

The next lemma shows that $f_{t,s}(x)$ is a concave function of x thus satisfying the condition of Theorem 1.5 (note that X is a convex set).

Lemma 1.2 $f_{t,s}(x)$ *is a concave function of x.*

Proof.   It suffices to prove that $l(s(t; x))$ is a convex function of x. Let u and v be two terminal points. We show that the distance $d(u, v)$ between u and v is a convex function of x. Find a path in $s(t; x)$ which connects points u and v. Suppose that the path has k edges with distinct lengths $x_{1'}, x_{2'}, \cdots, x_{k'}$ and directions $e_1, e_2, \cdots, e_k$, respectively, where each e_i is a unit vector. Then $d(u, v) = ||x_{1'}e_1 + x_{2'}e_2 + \cdots + x_{k'}e_k||$. Observe that norm functions are convex and the sum of terms inside the norm is linear with respect to x. Hence, $d(u, v)$ is a convex function with respect to x. Finally, we notice that the sum of convex functions is also a convex function. This proves that $l(s(t; x))$ is a convex function of x. □

By Theorem 1.3, determining the Steiner ratio is reduced to finding the minimax value at critical points. Note that the transformation between the Steiner ratio problem and the minimax problem is based on a mapping between sets of n points in the Euclidean plane and points in the $(2n - 3)$-dimensional space. Thus each critical point corresponds to a set of n points with a nice geometric structure, called a *critical structure*. Finally, we only need to verify Gilbert-Pollak conjecture on the terminal set with critical structures.

For a technical reason, we also need to modify the conjecture at the beginning. This modification is necessary because the critical structure obtained above is not nice enough to enable us to handle. The modified conjecture will make the critical structure much nicer. In the following subsections, we will refine the proof of Gilbert-Pollak conjecture [81] by applying Theorem 1.5. We will show how to modify the conjecture, how to determine the critical structure and how to verify the conjecture for the terminal set with critical structures.

1.2.2 *Characteristic Area*

Consider a full Steiner tree $t(x)$. Two terminal points are called *adjacent* if one can reach the other by always moving in a *clockwise* direction or always moving in a *counterclockwise* direction. Clearly, each terminal point has two adjacent terminal points.

Now consider two adjacent terminal points u and v with path $us_1s_2 \cdots s_k v$ connecting them as shown in Fig.1.6. It is easy to verify the following three facts:

(F1) Each angle $\angle s_{i-1}s_i s_{i+1}$ contains either u or v (maybe both).
(F2) If u (resp. v) lies inside of $\angle s_{i-1}s_i s_{i+1}$, then u (resp. v) lies inside of $\angle s_1 s_2 s_3, \cdots, \angle s_{i-1}s_i s_{i+1}$ (resp. $\angle s_{i-1}s_i s_{i+1}, \cdots, \angle s_{k-2}s_{k-1}s_k$).
(F3) There exists an index i such that both u and v lie inside of $\angle s_{i-1}s_i s_{i+1}$.

Choose an index i satisfying condition (F3), and connect u to s_2, $s_3, \cdots, s_i$ and connect v to $s_i, s_{i+1}, \cdots, s_{k-1}$. We will obtain triangles $\triangle us_1s_2, \triangle us_2s_3, \cdots, \triangle us_{i-1}s_i, \triangle us_i v, \triangle s_i s_{i+1}v, \cdots, \triangle s_{k-1}s_k v$. Pasting these triangle along with their edges such that every point between them has a neighborhood isometric to a neighborhood in the Euclidean plane, we obtain a simply connected region either in the plane or in a *multilayer Riemann surface* as shown in Fig.1.6(a). Call this region a *cell*. Observe that when the cell has more than one layer, there are two ways to place the layers. To be specific, we choose the one in the right hand screw rule as shown in Fig.1.6(b).

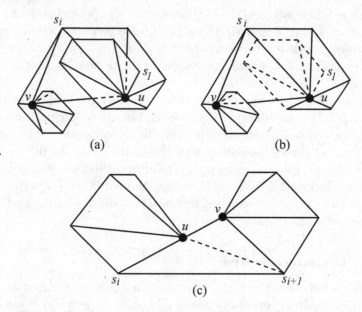

Fig. 1.6   Simple connected regions in multilayer Rimann surface.

It is worth mentioning that although there are more than one choices to choose such index i, we would obtain the same cell. The *characteristic area* of $t(x)$ is obtained by pasting all cells along all edges in $t(x)$ such that every point on $t(x)$ has a neighborhood isometric to a neighborhood in the Euclidean plane. Denote this area by $C(t;x)$. Clearly, the characteristic area is simply connected region in a multilayer Riemann surface satisfying the following properties:

(P1) Every interior point has a neighborhood isometric to a disc in the Euclidean plane.

(P2) All terminal points lie on the boundary of $C(t;x)$.

Note that the characteristic area $C(t;x)$ varies with x. For some x, $t(x)$

may have a self-intersection in the Euclidean plane as shown in Fig.1.7(a) but it has no self-intersection in $C(t; x)$ as shown in Fig.1.7(b). Let us put such x together with $C(t; x)$ into our consideration. Let $X(t; x)$ denote the set of all edge-length vectors $y \in X$ such that x together with $C(t; x)$ can be smoothly moved to y by varying the edge-lengths of all triangles which consist of the characteristic area.

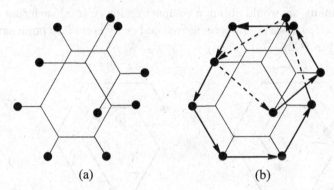

(a)   (b)

Fig. 1.7   (a) A self-intersection in the Euclidean plane, and (b) no self-intersection in the characteristic area.

Clearly, $X(t; x)$ is a compact set and $C(t; y)$ also has the properties (P1-2) for any $y \in X(t; x)$. If none of the triangles in the decomposition of $C(t; y)$ is degenerated, then y must be an interior point of $X(t; x)$. Thus, for every boundary point y of $X(t; x)$, $C(t; y)$ must have a degenerated triangle. This means that this triangle has either an angle of $180°$ or an edge of length zero.

Let us consider again the triangles $\triangle u s_1 s_2, \cdots, \triangle u s_{i-1} s_i, \triangle u s_i v$, $\triangle s_i s_{i+1} v, \cdots, \triangle s_{k-1} s_k v$. In each of them except $\triangle u s_i v$, every angle other than the angle at u or v is at most $120°$. So only the angle formed at u may become $180°$. Therefore, if a triangle other $\triangle u s_i v$ is degenerated, then one of the following two cases has to occur:

(C1) y has at least one zero-component.

(C2) $t(y)$ has a terminal point which lies on the path from the point to an adjacent terminal point.

Note that Case (C2) does not include the case of Fig.1.8(a) because the terminal point and the path which seem to overlap are in different layers. But, it includes the case of Fig.1.8(b).

If only $\triangle u s_i v$ is degenerated, then either (C1) or (C2) occurs except that $\angle s_i u v = 180°$ or $\angle u v s_i = 180°$. In the exceptional cases, for instance, $s_i u v = 180°$ as shown in Fig.1.6(c), we will replace $\triangle u s_i v$ and $\triangle v s_i s_{i+1}$ by $\triangle u s_i s_{i+1}$ and $\triangle u s_{i+1} v$, respectively. In this new set of triangles, no one is degenerated. We expand $X(t; x)$ by varying edge-lengths of this new set of triangles. In this expanded $X(t; x)$, y becomes an interior point. Since there are only finitely many possibilities to decompose each cell into triangles in the above form, only finitely many expanding operations are needed. Finally, we would obtain a compact region $X'(t; x)$ such that every point in $X'(t; x)$ has a characteristic area and every boundary point satisfies either (C1) or (C2).

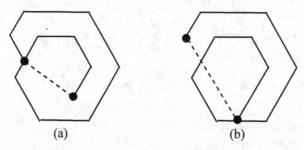

(a)   (b)

Fig. 1.8   (b) belongs to Case (C2) but (a) does not.

Note that for each pair of adjacent terminal points, there are several ways to construct their cell. However, they all yield the same cell. Thus, the characteristic area is unique for each $t(x)$. If for two full Steiner trees $t(x)$ and $t(y)$ every pair of corresponding cells are built up in the same way, then $X(t; x) = X(t; y)$ and hence $X'(t; x) = X'(t; y)$. It follows that there are only (finitely many) 2^n $X'(t; x)$'s for each full Steiner tree topology t. Let X_t denote the union of all $X'(t; x)$'s. Then X_t is a compact set and every boundary point of X_t must have the property (C1) or (C2).

1.2.3   *Inner Spanning Trees*

A spanning tree on $P(t; x)$ is called an *Inner Spanning Tree* (IST) with respect to $t(x)$ if it lies inside of $C(t; x)$. Let $I(t; x)$ denote the set of inner spanning tree topologies. We will prove the following theorem.

Theorem 1.10   *For every full Steiner tree topology t,*

$$\min_{x \in X_t} \max_{s \in I(t;x)} f_{t,s}(x) \geq 0. \tag{1.19}$$

Before we prove the above theorem at the end of this section, we make some observations and prove a series of lemmas. Let $l_{ist}(P(t;x))$ denote the length of minimum inner spanning tree with respect to $t(x)$. Then inequality (1.19) is equivalent to the following inequality

$$L_{smt}(P(t;x)) \geq \frac{\sqrt{3}}{2} l_{ist}(P(t;x)). \tag{1.20}$$

Since $l_{ist}(P(t;x)) \geq l_{mst}(P(t;x))$, the Gilbert-Pollak conjecture is a consequence of Theorem 1.10. To prove the theorem, define

$$g_t(x) = \max_{s \in I(t;x)} f_{t,s}(x) \quad \text{and} \quad M(t;x) = \{i \in I(t;x) \mid f_{t,s}(x) = g_t(x)\}.$$

In order to apply Theorem 1.5, we will need the following lemma.

Lemma 1.3 *For every $x \in X_t$, there is a neighborhood of x such that for any y in the neighborhood, $M(t;y) \subseteq M(t;x)$.*

Proof. First, we show that for any $m \in M(t;x)$ there exists a neighborhood $N(x)$ of x such that $m \in I(t;y)$ for any $y \in N(x)$. Suppose, by contradiction, that such a neighborhood does not exist. Then there is a sequence of points $\{y_k\}$ converging to x such that $m \notin I(t;y_k)$. Thus every $m(t;y_k)$ has at least one edge not in the characteristic area $C(t;y_k)$. Since the number of edges is finite, there exists a subsequence of $\{m(t;y_k)\}$ each of which contains an edge not in $C(t;x)$, but these edges converge to an edge uv in $m(t;x)$. It is easy to see that uv is on the boundary of the area $C(t;x)$ and that u and v are not adjacent. (Note that an edge between two adjacent terminal points always lies in the characteristic area.) Since all vertices in an inner spanning tree lie on the boundary of $C(t;x)$, there is a terminal point lying in the interior of the segment uv, contradicting the minimality of $m(t;x)$.

Now we prove the lemma by contradiction again. Suppose that there is a sequence of points $\{y_k\}$ converging to x such that for each y_k, a spanning tree topology m_k exists such that $m_k \in M(t;y_k) \backslash M(t;x)$. Since the number of spanning tree topologies is finite, there is a subsequence of points $\{y_{k'}\}$ each with the same $m_{k'}$, denoted by m. We can also also assume that this subsequence lies inside of the neighborhood $N(x)$ of x. Thus for every k', $l(m(t;y_{k'})) \leq l(m'(t;y_{k'}))$ for all $m' \in M(t;x)$ since $M(t;x) \subseteq I(t;y_{k'})$. Letting $k' \to \infty$, we obtain that $l(m(t;x)) \leq l(m'(t;x))$ for $m' \in M(t;x)$. Since $m \notin M(t;x)$, $m(t;x)$ must not be an inner spanning tree. It follows that there exists a neighborhood of x such that for any point y in the neighborhood, $m(t; y)$ is not an inner spanning tree for $t(y)$, contradicting the existence of the subsequence $\{y_{k'}\}$. This completes the proof.   □

An immediate consequence of above lemma is that $g_t(x)$ is continuous over X_t. Let

$$F(t) = \min_{x \in X_t} g_t(x).$$

By Theorem 1.5 and Lemmas 1.2-3, $F(t)$ is achieved at some critical point. Choose a full topology t^* such that $F(t^*) = \min_t F(t)$, where t is taken over all full Steiner tree topologies on n terminal points.

We prove Theorem 1.10 by contradiction argument. Suppose that the theorem is not true, i.e., $F(t^*) < 0$, and that n is the smallest number of terminal points such that $F(t^*) < 0$. From now on, a point x in X_{t^*} is called a *minimum point* if and only if $g_{t^*}(x) = F(t^*)$.

Lemma 1.4   *Every minimum point is an interior point of X_{t^*}.*

Proof.   Suppose to the contrary that there exists a minimum point x on the boundary of X_{t^*}. First, assume that Case (C1) occurs, that is, $t^*(t)$ has some edges vanished. If there is a vanished edge incident to a terminal point, then $t^*(x)$ can be decomposed into several edge-disjoint smaller Steiner trees. Since every smaller Steiner tree has fewer terminal points, we can apply Theorem 1.10 to them. Note that a union of inner spanning trees for the smaller Steiner trees is an inner spanning tree for $t^*(x)$. We find a contradiction to $F(t^*) < 0$ by summing over all inequalities. So every vanished edge is between two Steiner points. In this case, we can find a topology t satisfying the following conditions:

(C1.1) Two points are adjacent in t if and only if they are adjacent in t^*.
(C1.2) There is a tree T interconnecting n points in $P(t^*; x)$, with the
        topology t and with length less than $l(t^*(x))$.

We call topology t a *companion* of topology t^* if t satisfies condition (C1.1). To find the desired topology, consider a connected component Z of the subgraph consisting of edges in t^* which correspond to vanished edges in $t^*(t)$. Let Z' be the tree obtained from Z by adding all adjacent edges. Then Z' is a three-regular tree (i.e., a tree in which every vertex has degree on or three). Since all vertices of Z are Steiner points, all leaves of Z' are not vertices of Z. In fact, each leaf is associated with an edge in $Z' \setminus Z$. Two leaves are *adjacent* if they can reach each other by always moving in a clockwise direction or always moving in a counterclockwise direction.

Suppose that Z' has k leaves $u_1, u_2, \cdots, u_k$, where u_i and u_{i+1} are adjacent for each $i = 1, 2, \cdots, k$ (assume that $u_{k+1} = u_1$).

We use the same notation to denote the corresponding points in $t^*(x)$. Suppose that under $t^*(x)$, Z is vanished to the point s. Note that the point u has a neighborhood isometric to a disc in the Euclidean plane. We then have $\angle u_1 s u_2 + \angle u_2 s u_3 + \cdots + \angle u_k s u_1 = 360°$. Since Z has at least two points, Z' has at least four leaves. Thus, $k \geq 4$, and there exists an index i' such that $\angle u_{i'} s u_{i'+1} < 120°$. Now note that for four points, there exist exactly two full Steiner tree topologies. In a Steiner tree topology, if we change a subtree isomorphic to a full Steiner tree for four points to the one isomorphic to the other full topology for four points, then we obtain a companion. Through finitely many operations like this, we can obtain a companion t of t^* such that $su_{i'}$ and $su_{i'+1}$ correspond to two adjacent edges in t. Thus, t also satisfies condition (C1.2). Fig.1.9 demonstrates this process starting from (a) and ending up with (d).

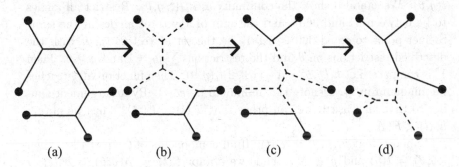

(a)   (b)   (c)   (d)

Fig. 1.9   Changing a subtree isomorphic to a full Steiner tree yields a companion.

In the following we study two cases.

Case 1. The Steiner tree of topology t for $P(t;x)$ exists. In this case, there exists a parameter vector y such that $P(t;y) = P(t^*;x)$. Let $h = 1/l(t(y))$. Then we have $h > 1$ since $l(t(y)) \leq l(T) < l(t^*(x)) = 1$. Note that $t(hy)$ is similar to $t(y)$. Hence, let s be an MST topology for the point sets $P(t;hy)$ and $P(t^*;x)$, we have the following equalities

$$f_{t,s}(hy) = 1 - \frac{\sqrt{3}}{2} l\big(s(t;hy)\big) = 1 - \frac{\sqrt{3}}{2} h\, l\big(s(t;y)\big)$$

$$= 1 - \frac{\sqrt{3}}{2} h\, l\big(s(t^*;x)\big) < g_{t^*}(x) = F(t^*),$$

Since $hy \in X_t$, we have $F(t) \leq g_t(hy) < F(t^*)$, contradicting the minimality of $F(t^*)$.

Case 2. The Steiner tree of topology t for $P(t^*; x)$ does not exist. In this case, we cannot use the above argument for Case 1 directly since $g_t(y)$ is undefined. (Remember that $F(t^*)$ is the minimum over all full Steiner topologies. So ever though T is a shorter tree, there is no contradiction to the minimality of $F(t^*)$.) Now, consider any tree of topology t. A tree under topology t can be determined by edge-lengths and angles at every Steiner point. Represent the lengths by a length vector y and the angles by an angle vector θ. Denote such a tree by $t(y, \theta)$. Two terminal points are said to be *adjacent* in $t(y, \theta)$ if in a Steiner tree of topology t, the corresponding two terminal points are adjacent.

Note that we can define an *inner spanning tree* and a minimum inner spanning tree for $t(y, \theta)$ in a similar way, and then construct the *characteristic area* for $t(y, \theta)$ by connecting every pair of adjacent terminal points. Let $l_{ist}(t; y, \theta)$ denote the length of a minimum inner spanning tree for $t(y, \theta)$. We can also show the continuity of $l_{ist}(t; y, \theta)$. Restrict all angles to be between $0°$ and $360°$, and the sum of any three angles at the same Steiner point to equal $360°$. Let Y_t be the set of vectors (y, θ) with the described restrictions on θ and the restrictions $\sum y_i = 1$ and $y \geq 0$. Then Y_t is compact. Let $h_t(y, \theta) = 1 - (\sqrt{3}/2)l_t(y, \theta)$. Then function $h(,)$ reaches its minimum in Y_t. Denote this minimum by $h^*(t)$. By an argument similar to that for Case 1, we can prove that $h^*(t) < F(t^*)$. Thus we obtain $h^*(t) < F(t)$.

Suppose that $h_t(y, \theta) = h^*(t)$. If all components of θ equal $120°$, then $t(y, \theta) = t(y)$ and $y \in X_t$. Thus we obtain $F(t) \leq h_t(y, \theta) = h^*(t)$, a contradiction. Therefore, we have $\theta < 120°$. Note that for an angle of less than $120°$ in $t(y, \theta)$, at least one edge of the angle must be vanished; Since otherwise, we can shorten the tree without changing the topology. Thus, $t(y, \theta)$ contains vanished edges. If there exists a vanished edge incident to a terminal point, we decompose $h(y, \theta)$ and find a full topology t' with fewer terminal points such that $h^*(t') < 0$. If there exists a vanished edge between two Steiner points, then we can find a new companion t' of t such that $h^*(t') < h^*(t)$. Repeating the above argument, we will obtain infinitely many full topologies with at most n terminal points, contradicting the finiteness of the number of topologies. Therefore, the assumption which is made at the beginning of the proof cannot hold, i.e., Case (C1) cannot occur.

In the end, assume that Case (C2) occurs. Then $t(x)$ (in its characteristic area) has a terminal point touching an edge or another terminal point. In the former case, we can decompose $t(x)$ at the touching point to obtain two trees each with less than n terminal points. In the latter case, we can reduce the number of terminal points by one. In either case, a contradiction will be achieved by an argument similar to the one used at the beginning of the proof. The proof is then finished. □

1.2.4 *Critical Structure*

In this subsection we will determine the geometric structure of $P(t^*, x)$ for every interior minimum point x in X_{t^*}. Let $\Gamma(t^*; x)$ denote the union of minimum inner spanning trees for $P(t^*; x)$. We will first show some properties of $\Gamma(t^*; x)$ in the following lemmas.

Lemma 1.5 *Two minimum inner spanning trees can never cross.*

Proof. We will show that edges of two trees meet only at vertices of the trees. Suppose, by contradiction argument, that uv and $u'v'$ are two edges crossing at the point x as shown in Fig.1.10, and that they belong to two minimum inner spanning trees T_1 and T_2, respectively. Without loss of generality, assume that xu is the shortest one among those four segments xu, xv, xu', and xv'. Removing the edge $u'v'$ from T_2, the remainder has two connected components containing u' and v', respectively. Without loss of generality, assume that u and v' are in the same component. Note that $l(uv') < l(xu) + l(xv') \leq l(u'v')$.

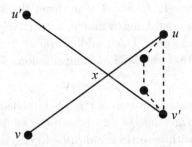

Fig. 1.10 Two edges uv and $u'v'$ cross each other at point x.

If uv' lies in the characteristic area, then connecting the two components by uv' results in a shorter inner spanning tree, contradicting the minimality of T_2. If uv' does not lie in the characteristic area, there must exist some terminal points lying inside the triangle $\triangle xuv'$. Consider the convex hull of those terminal points and the two points u and v'. The boundary of the convex hull other than the edge uv' must lie in the characteristic area. This boundary contains a path from u to v' as shown in Fig.1.10. In this path there exist two adjacent vertices which belong to different connected components of $T_2 \setminus \{u'v'\}$. Connecting two such adjacent vertices also results in an inner spanning tree shorter than T_2, a contradiction again. The proof is then finished. $\square$

Lemma 1.6   *Every polygon of* $\Gamma(t^*; x)$ *has at least two equal longest edges.*

Proof.   Suppose to the contrary that $\Gamma(t^*, x)$ has a polygon Q with the unique longest edge e. Let T^* be the minimum inner spanning tree containing e. For every edge e' of Q not in T^*, the union of T^* and e' contains a cycle. If this cycle contains e, then replacing e with e' in T^* yields an inner spanning tree shorter than T^*, a contradiction. Therefore, such a cycle does not contain e. Hence, for every e' in Q not in T^*, T^* has a path connecting two endpoints of e'. These paths and e form a cycle in T^*, a contradiction. $\square$

Lemma 1.7   *Let u, v, and w be three terminal points. Suppose that all three edges uv, vw, and wu lie in $C(t^*; x)$. If the edge uv is in $\Gamma(t^*; x)$, then $l(uv) \leq \max\{l(uw), l(vw)\}$. Moreover, if uv is in $\Gamma(t^*; x)$ and $l(uv) \geq \max\{l(uw), l(vw)\}$, then either vw or wu is in $\Gamma(t^*; x)$ and also has the same length as uv.*

Proof.   To prove the first half of the lemma, suppose to the contrary that $l(uv) > \max\{l(uw), l(vw)\}$. Removal of uv from the MST results in two connected components containing u and v, respectively. Assume that w is in one of the components. Thus, adding uw or vw would result in a spanning tree shorter than the MST, a contradiction. The second half can be proved in a similar way. $\square$

Note that the characteristic area of $t^*(x)$ is bounded by a polygon of n edges. Partitioning the area into $(n-2)$ triangles by adding $(n-3)$ edges, we will obtain a network with n vertices and $(2n - 3)$ edges. This network will be called a *triangulation* of $C(t^*; x)$. Let us first ignore the full Steiner tree $t^*(x)$ and consider the relationship between the vertex set and the length of edges. Note that in the previous discussion, when we say that a set P of points is given, we really mean that the distance between every two points in the set is given, that is, relative positions between those points have been given. Understanding in such a way, we make the following observations.

(O1) The vertex set of $t^*(x)$ (the set of terminal points, $P(t^*; x)$) can be determined by $(2n - 3)$ edge lengths of the network.

(O2) The $(2n - 3)$ edge-lengths are independent variables, that is, the network could be varied by changing any edge-length and fixing all others as long as the triangle inequality is preserved in each triangle.

Note that observation (O2) follows from Case (C2), from which the triangles can be built up sequentially.

Lemma 1.8 *Every $\Gamma(t^*; x)$ can be embedded in a triangulation of $C(t^*; x)$.*

Proof. Since $\Gamma(t^*; x)$ and the boundary of $C(t^*; x)$ partition $C(t^*; x)$ into polygons, it suffices to prove that every polygon lying in $C(t^*; x)$ has a triangulation. To show this, first note that the sum of all inner angles of any polygon in $C(t^*; x)$ equals $(n - 2) \times 180°$, where n is the number of vertices of the polygon. This fact can be proved by noting that the polygon can be obtained by pasting several polygons in the Euclidean plane and the pasting that we did in Section 1.2.1 preserves the formula. Now, consider a polygon of $n > 3$ vertices. Since the polygon satisfies the formula $(n - 2) \times 180°$, it must have an inner angle less than $180°$, say $\angle abc < 180°$. If the segment ac lies in $C(t^*; x)$, then ac partitions the polygon into two smaller polygons. If ac does not lie in $C(t^*; x)$, then we consider a segment ac', where c' moves from b along bc until it touches the boundary of the polygon. At the last position of ac', there must be a vertex of the polygon other than a lying on ac', say d. Then bd partitions the polygon into two smaller polygons. $\square$

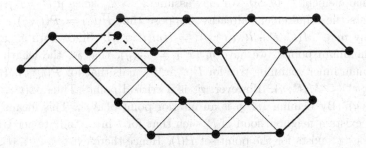

Fig. 1.11   A critical structure.

It follows from the above lemma that all edges in every $\Gamma(t^*; x)$ are independent. A $\Gamma(t^*; x)$ is said to have a *critical structure* if $\Gamma(t^*; x)$ partitions $C(t^*; x)$ into $(n - 2)$ equilateral triangles as shown in Fig.1.11. Such a structure has the property that any perturbation would change the set of topologies of a minimum inner spanning tree. The following lemma shows that every minimum point has $\Gamma(t^*; x)$ with a critical structure.

Lemma 1.9   *If x^* is a minimum point, then $\Gamma(t^*; x^*)$ divides $C(t^*; x^*)$ into $(2n - 3)$ equilateral triangles.*

Proof.   First, let us embed $\Gamma(t^*; x^*)$ into a triangulation of $C(t^*; x^*)$. Suppose to the contrary that the lemma is false. Then two cases must occur. We will show that in each case, the number of MSTs can be increased, i.e., we can find another minimum point y such that $M(t^*; x^*) \subset M(t^*; y)$ and $M(t^*; x^*) \neq M(t^*; y)$.

Case 1.   There is an edge in the triangulation which does not belong to $\Gamma(t^*; x^*)$. Let l_M be the length of the longest edge which is in the triangulation but is not in $\Gamma(t^*; x^*)$. We shrink all longest edges and keep other edge-lengths until a new MST is produced. Let l'_M be the length of the longest edge at the last minute during the shrink. Note that the triangular inequality is always preserved in every triangle if shrinking happens to all longest edges in the triangle or shrinking happens to the shortest edge in an isosceles. The latter is guaranteed by Lemma 1.7. Thus, during the shrinking from l_M to l'_M, we do not need to worry about the condition on the triangular inequality.

Now, for each $l \in [l_M, l'_M]$, denote by $\overline{P}(l)$ the corresponding set of terminal points. Then $\overline{P}(l_M) = P(t^*; x^*)$. Consider the set S_l of all $l \in [l_M, l'_M]$ satisfying the condition that there is a minimum point y in X_{t^*} such that $\overline{P}(l) = P(t^*; y)$. Since $l_M \in S_l$, S_l is nonempty. Moreover, S_l is a closed set since all minimum points form a closed set. Consider now the minimal element l^* of S_l. We may assume $l^* > l'_M$ since if $l^* = l'_M$, then y meets the requirement already. Suppose that $\overline{P}(l^*) = P(t^*; y)$. Then for any $m \in M(t^*; x^*)$, $l(m(t^*; y)) = l(m(t^*; x^*))$. Since both x^* and y are minimum points, we have $g_{t^*}(x^*) = g_{t^*}(y)$, that is, the length of a minimum inner spanning tree for $P(t^*; x^*)$ equals that for $P(t^*; y)$. Hence $M(t^*; x^*) \subseteq M(t^*; y)$. However, x^* is a critical point. Thus $M(t^*; x^*) = M(t^*; y)$. By Lemma 1.4, y is an interior point of X_{t^*}. This means that there exists a neighborhood of l^* such that for l in it, the Steiner tree of topology t^* exists for the point set $\overline{P}(l)$. Hence there exists an $l \in (l'_M, l^*)$ such that $\overline{P}(l) = P(t^*; z)$ for some vector z (not necessarily in X_{t^*} but

$hz \in X_{t^*}$ for some $h > 0$). Since $l(m(t^*; x))$ is continuous with respect to x, there is a neighborhood of y such that for every point y' in the neighborhood, $M(t^*; y') \subseteq M(t^*; y)$. So l can be chosen to make z satisfy $M(t^*; z) \subseteq l(m(t^*; y))$, too. Note that $M(t^*; x^*) = l(m(t^*; y))$ and for every $m \in M(t^*; x^*)$, $l(m(t^*; z)) = l(m(t^*; x^*))$. It follows that for every $m \in M(t^*; x^*)$, $m(t^*; z)$ is a minimum inner spanning tree for $P(t^*; z)$. Thus we have $M(t^*; z) = M(t^*; x^*)$ and $g_{t^*}(x^*) = g_{t^*}(z)$. Suppose that $hz \in X$ where h is a positive number. By the second half of Theorem 1.5, we obtain $g_{t^*}(x^*) = g_{t^*}(hz) = h\,g_{t^*}(z)$. So $h = 1$, i.e., $z \in X$. Hence, z is a minimum point, contradicting the minimality of l^*.

Case 2. Every edge in the triangulation belongs to $\Gamma(t^*; x^*)$. But, $\Gamma(t^*; x^*)$ has a nonequilateral triangle. In this case, we can give a similar proof by increasing the length of all shortest edges in $\Gamma(t^*; x^*)$ and considering the ratio of lengths between the shortest edge and the longest edge in $\Gamma(t^*; x^*)$. The proof is then finished. □

1.2.5 *Hexagonal Trees*

In this subsection, we will prove $g_{t^*}(x^*) \geq 0$ where x^* is a minimum point. For this purpose, we first study a different kind of trees [145]. A tree in $C(t^*; x^*)$ is called a *hexagonal tree* if every edge of the tree is parallel to some edge in $\Gamma(t^*; x^*)$. The shortest hexagonal tree interconnecting the terminal set P is called a *Minimum Hexagonal Tree* (MHT) on P. Let $l_{mht}(P)$ denote the length of the minimum hexagonal tree on P. Weng [267] discovered the following property about the minimum hexagonal tree.

Lemma 1.10 *For any terminal set P, $l_{smt}(P) \geq \frac{\sqrt{3}}{2} l_{mht}(P)$.*

Proof. Note that if a triangle $\triangle abc$ has the angle $\angle bac \geq 120°$, then $l(bc) \geq \frac{\sqrt{3}}{2}(l(ab) + l(ac))$. Thus each edge of an SMT can be replaced by two edges meeting at an angle of $120°$ and parallel to the given directions.

A point on a hexagonal tree but not in P is called a *junction* if the point is incident to at least three lines. A hexagonal tree for n points is said to be *full* if all terminal points are leaves. Any hexagonal tree can be decomposed into a union of edge-disjoint smaller full hexagonal trees. Such a smaller full hexagonal tree will be said to be a *full component* of the hexagonal tree.

In a hexagonal tree, an *edge* is referred to as a *path* between two vertices (terminal points or junctions). Thus an edge can contain several straight segments. An edge is called a *straight edge* if it contains only one straight segment, and is called a *nonstraight edge* otherwise. Any two segments adjacent to each other in a nonstraight meet at an angle of 120° since if they meet at an angle of 60° then we can shorten the edge easily.

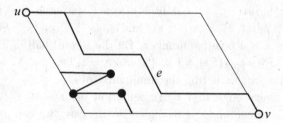

Fig. 1.12   All shortest hexagonal paths from u to v form a parallelogram.

In any minimum hexagonal tree, an edge with more than two straight segments can be replaced by an edge with at most two segments. To see this, consider a nonstraight edge e in a minimum hexagonal tree T_{mht}. Suppose that u and v are two endpoints of edge e. The all shortest hexagonal paths from a to b form a parallelogram as shown in Fig.1.12. This parallelogram must lie in $C(t^*; x^*)$ since otherwise, the part of this parallelogram which is inside of $C(t^*; x^*)$ must contain a piece of the boundary of $C(t^*; x^*)$. This boundary must have at least two consecutive segments in the different directions to ensure that it passes through the parallelogram without crossing edge e. The common endpoint of the two segments is a terminal point lying in the parallelogram. Consider all such terminal points and all shortest hexagonal paths from u to v in $C(t^*; x^*)$. One of the paths must pass through one of the terminal points, say w as shown in Fig.1.12. Replacing e by this path and deleting an edge incident to w would yield a shorter hexagonal tree, contradicting the minimality of T_{mht}. Now since the parallelogram lies in $C(t^*; x^*)$, we can use a path with at most two straight segments to replace edge e.

From now on, we make the convention that any edge in a minimum hexagonal has at most two straight segments. In addition, when we talk about an edge of a junction, its *first segment* is the segment incident to the junction. The other segment, if it exists, is the *second segment* of the edge. Note that the junctions as shown in Fig.1.13 can result in a shorter tree. Thus those kinds of junctions cannot exist in a minimum hexagonal tree. The following facts then follow.

(F1) If a minimum hexagonal tree has a nonstraight edge, then the parallelogram obtained by flipping this edge lies inside of $C(t^*; x^*)$.

(F2) If a minimum hexagonal tree has an angle equal to 60°, then the isosceles trapezoid whose two edges are identical to the two edges of the angle lies inside of $C(t^*; x^*)$.

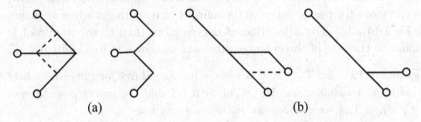

(a)   (b)

Fig. 1.13   Junctions that can not exist in a minimum hexagonal tree.

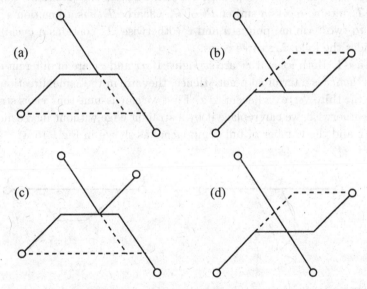

Fig. 1.14   Decrease the number of nonstraight edges: (a,b) a junction of degree three, and (c,d) a junction of degree more than three.

Lemma 1.11   *Let T_{mht} be a minimum hexagonal tree for the terminal set P with the maximum number of full components. Then T_{mht} can be chosen to have the property that every junction of degree three in T_{mht} having at most one nonstraight edge.*

Proof. First, consider a junction of degree three has two nonstraight edges. Then these two edges have segments in the same direction. Flip the edges if necessary to line up these two segments, then the second segments of these two edges as well as the first segment of the third edge comprise three segments each lying completely on one side of the line just constructed. Therefore, one side has the majority of the three segments, and as a result, we can move the line to decrease the number of nonstraight edges as shown in Fig.1.14(a,b). For a junction of degree more than three, the proof is similar to the case of three-degree junctions as shown in Fig.1.14(c,d).   □

Lemma 1.12   *Let T_{mht} be a minimum hexagonal tree for the terminal set P with the maximum number of full subtrees and the property in Lemma 1.11. Then T_{mht} is a minimum inner spanning tree.*

Proof.   Suppose, by contradiction argument, that the lemma is false. Then T_{mht} has a full component T' with at least one junction. Suppose that T' interconnects a subset P' of P. Clearly, T' has a junction x adjacent to two terminal points u and v (otherwise, T' contains a cycle). We consider the following three cases.

Case 1. Both ux and xv are straight. If ux and xv are in different directions then x is a terminal point. Hence, they are in the same direction. Let w be the third vertex adjacent to x. First we can assume that xw is straight since otherwise, we can replace it by a straight edge without increasing the length and the number of full components as shown in Fig.1.15(a).

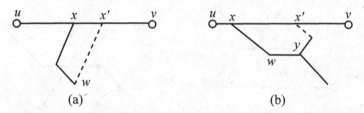

Fig. 1.15   (a) Replace a nonstraight edge with a straight one, and (b) shorten T'.

Since w being a terminal point implies that x is a terminal point, we deduce that w is a junction. We will show that one of the followings occurs:

Case 1.1. x is a terminal point.

Case 1.2. w can be moved further away from x.

Because the movement of w cannot go on forever, x is a terminal point, this contradicts the definition of the junction.

Let L_w be a line through w, which is parallel to uv. If w has a straight edge overlapping L_w on the right of w, then we go from w along the edges of T' to the right as far as possible. Suppose that we end at a point y. Then $l(wy) < l(xv)$ since if $l(wy) \geq l(xv)$, then xw can be moved to the right until x and v are identical so that the number of full components is increased. Since $l(wy) < l(xv)$, y cannot be a terminal point since otherwise, we can move xw to touch y which increases the number of full components. However, y cannot be a junction either since otherwise, T' could be shortened as shown in Fig.1.15(b). Thus y is a corner of a nonstraight edge. A similar situation happens to the left hand side of w.

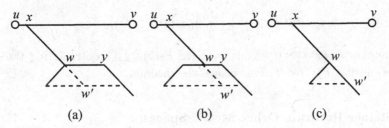

(a)   (b)   (c)

Fig. 1.16 w is moved further away from x to w'.

Now we can move w further away from x as shown in Fig.1.16. If w has no edge with segment overlapping L_w, then w can also be moved further away from x. As this movement cannot last forever, eventually w will become a terminal point and x is a terminal point, that is Case 1.1.

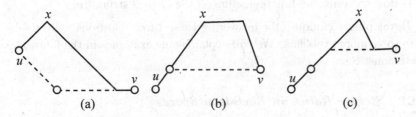

(a)   (b)   (c)

Fig. 1.17 x is moved to a terminal point.

Case 2. ux is a straight edge and xv is a nonstraight edge with a segment in the same direction as ux. In this case, flip xv, if necessary, to line up the two first segments of ux and xv. Let vy be the first segment of xv. Then y must be a terminal point. If y is not identical to v, then we can shorten

T_{mht} by deleting an edge incident to y. If y is identical to v, then we go back to Case 1.

Case 3. ux is a straight edge and xv is a nonstraight edge without a segment in the same direction as ux. In this case, x can be moved either to u or to a terminal point as shown in Fig.1.17. Such a movement increases the number of full components. It follows from facts (F1-2) that the new location of x belongs to $C(t^*; x)$.   □

Finally, we are ready to prove the main theorem, Theorem 1.10.

Proof of Theorem 1.10.   It follows from Lemma 1.10 and Lemma 1.12 that for any minimum point x^*,

$$l_{smt}\big(P(t^*; x^*)\big) \geq \frac{\sqrt{3}}{2} l_{mht}\big(P(t^*; x^*)\big),$$

which implies $g_{t^*}(x^*) \geq 0$. Thus we have $F(t^*) \geq 0$, contradicting the assumption that $F(t^*) < 0$. This proves the theorem.   □

1.3   Steiner Ratios in Other Metric Spaces

From discussions of Section 1.3, we can see that to apply the minimax approach for solving an (optimization) problem, the following three questions should be addressed:

(Q1) How to transfer the problem to such a minimax problem satisfying the condition that the corresponding functions are concave.
(Q2) How to determine the critical geometric structure.
(Q3) How to verify the function value on the critical structure.

Developing techniques for answering these three questions will enable us to solve more problems. We will explain some examples in the following discussions.

1.3.1   *Steiner Ratios in Euclidean Spaces*

Gilbert and Pollak [114] also conjectured that in any Euclidean space the Steiner ratio is achieved by the vertex set of a terminal simplex. Chung and Gilbert [62] constructed a sequence of Steiner trees on regular simplices whose lengths go decreasingly to $\sqrt{3}/(4 - \sqrt{2})$. Although the constructed trees are not known to be SMTs, they conjectured that $\sqrt{3}/(4 - \sqrt{2})$ is the best lower bound for Steiner rations in Euclidean spaces. Clearly, if

$\sqrt{3}/(4 - \sqrt{2})$ is the limit of Steiner ratios in d-dimensional Euclidean space $\Re^d$ as d goes to infinity, then Chung-Gilbert conjecture is a corollary of Gilbert and Pollak's general conjecture. However, the general conjecture of Gilbert and Pollak has been disproved by Smith [247] for dimension from three to nine and by Du and Smith for dimension larger than two. Now interesting questions about Chung-Gilbert conjecture arise: Could Chung-Gilbert conjecture also be false? If the conjecture is not false, can we prove it by the minimax approach?

First, we claim that Chung-Gilbert conjecture could be true. In fact, we could get rid of Gilbert-Pollak general conjecture, and use another way to reach the conclusion that the limit of Steiner ratios for regular simplices is the best lower bound for Steiner ratios in Euclidean spaces. To support our claim, let us analyze a possible proof of such a conclusion as follows.

Consider n terminal points in $(n-1)$-dimensional Euclidean space. Then all of $n(n - 1)/2$ distances between the n points are independent. Suppose that we could do a similar transformation and the minimax theorem could apply to these n points to obtain a similar result in the proof of Gilbert-Pollak conjecture for the case of Euclidean plane, i.e., a terminal set with critical geometric structure has the property that the union of all MSTs contains as many equilateral triangles as possible. Then such a critical structure must be a regular simplex. This observation tells us two facts:

(1) Chung-Gilbert conjecture can follow from the following two conjectures (the second one due to Smith [247]).
(2) It may be possible to prove Conjecture 1 by the minimax approach if we could find a right transformation.

Conjecture 1 *The Steiner ratio for n terminal points in an Euclidean space is not smaller than the Steiner ratio for the vertex set of $(n-1)$-dimensional regular simplex.*

Conjecture 2 *The limit of Steiner ratios in simplices is $\sqrt{3}/(4 - \sqrt{2})$.*

One may wonder why we need to find another transformation. What happens to the transformation used in the proof of Gilbert-Pollak conjecture in the Euclidean plane. The reason is that the transformation we used in Section 1.3.1 turns out not to be applicable for Conjecture 1. In fact, in the Euclidean plane, with a fixed graph structure, all edge-lengths of a full Steiner tree can determine the set of original terminal points, and furthermore the length of a spanning tree for a fixed graph structure is a convex function of the edge-lengths of the Steiner tree. However, in an

Euclidean space of dimension more than two, edge-lengths of a full Steiner tree are insufficient for determining the set of given terminal points. Even worse than that, adding other parameters may destroy the convexity of the length of a spanning tree as a function of the parameters.

Smith [247] showed, by an exhaustive computing, that for $d = 3, 4, \cdots, 7$, the Steiner trees constructed by Chung and Gilbert are actually SMTs; But, for $d = 8$, their Steiner tree is not an SMT. Smith also conjectured that the trees of Chung and Gilbert are SMTs if $d = 3 \cdot 2^p$.

From the above discussion, we see that proving Chung-Gilbert conjecture requires a further development of the minimax approach.

1.3.2 *Steiner Ratio in Rectilinear Spaces*

Although Hwang [142] determined the Steiner ratio in the rectilinear plane in the very early stage of the study of Steiner trees, there is still no progress on the Steiner ratio in rectilinear spaces by now. In fact, Graham and Hwang [119] conjectured that the Steiner ratio in a d-dimensional rectilinear space is $d/(2d-1)$ for any $d \geq 2$. The difficulty for extending Hwang's approach to proving Graham-Hwang conjecture is due to the lack of knowledge on the full rectilinear Steiner trees in high dimensional spaces. Note that for a full rectilinear Steiner tree in the plane, all Steiner points lie on a path of the tree. However, it is not known whether a similar result holds for full rectilinear Steiner trees in a space of dimension more than two.

Graham-Hwang conjecture can be easily transferred to a minimax problem that is needed for applying our minimax approach as follows: Choose lengths of all straight segments of a Steiner tree. When the topology of the Steiner tree is fixed, the terminal set can be determined by such segment-lengths, the length of the Steiner tree is a linear function and the length of a spanning tree is a convex function of such segment-lengths, thus $f_i(\cdot)$ is a concave function of segment-lengths. Unfortunately, under such a transformation, it is hard to determine the *critical structure*. To explain the difficulty, we notice that in general the critical points could exist in both the boundary and interior of the polytope. In the proof of Gilbert-Pollak conjecture in the Euclidean plane, the crux is that only interior critical points need to be considered in a contradiction argument. The critical structure of interior critical points are relatively easy to be determined. However, for the current transformation on Graham-Hwang conjecture, we have to consider some critical points on the boundary. It requires a new technique to either determine critical structure for such critical points or eliminate them from our consideration.

One option is to combine the minimax approach and Hwang's method. In fact, using the minimax approach, we may get a useful condition on the set of terminal points. With such a condition, the terminal set can have only certain type of full Steiner trees. This may reduce the difficulty of extending Hwang's method to higher dimension.

The significance of developing new techniques for determining critical structures corresponding to critical points on the boundary is not only for solving Graham-Hwang's conjecture, but also for solving some other interesting problems. For example, it can be immediately applied to some packing problems. One of them is to find the maximum number of objects which can be put in a certain container. When the objects are discs or spheres, the problem can be transferred to a minimax problem that meets our requirement. To determine such a number exactly, we have also to deal with critical points on the boundary of the polytope.

1.3.3 Steiner Ratio in Banach Spaces

Examining the proof of Gilbert-Pollak conjecture in the Euclidean plane, we observe that the whole proof does not use the property of the Euclidean norm except the last part when verifying the conjecture on terminal sets of critical structure. This means that using the minimax approach to determine the Steiner ratio in Minkowski plane[3] (2-dimensional Banach space[4]), we would have no technical difficulty finding a transformation and determining critical structures, but only meet a difficulty verifying the conjecture for terminal sets with critical structure.

SMTs in the Minkowski planes have been studied in [249; 3; 66; 193; 77; 82]. In these papers, some fundamental properties of SMTs in Minkowski planes have been established. The following two nice conjectures about the Steiner rations in Minkowski planes were proposed in [66; 77] and [77], respectively.

Conjecture 3 *In any Minkowski plane, the Steiner ratio is between 2/3 and $\sqrt{3}/2$.*

Conjecture 4 *The Steiner ratio in a Minkowski plane equals that in its*

---

[3]The *Minkowski plane*, named after H. Minkowski (1864-1909), is a two-dimensional affine space provided with a metric that is invariant under translations.

[4]A *Banach space*, named after S. Banach (1892-1945), is a complete vector space with a norm $|| \cdot ||$.

dual plane.

With some new techniques in the critical structures, Gao et al. [108] proved the first half of Conjecture 3 that in any Minkowski plane, the Steiner ratio is at least 2/3, and Wan et al. [259] showed that Conjecture 4 is true for three, four, and five points. With a different approach, Du et al. [77] also proved that in any Minkowski plane, the Steiner ratio is at most 0.8766.

Chung-Gilbert conjecture and Conjecture 4 can be naturally extended to high dimensional Banach spaces as follows.

Conjecture 5 *In any infinite dimensional Banach space, the Steiner ratio is between 1/2 and $\sqrt{3}/(2 - \sqrt{2})$.*

Conjecture 6 *The Steiner ratio in any Banach space equals that in its dual space[5].*

Significant results on these two conjectures could be produced by further developments of minimax approach from successful application in two-dimensional problems to d-dimension for $d \geq 3$.

1.4   Discussions

The proof of Gilbert-Pollak conjecture implies that minimum spanning tree algorithm is a $2/\sqrt{3}$-approximation algorithm for the Steiner tree problem in the Euclidean plane. A natural problem is to design a better algorithm with approximation ratio smaller than $2/\sqrt{3}$. A simple method is to start from an MST for given terminal set and then improve it by adding some Steiner points. Clearly, every solution obtained in such a way would have an approximation ratio no worse (greater) than $2/\sqrt{3}$. The problem is how much better than $2/\sqrt{3}$ that one can achieve.

It was a long-standing problem whether there exists a polynomial time approximation algorithm whose performance ratio is smaller than the inverse of the Steiner ratio or not. Over more than thirty years numerous heuristics (e.g., [26; 35; 44; 97; 175; 178; 179; 251; 271]) for Steiner minimum tree problem have been proposed for terminal points in various metric spaces. Their superiority over MST-based algorithms were often claimed by computation experiments without theoretical proof.

---

[5]The *dual* of a real vector space $\mathcal{S}$ is the vector space of linear functions $f : \mathcal{S} \to \Re$.

In particular, Chang [44; 45] proposed an approximation algorithm for Steiner minimum tree problem in the Euclidean plane as follows: Start from an MST and at each iteration choose a Steiner point such that using this Steiner point to connect three vertices in the current tree could replace two edges in the MST and this replacement achieves the maximum saving among such possible replacements.

Smith et al. [248] also adopted the idea of greedy improvement. But, they start with Delaunay triangulation instead of an MST. Since every MST is contained in Delaunay triangulation, the performance ratio of their approximation algorithm can also be bounded by the inverse of the Steiner ratio. The advantage of this algorithm over Chang's algorithm is on the running time, it runs only in time of $O(n \log n)$ while Chang's algorithm runs in time of $O(n^3)$.

Kahng and Robin [160] proposed an approximation algorithm for the Steiner tree problem in the rectilinear plane by adopting the same idea as that of Chang. For these three algorithms, it can be proved that for any particular set of terminal points, the ratio of lengths of the approximation solution and that of the SMT is smaller than the inverse of the Steiner ratio. Some experimental results also show that the approximation solutions obtained by these algorithms are very good. However, no theoretical proof has been found to show any one of them being a better approximation algorithm.

The first significant work on finding a better approximation algorithm was made by Bern [31]. He proved that for the rectilinear metric and Poisson distributed terminal points, a greedy approximation obtained by a very simple improvement over an MST has a shorter average length. Later, Hwang and Yao [146] extended this result to the case when the number of terminal points is fixed.

In 1991, Zelikovsky [276] made the first breakthrough to the problem by proposing a better approximation algorithm for the Steiner tree problem in graphs. In 1996, Arora [10] made another breakthrough by proving that the Steiner tree problem in the Euclidean plane admits a series of approximation algorithms whose performance ratios could arbitrarily approach to 1. These are two important developments on Steiner tree problems in 1990s. In the next two chapters, we will study their works in details, respectively.

Chapter 2

k-Steiner Ratios and Better Approximation Algorithms

In this chapter we will study the Steiner tree problem in graphs as well as in metric spaces. The Steiner tree problem in graphs is also NP-hard [163] and it can be defined as follows.

Problem 2.1 *Steiner Tree Problem in Graphs*

Instance A connected graph $G(V, E)$ with a cost[1] $l(e)$ on each edge $e \in E$, a terminal set $P \subset V$.

Solution A Steiner tree T for P that interconnects all terminals in P.

Objective Minimizing the total length of the edges in T, $l(T) \equiv \sum_{e \in T} l(e)$.

A *k-Steiner tree* is a Steiner tree with all full components of size at most k. The *k-Steiner minimum tree* (*k*-SMT) is a *k*-Steiner tree with the shortest length. Note that a 2-SMT is clearly an MST; But *k*-Steiner tree for $k \geq 3$ may not exist in some cases (e.g. a star graph with more than k rays and terminal set containing all vertices except the central vertex), in such cases it is assumed that a *k*-Steiner tree is allowed to use the edges and Steiner vertices in more than one full components. As a result, we can consider a *k*-Steiner tree as a collection of full components with at most k terminals such that all terminals are connected.

Given a terminal set P, by the definition of *k*-Steiner trees, a *k*-SMT is an SMT for $k \geq |P|$, so it is reasonable to expect that a *k*-SMT gives a good approximation of SMT for sufficiently large k. To measure the difference in lengths between *k*-SMTs and SMTs, let $l_{smt}(P)$ be the length of the SMT of P, and let $l_{k-smt}(P)$ be the length of the *k*-SMT of P, we define the *k-Steiner ratio* in a metric space M as the following

---

[1] To keep consistence we will still use $l(\cdot)$ to represent the cost function on edges.

$$\rho_k(M) \equiv \inf \left\{ \frac{l_{smt}(P)}{l_{k-smt}(P)} \,\Big|\, P \subset M \right\}. \tag{2.1}$$

Note that the Steiner ratio ρ_M (1.1) is equal to k-Steiner ratio $\rho_k(M)$ of (2.1) with $k = 2$. In particular, in the rectilinear plane L_1, $\rho_2(L_1) = 2/3$ [142], and in the Euclidean plane L_2, $\rho_2(L_2) = \sqrt{3}/2$ [81].

Every edge-weighted graph $G(V, E)$ can be considered as a metric space M where the distance between two vertices[2] in V equals to the length of the shortest path between them in G. Thus we can define k-*Steiner ratio in graphs* as follows

$$\rho_k \equiv \inf_G \rho_k(G) = \inf_M \rho_k(M),$$

which means that the k-Steiner ratio in graphs is the same as the k-Steiner ratio over all metric spaces. It is well-known that $\rho_2 = 1/2$ [179; 60; 253].

Unfortunately, computing k-SMTs for $k \geq 4$ turned out to be NP-hard in general and the complexity of $k = 3$ is unknown [29]. Indeed, the NP-hardness proofs for computing SMTs in the Euclidean and rectilinear planes [109; 111] are applicable for computing k-SMTs in the planes when $k \geq 4$ is fixed, respectively. For the case of graphs, the bounded degree *independent set problem*[3] can be reduced to computing k-SMTs for $k \geq 4$ by using the reduction of Hakimi [126]. As a result, k-Steiner ratio ρ_k can not serve as an approximation performance ratio as classical ratio ρ.

However, studying k-Steiner ratio for $k \geq 3$ is still very important since it plays a key role in the design of better approximation algorithms for Steiner tree problems. In fact, it was a long-standing open problem whether there exits a polynomial time approximation for the Steiner tree problem in each metric space with a performance ratio smaller the the inverse of the Steiner ratio, i.e., $\rho^{-1} = \rho_2^{-1}(M)$. Zelikovsky [276] made the first breakthrough in the Steiner tree problem in graphs. He proposed a polynomial time algorithm using 3-Steiner trees whose performance ratio is upper bounded by

$$\frac{\rho_2^{-1} + \rho_3^{-1}}{2}.$$

---

[2]When studying Problem 2.1 we will use graph term "vertex/vertices" instead of geometry term "point/points".

[3]Given a graph $G(V, E)$, the problem asks for the subset of V with maximal size such that there is no edge in E between any two vertices in the set.

By bounding ρ_2 and ρ_3, he proved that the ratio is $11/6 < \rho_2^{-1} = 2$. Later Du et al. [91] and Berman and Ramaiyer [29] generalized Zelikovsky's idea to k-Steiner trees. Du et al. [91] showed that a generalized Zelikovsky's algorithm has a performance ratio upper bounded by

$$\frac{(k-2)\rho_2^{-1} + \rho_k^{-1}}{k-1}.$$

Berman and Ramaiyer [29] proposed a better algorithm that has performance ratio at most

$$\rho_2^{-1} - \frac{\rho_2^{-1} - \rho_3^{-1}}{2} - \frac{\rho_3^{-1} - \rho_4^{-1}}{3} - \cdots - \frac{\rho_{k-1}^{-1} - \rho_k^{-1}}{k-1}$$

In addition, Zelikovsky [277] also obtained a better approximation algorithm with performance ratio no more than

$$\frac{1 - \ln \rho_2 + \ln \rho_k}{\rho_k}.$$

A main part of the above mentioned works was to establish the lower bound for the k-Steiner ratio ρ_k since a better lower bound will give a better performance ratio for their approximations. Zelikovsky [276] first showed that $\rho_3 \geq 3/5$ and later Du [76] showed that $\rho_3 \leq 3/5$, which means that $\rho_3 = 3/5$. Du et al. [91] proved that $\rho_k \geq \lfloor \log_2 k \rfloor / (1 + \lfloor \log_2 k \rfloor)$. Berman and Ramaiyer [29] proved that in the rectilinear plane $\rho_3(L_1) = 4/5$ and $\rho_k(L_1) = (2k-1)/(2k)$ for $k \geq 4$.

In this chapter we shall first prove that for $k = s + 2^r$ with $0 \leq s < 2^r$, the k-Steiner ratio in graphs is

$$\rho_k = \frac{s + r2^r}{s + (r+1)2^r}. \tag{2.2}$$

And then we shall present some approaches proposed for designing better approximation algorithms for Steiner tree problems. In the end we will discuss some related open problems.

2.1 k-Steiner Ratio

To determine the k-Steiner ratio, we will need some concepts. Let T be a Steiner tree in a graph $G(V, E)$, the edges of T are actually shortest paths between vertices of the graph. In fact, we may think of the vertices along such a path as 2-degree Steiner vertices. The Steiner vertices all belong to

V, but a vertex in V can be used more than once as a Steiner vertex in different components of the same Steiner tree.

Recall that a *binary tree* is such a rooted tree that each vertex has at most two children. We call a binary tree where every internal vertex has exactly two children a *terminal binary tree*. A *complete binary tree* is a terminal binary tree where all leaves have the same depth. A binary tree that is *complete* except perhaps at the bottom level is a terminal binary tree where all leaves have depth l or $(l+1)$ for some l.

2.1.1   *Upper Bound for k-Steiner Ratio*

To prove that the term on the right side of equality (2.2) is a upper bound for the k-Steiner ratio, we consider a special metric space M_n based on a weighted tree B_n as shown in Fig.2.1. Let B_n be a complete binary tree with n levels of edges and a final bottom level where each internal vertex has only one child. The edges at the i-th level have length 2^{n-i} and the edges at the $(n+1)$-th level, the bottom level, have length 1. The points of M_n are the vertices of B_n. The distance between two points is the length of the shortest path between them in B_n, we can think of edges between two points in M_n as paths between those points in B_n. It is easy to verify that this defines a *metric space*.

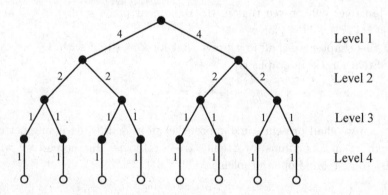

Fig. 2.1   The metric space M_n based on a weighted binary tree B_n.

Let the leaves of B_n be the set P_n of terminal points. We will compute the length of SMT for P_n, $l_{smt}(P_n)$. Let $l_{k-smt}(P_n)$ be the length of the k-SMT of P_n. The ratio of these two values will yield a upper bound for ρ_k as $n \to \infty$.

Lemma 2.1   *The length of the Steiner minimum tree for P_n is the total length of B_n, $l_{smt}(P_n) = (n+1)2^n$.*

Proof.   Note that in B_n the total length of the edges at any fixed is 2^n, this is true at the bottom two levels, and each level above has half as many edges each with twice the length. As there are $(n+1)$ levels, we obtain $l_{smt} = (n+1)2^n$.   □

Lemma 2.2   *There is a k-Steiner minimum tree for P_n such that each Steiner point has degree exactly 3.*

Proof.   Note that the terminal points of a component in any k-SMT can be interconnected by a binary tree embedded in B_n, and it follows from Lemma 2.1, this must be a shortest tree interconnecting these points. Note that Steiner points of degree 2 can be removed and the two adjacent edges replaced by a single edge, which is a path in B_n, between the two adjacent vertices. Therefore, we can assume all Steiner points in this component tree have degree 3. Components obtained in such a way yield the desired k-SMT.   □

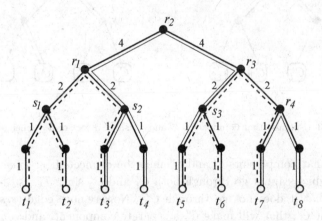

Fig. 2.2   A k-Steiner tree and roots of its components.

We could consider a component in a k-SMT for P_n as a regular binary tree with at most k leaves embedded in B_n, where the edges are paths in B_n. The root of a tree component is defined as the 2-degree Steiner point at the highest level, which will be called the *component root*. In the case of $n = 3$ as shown in Fig.2.2, one component consists of one Steiner point

s_1, three terminals t_1, t_2, and t_3 with root at the degree-2 Steiner point r_1. Observe that r_1 is a degree-3 point in B_n and the edge between s_1 and t_3 is a path of length 6 in B_n.

Lemma 2.3 *There is a k-Steiner minimum tree for P_n such that each Steiner point has degree 3 and no point of B_n is used more than once as a Steiner point or component root of any component.*

Proof. Let T_{k-smt} be a k-SMT as described in Lemma 2.2. Suppose that there exists a vertex s at the highest level of B_n that is used as a root or Steiner point of two components, C and C', in T_{k-smt}. Let s_1 and s_2 be the two children of s. Then edges (s, s_1) and (s, s_2) are used in both components C and C'. Let C_1 and C_1' be the parts of components C and C', respectively, that lie below s, and let C_2 and C_2' be the parts that lie below s_2. See Fig.2.3(a) where C consists of the thick links where C' the thin links.

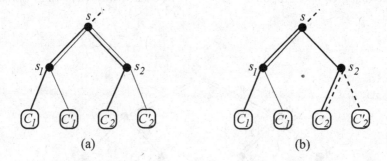

(a)   (b)

Fig. 2.3   Suppose that components C and C' share a root or a Steiner point.

Note that components C and C' must be connected by some path in T_{k-smt}, which cannot go through both C_1' and C_2' since T_{k-smt} is a tree. Assume that it does not go through C_2'. Now remove edge (s, s_2) from component C', that will make C_2' a separate component, and connect C_2' to component C by a path from s_2 to a terminal point in C_2, which is a leaf in B_n. See Fig.2.3(b) where the new component consists of the dashed links.

Observe that in B_n the length of edge (s, s_2) is the same as the length of the path from s_2 to a leaf since the length of the edges doubles at each level up the tree. Therefore, the above modification will not increase the length of the Steiner tree; Moreover, it will not increase the size of any component and add any Steiner points. Thus all Steiner points are still of degree 3. Vertex s_2 has become the root of a new component, so we might have added points that are used as roots or Steiner points of more than one component; However, we have only added them below s.

If we repeat the above described process, by mathematical induction we can successively remove the Steiner points or component roots that overlap at vertices of B_n from level to level going down towards the bottom of the tree until no such points exist. The proof is then finished. $\square$

Lemma 2.4 *There is a k-Steiner minimum tree for P_n such that each Steiner point has degree 3 and every internal vertex of B_n, except at the lowest level, is used exactly once as a Steiner point or component root of some component.*

Proof. It follows from Lemma 2.3 that there is a k-SMT T_{k-smt} such that every internal vertex of B_n is used at most once. The internal vertices at the lowest level cannot be component roots or Steiner points because they have only one child. We will show that $(2^n - 1)$ Steiner points and roots are needed to interconnect all 2^n points in P_n, and since there are only $(2^n - 1)$ internal vertices of B_n not at the lowest level, every point must be used exactly once. Fig.2.2 shows such a k-SMT T_{k-smt} with $k = 3$.

Now remove the component roots and Steiner points one by one from T_{k-smt}. Notice that when a component root is removed, the tree will be split into two pieces at the root, and when a Steiner point is removed, the tree will be split into two pieces by disconnecting one of the two children of the Steiner point. After all points are removed, the 2^n terminal points must be completely disconnected. Since removing one point adds only one new piece, we just have removed $(2^n - 1)$ points to yield the 2^n disconnected pieces. The proof is then finished. $\square$

Let $T_{k-smt}(P_n)$ be a k-SMT on P_n that satisfies Lemma 2.4. We then can think of each component C in $T_{k-smt}(P_n)$ as a regular binary tree whose internal vertices are the component root and the Steiner points. Below each of these internal vertices are two edges of B_n. By Lemma 2.2, all Steiner points have two edges below them and the component root must have two edges below it in $T_{k-smt}(P_n)$. We call these edges of B_n the *peak edges* of component C, and denote them by P_C, and we call the rest of the edges used by C the *connecting edges* of C, and denote them by C_C. In addition, denote all of the peak edges of $T_{k-smt}(P_n)$ by $P_{k,n}$ and all the connecting edges by $C_{k,n}$. When we refer to a *peak*, we mean those two peak edges and the vertex they incident to.

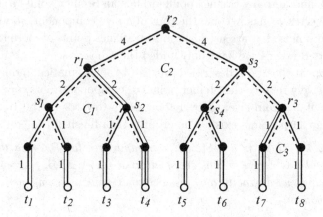

Fig. 2.4   The peak edges and connecting edges of components.

Fig.2.4 shows a k-SMT $T_{k-smt}(P_n)$ for $k = 4$ and $n = 3$. It consists of three components C_1, C_2, and C_3 each has some peak (dashed) edges and connecting (thin) edges. In particular, the component C_1 that contains four terminals t_i for $i = 1, 2, 3, 4$ and has three pairs of peak edges with total length of 8 and four connecting edges with total length of 4. For the component C_3 that contains t_7 and t_8, these two values are both equal to 2. The following lemma gives a relationship between these two values in general.

Lemma 2.5   *For any k with $k = s + 2^r$, where $0 \leq s < 2^r$,*

$$\sum_{e \in C_C} l(e) \geq \left(\frac{2^r}{s + r \cdot 2^r} \right) \sum_{e \in P_C} l(e). \tag{2.3}$$

Proof.   We will prove the lemma for components of size k in the k-SMT by mathematical induction on the sum of the depths of the peaks from the root of the component. We compute the the depths of the peaks by counting edges in B_n, not edges in M_n; and we count each edge the same, ignoring the lengths of the edges. Note that the sum of the peak depths is minimum when the peaks form a complete binary tree, except perhaps at the lowest level, so we will consider this case first.

In a component C of size k whose peaks form a complete binary tree, except perhaps at the lowest level, we have 2^r peak edges at the second to the lowest level and $2s$ peak edges at the lowest level. The components C_1 and C_2 in Fig.2.4 are such two components. Suppose that a peak edge at the lowest level has length w. Then the peak edges at the lowest level have total length of $2s \cdot w$ and at the r higher levels each level has total length of $w \cdot 2^{r+1}$. Thus we have

$$\sum_{e \in P_C} l(e) = 2w(r \cdot s + 2^r).$$

The connecting edges of C form paths from the peak edges at the lowest level down to the terminal points at the leaves of B_n. By the construction of B_n, these paths have the same length as the peak edges at the lowest level where they originate. There are $2s$ such paths of length w at the lowest level and $(2^r - s)$ such paths of length $2w$ one level higher. Thus we have

$$\sum_{e \in C_C} l(e) = 2w \cdot 2^r.$$

In this case, inequality (1.3) is satisfied. In fact, it is an equality.

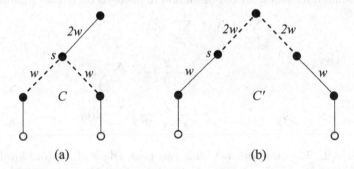

Fig. 2.5   Case 1: (a) C has a peak with a nonpeak edge above it, and (b) C' after moving the peak of C up.

We next will show that the ratio of total length of the connecting edges to the total length of the peak edges decreases as the sum of the depths of the peak edges decreases. In fact, as we change the configuration of the component by moving a peak up, the total length of the connecting edges remains unchanged, while the total length of the peak edges increases. When the peaks cannot move up any further, the sum of the peak depths achieves the minimum, and as we just showed, inequality (2.3) holds. Hence it will hold for any configuration of a component. To show this fact, we will consider two cases in which the sum of the peak depths might not be minimum: either a peak has a nonpeak edge above it as shown in Fig.2.5(a) or else there are two peak edges at the bottom level whose depths differ by more than 1 as shown in Fig.2.6(a). We assume that inequality (2.3) has been established for all possible components where the sum of the peak depths is at most m and that the sum of the peak depths of component C is $(m + 1)$, so it is not minimum.

Case 1. The component C has a peak with a nonpeak edge above it as shown in Fig.2.5(a). Suppose that the peak edges with the nonpeak edge above have length w. Consider the component C' obtained by moving the peak with the nonpeak edge above it up one level as shown in Fig.2.5(b). This decreases the sum of the peak depths by 1, so the induction hypothesis applies to C'.

Note that the total length of the connecting edges of C and C' are the same since the connecting edge of length $2w$ above the peak in C has been replaced by two connecting edges of length w in C'. In addition, the total length of the peak edges of C is less than that in C' since the two peak edges of length w in C have been moved up so they each have length $2w$ in C'. Combining this with the induction hypothesis on C', we obtain

$$\sum_{e \in C_C} l(e) = \sum_{e \in C_{C'}} l(e)$$

$$\geq \left(\frac{2^r}{s + r \cdot 2^r}\right) \sum_{e \in P_{C'}} l(e)$$

$$\geq \left(\frac{2^r}{s + r \cdot 2^r}\right) \sum_{e \in P_C} l(e) \tag{2.4}$$

Case 2. The component C has two peak edges at bottom level whose depths differ by more than 1 as shown in Fig.2.6(a). Consider the peak edges at the lowest and highest bottom level in C. The lowest peak must have two paths to two leaves below t; the highest peak must have at least one path to a leaf below it. Suppose that the highest peak edge has length $2w'$ and the lowest peak edges have length w. Then by the construction of B_n, the paths below these peak edges also have length $2w'$ and w, respectively.

Now consider the component C' obtained from C by moving the lowest peak just below the peak edge at the highest bottom level as shown in Fig.2.6(b). This operation decreases the sum of the peak depths by at least 1, so our induction hypothesis applies to C'.

The total length of the connecting edges of C and C' are again the same since the connecting path of length $2w'$ below the highest peak in C has been replaced by two connecting paths of length w' in C', and the two connecting paths of length w below the lowest peak in C have been replaced by one connecting path of length $2w$ in C'. The total length of the peak edges in C is less than that in C' since the two lowest peak edges of length w in C have been moved up, so they have length w' in C'. As the peaks at the lowest and highest bottom level differ in depth by at least, we have that $w < w'$. Combining this inequality with the induction hypothesis on C', we get inequality (2.4) again.

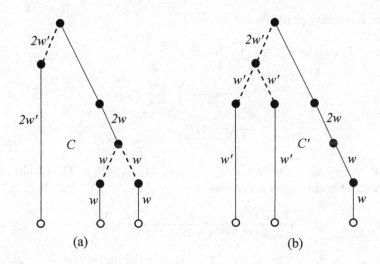

Fig. 2.6   Case 2: (a) C, and (b) C' after moving the peak of C up.

The above analysis for two cases complete the induction step, and the proof is then finished.   □

Theorem 2.1   *For any k with $k = s + 2^r$, where $0 \le s < 2^r$, the k-Steiner ratio*

$$\rho_k \le \frac{s + r2^r}{s + (r+1)2^r}.$$ (2.5)

Proof.   First consider the case where each component C has size k. By Lemma 2.5, summing the inequality (2.3) over all components of $T_{k-smt}(P_n)$ yields

$$\sum_{e \in C_{k,n}} l(e) \ge \left(\frac{2^r}{s + r2^r} \right) \sum_{e \in P_{k,n}} l(e).$$ (2.6)

By Lemma 2.4, the Steiner points and component roots of $T_{k-smt}(P_n)$ cover all of the internal vertices of B_n exactly once, except at the lowest level. Thus the peak edges of $T_{k-smt}(P_n)$ cover all of the edges of B_n exactly once, except for the edges at the lowest level. The total length of the edges at the lowest level is 2^n, and the total lengths of all edges of B_n is $l_{smt}(P_n)$ by Lemma 2.1. Therefore, we have

$$l_{smt}(P_n) = 2^n + \sum_{e \in P_{k,n}} l(e).$$

Then by inequality (2.6), we obtain

$$l_{smt}(P_n) = \sum_{e \in P_{k,n}} l(e) + \sum_{e \in C_{k,n}} l(e)$$

$$\geq \left(1 + \frac{2^r}{s + r2^r}\right) \sum_{e \in P_{k,n}} l(e)$$

$$= \frac{s + (r+1)2^r}{s + r2^r} \left(l_{smt}(P_n) - 2^n\right). \tag{2.7}$$

Remember that ρ_k is the infimum of $l_{smt}(P_n)/l_{k-smt}(P_n)$. Then from inequality (2.7) and Lemma 2.1 we deduce

$$\rho_k \leq \frac{l_{smt}(P_n)}{l_{k-smt}(P_n)} \leq \frac{s + r2^r}{s + (r+1)2^r} + \frac{1}{n+1}.$$

Letting $n \to \infty$ implies inequality (2.5) yielding the upper bound for ρ_k.

Now consider the case where a component C has size $k' < k$ (as the component C_3 in Fig.2.4). In this case, it is easy to verify that for $k' = 2^{r'} + s'$, where $0 \leq s' < s^{r'}$, we have

$$\frac{2^{r'}}{r'2^{r'} + s'} \geq \frac{2^r}{s + r \cdot 2^r},$$

from which we deduce that inequality (2.3) also holds for any component of size less than k in a k-Steiner tree. Therefore, the above argument is also applicable to this case. The proof is then finished.   □

2.1.2   Lower Bound for k-Steiner Ratio

The proof for the lower bound of k-Steiner ratio ρ_k applies the same approach as the one used in the lower bound proof in [91]. We first convert a Steiner tree into a weighted regular binary tree. Then by labelling the vertices of this binary tree, we construct $(s + r \cdot 2^r)$ different k-Steiner trees and show that one of these trees has small enough length to enable us to get the lower bound.

Lemma 2.6 *For any regular binary tree T, there exists an one-to-one mapping $f(\cdot)$ from internal vertices to leaves such that for any internal vertex u,*
(i) *$f(u)$ is a descendant of u;*
(ii) *all paths $p(u)$ in T from u to $f(u)$ are edge-disjoint.*

Proof. To prove the lemma, we add the following additional requirement:
(iii) There is a leaf v so that the path from the root to v is edge-disjoint
     from all other paths $p(u)$.
We will prove by mathematical induction on the height of the tree that all three conditions can be satisfied. Fig.2.7 shows a desired mapping where $f(u_i) = v_i$ for each internal vertex u_i.

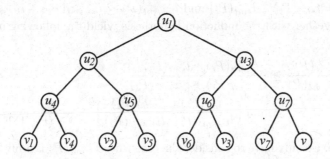

Fig. 2.7   A mapping that satisfies all three conditions.

When the tree has height 1, this is trivially true. Let T be a tree of height $h \geq 2$. Then T has two subtrees T_1 and T_2, each of them has height at most $(h - 1)$ and rooted at the two children of the root of T. By induction hypothesis, there are two mappings $f_1(\cdot)$ and $f_2(\cdot)$ on the internal vertices of T_1 and T_2 satisfying conditions (i) and (ii). In addition, there are two vertices v_1 and v_2 in T_1 and T_2, respectively, that satisfy condition (iii). Now define $f(\cdot)$ as the union of $f_1(\cdot)$ and $f_2(\cdot)$, and define $f(\cdot)$(root of T)= v_1. Clearly, $f(\cdot)$ satisfies conditions (i,ii) and v_2 satisfies condition (iii) for T. This completes the induction process and the proof. □

Theorem 2.2 *For any k with $k = s + 2^r$, where $0 \leq s < 2^r$, the k-Steiner ratio*

$$\rho_k \geq \frac{s + 2^r}{s + (r+1)2^r}. \tag{2.8}$$

Proof. We will prove that for any metric space M and any set P of terminal points in M,

$$\frac{l_{smt}(P)}{l_{k-smt}(P)} \geq \frac{s + 2^r}{s + (r+1)2^r}, \tag{2.9}$$

from which inequality (2.8) follows immediately since ρ_k is the infimum of these ratios.

We will prove inequality (2.9) by mathematical induction on n, the number of terminal points in P. If $n \leq k$, then the inequality is trivially true since $l_{smt}(P) = l_{k-smt}(P)$. For $n > k$, consider an SMT $T_{smt}(P)$ on P. If $T_{smt}(P)$ is not a full Steiner tree, then we can split it at a terminal point into two smaller Steiner trees each with fewer than n terminal points. Denote the terminal points of these two trees by sets P_1 and P_2. Then we have $l_{smt}(P) = l_{smt}(P_1) + l_{smt}(P_2)$ and $l_{k-smt}(P) \leq l_{k-smt}(P_1) + l_{k-smt}(P_2)$, which, together with the induction hypothesis, yield the following inequalities

$$\frac{l_{smt}(P)}{l_{k-smt}(P)} \geq \frac{l_{smt}(P_1) + l_{smt}(P_2)}{l_{k-smt}(P_1) + l_{k-smt}(P_2)}$$

$$\geq \min\left\{\frac{l_{smt}(P_1)}{l_{k-smt}(P_1)}, \frac{l_{smt}(P_2)}{l_{k-smt}(P_2)}\right\} \geq \frac{s + 2^r}{s + (r+1)2^r}.$$

Therefore, we only have to consider the case where $T_{smt}(P)$ is a full Steiner tree, that is, all terminal points are leaves of T.

Now by adding some edges of length zero and Steiner points, we first modify $T_{smt}(P)$ to be a tree where every Steiner point has degree exactly 3. And then we choose a root in the middle of an edge to convert $T_{smt}(P)$ into a weighted regular binary tree $T'_{smt}(P)$, where the weight of each edge in $T'_{smt}(P)$ is the length of the edge in the metric space. Fig.2.8 illustrates this process. A full Steiner tree $T_{smt}(P)$ interconnects seven terminal points $\{t_i | i = 1, 2, \cdots, 7\}$ via two Steiner points s_1 and s_2 as shown in Fig.2.8(a). After modification the new Steiner tree $T'_{smt}(P)$, as shown in Fig.2.8(b), interconnects $\{t_i | i = 1, 2, \cdots, 7\}$ via five Steiner points with three added (black) ones each of degree three where three (dashed) edges of length zero are added.

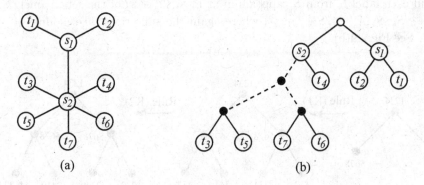

Fig. 2.8   A Steiner tree is modified into a weighted regular binary tree.

Next we label all internal vertices of $T'_{smt}(P)$ with sets of size 2^r chosen from the numbers in $\{1, 2, \cdots, s + r \cdot 2^r\}$. The labelling of a vertex is determined inductively by the labelling of the r vertices above it on a path to the root, its r immediate ancestors. Fig.2.9 illustrates the labelling process for $k = 5$, that is, $r = 2$ and $s = 1$.

We first show how to label the internal vertices on the i-th level in $T'_{smt}(P)$ for $1 \leq i \leq r$. Label the vertex on the first level (the root) with the set $\{1, 2, \cdots, 2^r\}$, and then label the two vertices on the second level with the set $\{1 + 2^r, 2 + 2^r, \cdots, 2 \cdot 2^r\}$; See Fig.2.9(a). In general, label all vertices on the i-th level, for $1 \leq i \leq r$, with the set $\{1 + (i-1)2^r, 2 + (i-1)2^r, \cdots, i \cdot 2^r\}$.

We then describe how to inductively label the internal vertices on the i-th level in T'_{smt} for $i > r$. The labelling is made in such a way that the label sets of up to r consecutive vertices on a path up the tree are disjoint, which is called the *disjointness property*. Clearly, the labelling of the first r levels satisfies the disjointness property.

Assume that the internal vertices on the first i levels have been labelled for $i \geq r$, and that the disjointness property holds up to the i-th level. We label the vertices at the $(i+1)$−st level by the following two rules.

(R1) Let v be a vertex at the $(i+1-r)$-level with label set $S_v = \{l_1, l_2, \cdots, l_{2^r}\}$. Label the j-th descendant of v on the $(i+1)$-st level with the set $S_j = \{l_j, l_{j+1}, \cdots, l_{j-s-1+2^r}\}$, where the subscripts take module of 2^r so that they are all in $\{1, 2, \cdots, 2^r\}$.

Vertex v has at most 2^r descendants on the $(i+1)$−st level, so we need at most the sets $S_1, S_2, \cdots, S_{2^r}$. Each of the sets S_j's has $(2^r - s)$ elements, and each label l_k from S_v appears in at most $(2^r - s)$ of the sets, namely $l_k \in S_k, S_{k-1}, \cdots, S_{k+s+1-2^r}$, where again the subscripts take module of 2^r. See Fig.2.9(b).

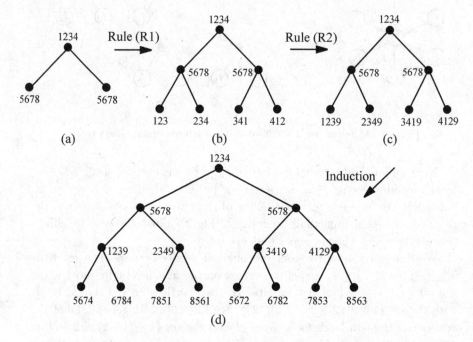

Fig. 2.9   Labelling process that guarantees the disjointness property.

(R2) For each vertex at the $(i+1)$-st level, add to its label set those s labels that are not in the label sets of any of its immediate r ancestors.

By the disjointness property, the r immediate ancestors of a vertex at the $(i+1)$-st level are labelled by r disjoint sets of size 2^r, so there must be exactly s numbers from $\{1, 2, \cdots, s + r \cdot 2^r\}$ unused. Also, the labels added by rule (R2) will be different from the labels added by rule (R1), so all vertices at the $(i+1)$-st level are given exactly 2^r labels by these two rules. See Fig.2.9(c).

The disjointness property now holds up to the $(i+1)$-st level, since a vertex v at the $(i+1)$-st has labels taken from its r-th ancestor's label set, which by the disjointness property at the i-th level are unused by v's $(r-1)$ immediate ancestors, and s other labels, which are also unused by its $(r-1)$ immediate ancestors. By the induction, we can label the entire tree as shown in Fig.2.9(d).

We now can use this labelling to construct $(s + r \cdot 2^r)$ k-Steiner trees. Each label l determines the k-Steiner tree T_l for $l = 1, 2, \cdots, s + r \cdot 2^r$. Each vertex labelled l becomes the component root of a component in T_l. In addition, we always have a component in T_l whose component root is the root of T'_{smt}, even if the root of T'_{smt} is not labelled by l. A component that roots at vertex v is connected, by paths in T'_{smt}, with the first vertices below v that are also labelled by l, which are called *intermediate leaves* of the component. From the intermediate leaf u, the component then follows the path $p(u)$ to the tree leaf $f(u)$ as given by Lemma 2.6. If there are no vertices labelled l on a path below v, then the component extends along that path all the way down to a tree leaf.

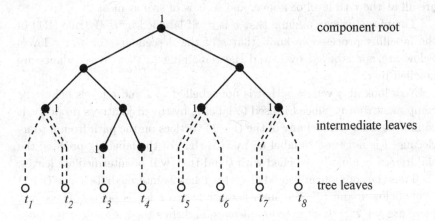

Fig. 2.10 k-Steiner tree T_1 construction based on node labelling given in Fig.2.9(d).

Fig.2.10 illustrates the construction of k-Steiner tree T_l based on the node labelling given in Fig.2.9(d). T_1 has four components, one of them consists of solid edges containing five terminals $\{t_2, t_3, t_4, t_5, t_7\}$ while all other three each consist of dashed edges containing two terminals.

In the following we first verify that T_l is in fact a tree that spans the terminal set P. (Remember that all points in P are the leaves of T'_{smt}.) This can be proved by mathematical induction on the height of the tree T'_{smt}. It is trivially true when the height is zero. Let C be the component at the top of T_l, the component whose component root is the root of T'_{smt}. Then consider a subtree below an intermediate leaf of C. By induction hypothesis, T_l restricted to this subtree is a tree spanning those points in

P which are leaves of the subtree. As component C is itself a tree, it joins one tree leaf from each of the subtrees below its intermediate leaves (and any tree leaves of T'_{smt} that are not below any intermediate leaves) into one large tree.

We next verify that every components in tree T_l has size at most k. If a component stops at a tree leaf before reaching an intermediate leaf, its size will be smaller, so we only look at the maximum-sized components that have all possible intermediate leaves.

Suppose that a component root v is not labelled by l. Then v must be the root of tree T'_{smt} and $l \geq 1 + 2^r$. By the initial labelling, all numbers in $\{1 + 2^r, 2 + 2^r, \cdots, r \cdot 2^r\}$ appears on all vertices of one of the $(r-1)$ levels below the root, and by rule (R2), the remaining s labels appear on all vertices of the r-th level below the root. Thus the intermediate leaves are all at the r-th level or above, and so it is of size at most $2^r \leq k$.

Therefore, we can assume that v is itself labelled by l. By rule (R1) of the labelling process, we know that s of the descendants of v at r levels below are not labelled by l, and the remaining $(2^r - s)$ descendants are labelled by l.

Now look at a vertex w that is not labelled by l and r levels below the component root v. Since l is used to label v, by the disjointness property, it cannot be used to label any of the $(r-1)$ vertices on the path from v to w. Because l is not used to label w, by rule (R2) of the labelling process, the children of w must be labelled with l and they will be intermediate leaves.

Thus the component rooted at vertex v has $2s$ intermediate leaves $(r+1)$ levels below v and $(2^r - s)$ intermediate leaves r levels below. Therefore, there are $s + 2^r = k$ intermediate leaves, and thus the component is of size k. Hence all components are of size at most k, and T_l is a k-Steiner tree on terminal set P.

In the end, let l_l be the sum of the lengths of the paths $p(u)$ from intermediate leaves u in T_l to tree leaves. We will estimate the value of $\sum_{l=1}^{s+r2^r} l_l$, from which we obtain an upper bound on the length of k-SMT. Since each internal vertex u in T'_{smt} is an intermediate leaf in exactly 2^r of the k-Steiner trees, namely T_l for each l in the label set of u, the length of path $p(u)$ will be counted exactly 2^r times in the sum. Moreover these paths are disjoint for different intermediate leaves, the sum of all of the paths in all of the terms of the sum will be at most 2^r times the length of the entire tree T'_{smt}, which is the length of the SMT T_{smt}. Therefore, we have

$$l_1 + l_2 + \cdots + l_{s+r2^r} \leq 2^r l_{smt}(P). \qquad (2.10)$$

Then there must be an index i such that

$$l_i \leq \frac{2^r}{s + r \cdot 2^r} \, l_{smt}(P). \qquad (2.11)$$

Note that the total length of edges in T_i is the length of the components from the component roots to the intermediate leaves plus the length from the intermediate leaves to the tree leaves. The components from the component roots to the intermediate leaves cover $T'_{smt}(P)$ exactly once, so this part has length equal to $l_{smt}(P)$ (the other part has length l_i). This, together with inequality (2.11), leads to

$$l(T_i) = l_{smt}(P) + l_i \leq \left(1 + \frac{2^r}{s + r \cdot 2^r}\right). \qquad (2.12)$$

Since $l_{k-smt}(P) \leq l(T_i)$, we get

$$\frac{l_{smt}(P)}{l_{k-smt}(P)} \geq \frac{l_{smt}(P)}{l_i} \geq \frac{1}{1 + \frac{2^r}{s+r \cdot 2^r}} = \frac{s + r2^r}{s + (r+1)2^r},$$

which implies inequality (2.9). The proof is then finished. $\qquad \square$

2.2 Approximations Better Than Minimum Spanning Tree

In the past decades numerous heuristics for Steiner tree problem have been proposed for terminal points in various metric spaces (and graphs as well). Their superiority over MST-based algorithms[4] [179; 60; 253], which produce $2/\sqrt{3}$-approximation and 2-approximation solutions in the Euclidean plane and graphs, respectively, were claimed and demonstrated only by simulations (no mathematical proof of superiority was ever obtained). In 1990, Zelikovsky [276] finally brought a breakthrough by giving a $\frac{11}{6}$-approximation algorithm for the Steiner tree problem in graphs. After that many better approximation algorithms were proposed, some of them are listed in the following tables.

---

[4]This kind of algorithms was already known in 1968 [114; 120].

Approximation ratios	References
$11/6 \approx 1.833$	Zelikovsky [276]
1.746	Berman and Ramaiyer [29]
1.734	Borchers and Du [39]
$1 + \ln 2 \approx 1.693$	Zelikovsky [277]
$5/3 \approx 1.666$	Prömel and Steger [224]
1.644	Karpinski and Zelikovsky [165]
1.598	Hougardy and Prömel [136]
$1 + \frac{1}{2}\ln 3 \approx 1.55$	Robins and Zelikovsky [239]

Table 2.1 Better algorithms for Steiner tree problem in graphs.

Approximation ratios	References
1.1546	Du et al. [91]
$1 + \ln(2/\sqrt{3}) \approx 1.1438$	Zelikovsky [277]

Table 2.2 Better algorithms for Steiner tree problem in Euclidean plane.

Approximation ratios	References
$11/8 \approx 1.375$	Zelikovsky [275]
1.323	Berman and Ramaiyer [29]
1.267	Karpinski and Zelikovsky [165]

Table 2.3 Better algorithms for Steiner tree problem in rectilinear plane.

All these *better approximation algorithms*, except for the randomized algorithm of Prömel and Steger [224], are based on greedy strategies originally due to Zelikovsky [276]. In particular, Du et al. [91] generalized Zelikovsky's idea by extending 3-Steiner trees to k-Steiner trees. They obtained a lower bound for the k-Steiner ratio in any metric space which achieves one as k goes to infinity. Thus in any metric space with the Steiner ratio less than one, there exists a k-Steiner ratio bigger than the Steiner ratio. Thus, they proved that a better approximation exists in any metric space satisfying the following conditions:

(C1) The Steiner ratio is smaller than one.

(C2) The Steiner minimum tree on any fixed number of points can be computed in polynomial time.

Note that the Euclidean space (particularly the Euclidean plane) satisfies the above two conditions.

In the next three subsections we will study some of the approaches proposed for designing approximation algorithms with performance ratios better than MST-based algorithms.

2.2.1 Greedy Strategy

Recall that a k-Steiner tree is a tree with a Steiner topology and each of its full components contains at most k terminals, and the k-Steiner minimum tree is a k-Steiner tree of the shortest length, denoted it by k-SMT and its length by $l_{k-smt}(P)$ for given terminal set P. As we have mentioned in the beginning of this chapter, no polynomial time algorithm is known for computing k-SMT for $k \geq 3$, so one cannot use a k-SMT as a better approximation for SMT. The key idea of Zelikovsky's algorithm [276] is to use 3-Steiner trees instead.

Zelikovsky's algorithm [276] essentially works as follows: Start from an MST and at each iteration choose a Steiner vertex such that using this Steiner vertex to connect three terminal vertices could replace two edges in the MST and such a replacement achieves the maximum length reduction among all possible replacements. Although Zelikovsky started from a point different from Chang's [44; 45], the two approximations algorithms are very similar. Indeed, they both start from an MST and improve it step by step by using a greedy principal to choose a Steiner vertex to connect a triple of vertices. The difference lies in the way of choosing triples of vertices, Chang's algorithm may choose some Steiner vertices while Zelikovsky's algorithm only chooses terminal vertices. This difference makes Chang's algorithm hard to be analyzed. However, it is an interesting problem to know which algorithm yields a better approximation solution. In the following we will present the general version of Zelikovsky's algorithm [91; 83].

The greedy strategy for computing an MST for given terminal set can be extended to the case where some of the terminal points are actually point-sets. For a given set P of n terminal points and subset $P_i \subset P$, $i = 1, 2, \cdots, m$, let $P' = \bigcup_{i=1}^{m} P_i$. An MST $T(P; P_1, P_2, \cdots, P_m)$ is a set of $(m + |P| - |P'| - 1)$ edges connecting the m point-sets and the $(|P| - |P'|)$ singletons where a point-set is considered connected if any of its points is connected. In fact, P_i may be considered as a *contracted* terminal point of all terminal points in P_i. The original MST $T_{mst}(P)$ can be considered as the special case of $m = 0$ (no contraction).

Fig.2.11(a) shows an MST for nine terminals in the Euclidean plane, and

(b) shows an MST $T_{mst}(P; P_1, P_2, \cdots, P_m)$ is a set of $(m + |P| - |P'| - 1)$ edges connecting the m point-sets and the $(|P| - |P'|)$ singletons where $m=2$ and $P_1 = \{t_1, t_2, t_4\}$ and $P_2 = \{t_3, t_6, t_7\}$.

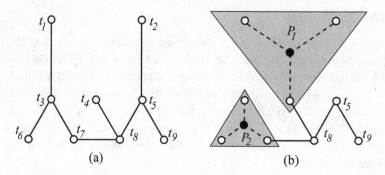

(a)   (b)

Fig. 2.11   (a) An MST for nine terminals, and (b) an MST for two point-sets and three singletons.

To assure that the connected graph is a tree, it is required that no set (including singleton) is contained in another set, i.e., $|P_i \cap P_j| \leq 1$, and that there does not exist a cyclic sequence of subscripts such that $|P_i \cap P_j| = 1$ for all i and j that are consecutive in the cyclic sequence. Such a set of terminal-sets $\{P_i\}$ is called a *cycle-free set*. In Fig.2.11, $\{P_1, P_2\}$ is cycle-free, but P_2 and $\{t_3, t_4, t_7\}$ do not make a cycle-free set.

Let S_i denote the set of terminals in a full SMT $T_{smt}(S_i)$. Then define the *gain* of cycle-free set $\{P_i\}$ as

$$gain(P; P_1, P_2, \cdots, P_m)$$
$$\equiv l_{mst}(P) - l(T_{mst}(P; P_1, P_2, \cdots, P_m)) - \sum_{i=1}^{m} l(T_{smt}(S_i)). \quad (2.13)$$

When $gain(;) > 0$, it can be considered as the saving of length over an MST obtained by using $T_{smt}(S_i)$ to connect points in P_i. For the example of Fig.2.11, the gain is the difference between the length of MST of Fig.2.11(a) and the length of Steiner tree of Fig.2.11(b).

Du and Zhang [90], extending the idea of Zelikovsky [276] for computing an approximate solution of 3-SMT in graphs, gave the following greedy algorithm for computing an approximate solution of k-SMT in an arbitrary metric space M. It computes an MST of terminal-set P in the *distance* graph $G_D(P)$[5] as an initial solution, then it improves the current solution by using i-trees with the maximal positive gain, where for $3 \leq i \leq k$, an i-tree is a full Steiner tree with i terminals.

Algorithm 2.1 *Greedy Algorithm*

Step 1 For each $i = 1, 2, \cdots, k$, compute an i-SMT for each $\binom{n}{i}$ subsets of i terminals. Store the obtained i-SMT in list L if it is a full SMT.

Step 2 For $j = 1, 2, \cdots, n - 1$, select T_j from L such that $T_1, \cdots, T_j$ is cycle-free and $gain(P; P_1, \cdots, P_j)$ maximizes $gain(P; P_1, \cdots, P_j^*)$ over all $T_{mst}(S_j)$ from L, where P_j is the terminal-set of T_j. Delete P_j^* from L if $P_1, P_2, \cdots, P_j^*$ is not cycle-free.

Step 3 Stop when L is empty and return $T_{k-G} := T_{mst}(P; P_1, \cdots, P_{j'})$.

Since the edge between any pair of two terminals is in L, the returned solution of Algorithm 2.1 is always a Steiner tree for terminal set P. However, the efficiency of Algorithm 2.1 depends on the existence of an efficient construction of minimum i-trees in metric space M. Given that there exists such a procedure which requires time of $f(k)$, then Step 1 takes time of $O(kn^k f(k))$ time to generate the $O(kn^k)$ SMTs where $n = |P|$. In Step 2, it takes time of $O(n \log n)$ to generate the MST $T_{mst}(P; P_1, P_2, \cdots, P_j^*)$ for each SMT $T_{mst}(S_j) \in L$ and there are $O(kn^k)$ SMTs in L. As Step 2 is repeated $O(n)$ times, Algorithm 2.1 requires total time of $O(kn^k(n^2 \log n + f(k)))$.

For a cycle-free set $\{P_1, P_2, \cdots, P_m\}$, we define

$$gain(P; P_{i+1}, \cdots, P_m | P_1, \cdots, P_i)$$
$$\equiv gain(P; P_1, \cdots, P_m) - gain(P; P_1, \cdots, P_i) \qquad (2.14)$$
$$= l(T_{mst}(P; P_1, \cdots, P_i)) - l(T_{mst}(P; P_1, \cdots, P_m)) - \sum_{j=i+1}^{m} l(T_{smt}(S_j)).$$

For an edge $e = (u, v)$ between two points u and v, let $P(e) = \{u, v\}$.

Lemma 2.7 *Let $\{P_j\}$ be a cycle-free set. For each edge e, either*

$$gain(P; \{P_j\} \mid P(e)) = gain(P; \{P_j\}) \qquad (2.15)$$

or there exists a $P_e \in \{P_j\}$ such that

$$gain(P; \{P_j\} \setminus P_e \mid P(e)) \geq gain(P; \{P_j\} \setminus P_e \mid P_e) \qquad (2.16)$$

---

[5]Given a graph $G(V, E)$ with $P \subset V$, the distance graph $G_D(P)$ is a complete graph of P where the distance between two vertices u and v in P is the length of the shortest path between u and v in the original graph $G(V, E)$.

Proof. Consider an edge e. Clearly, $T_{mst}(P) \cup \{e\}$ contains a circuit, and $gain(P; p(e))$ is simply the length of the longest edge e^* in this circuit. Let U and V be the two components of $T_{mst}(P)$ after e^* is deleted from $T_{mst}(P)$. Since the two endpoints of e^* must be connected in $T_{mst}(P; \{P_j\})$, one of the following two cases must occur.

Case 1. $T_{mst}(P; \{P_j\})$ contains an edge e' connecting U and V. Then by definitions (2.13-14), we have

$$gain(P; P(e) \mid \{P_j\}) = gain(P; P(e)). \tag{2.17}$$

From equality (2.17) we deduce

$$
\begin{aligned}
gain(P; \{P_j\} \mid P(e)) &= gain(P; \{P_j\}, P(e)) - gain(P; P(e)) \\
&= gain(P; \{P_j\}) + gain(P; P(e) \mid \{P_j\}) - gain(P; P(e)) \\
&= gain(P; \{P_j\}),
\end{aligned}
$$

which yields equality (2.15).

Case 2. There exists a $P_e \in \{P_j\}$ which contains one point $u \in V$ and one point $v \in V$. Let P'_e denote the set of terminal points in $T_e \setminus \{u, v\}$. Then we have

$$
\begin{aligned}
gain(P; \{P_j\} \setminus P_e \mid P_e) &= gain(P; \{P_j\}) - gain(P; P_e) \\
&= gain(P; \{P_j\} \setminus P_e, \{u, v\}) + gain(P; P'_e \mid \{P_j\}, \{u, v\}) \\
&\quad - gain(P; \{u, v\}) - gain(P; P'_e \mid \{u, v\}) \\
&\leq gain(P; \{P_j\} \setminus P_e, \{u, v\}) - gain(P; \{u, v\}) \\
&= gain(P; \{P_j\} \setminus P_e \mid \{u, v\}) \\
&= gain(P; \{P_j\} \setminus P_e \mid P(e)),
\end{aligned}
$$

which yields equality (2.16). The proof is then finished. $\qquad\square$

By processing edge e at a time, we can deduce the following corollary from the above lemma.

Corollary 2.1 *For any cycle-free set* $\{P_1, P_2, \cdots, P_m\}$,

$$gain\left(P; \{P_j\} \setminus \bigcup_e P_e \;\Big|\; \bigcup_e P_e\right) \leq gain\left(P; \{P_j\} \setminus \bigcup_e P_e \;\Big|\; \bigcup_e P(e)\right).$$

Theorem 2.3 *Let* $l_{k-G}(P)$ *be the length of Steiner tree* T_G *for* P *returned by greedy algorithm 2.1. Then for any* $k \geq 3$

$$l_{k-G}(P) \leq \frac{k-2}{k-1}\, l_{mst}(P) + \frac{1}{k-1}\, l_{k-smt}(P). \tag{2.18}$$

Proof. We will prove the theorem by mathematical induction on m, where m is the number of P_i in the list L that satisfies $|P_i| \geq 3$. Let $L_k = \bigcup_{i=1}^{j} T_{k-smt}(S_i)$ where each $T_{k-smt}(S_i)$ is a full SMT, and let $\{T_{k-smt}(S_1), T_{k-smt}(S_2), \cdots, T_{k-smt}(S_m)\}$ be the full SMTs chosen by the algorithm. If $m = 0$, then each $T_{k-smt}(S_i)$ must be an edge for otherwise $gain(P; P_1) \geq gain(P; P_i^*) > 0$ and $m \geq 1$. Thus $gain(P; P_1, \cdots, P_m) = 0 = gain(P; P_i^*)$, and inequality (2.18) is trivially true. So, we assume $m \geq 1$.

Now subtracting $l_{mst}(P) = l_{2-G}(P)$ from both sides of inequality (2.16), we obtain

$$gain(P; P_1, P_2, \cdots, P_m) \geq \frac{1}{k-1} gain(P; P_1^*, P_2^*, \cdots, P_j^*). \quad (2.19)$$

Suppose that T_1 contains t_1 terminal points. Then $T_{mst}(P_1)$ has $(t_1 - 1)$ edges. Without loss of generality, assume that

$$\bigcup_{e \in T_{mst}(P_1)} P_e = \{T_1^*, T_2^*, \cdots, T_t^*\}, \quad 1 \leq t \leq t_1 - 1.$$

Then by the way of choosing T_1 in the algorithm, we have

$$(k-1)gain(P; P_1) \geq (t_1 - 1)gain(P; P_1)$$
$$\geq \sum_{i=1}^{t} gain(P; P_i^*) \geq gain(P; P_1^*, P_2^*, \cdots, P_t^*).$$

Note that

$$gain(P; P_1, \cdots, P_m) = gain(P; P_1) + gain(P; P_2, \cdots, P_m | P_1) \quad \text{and}$$

$$gain(P; P_1^*, \cdots, P_j^*) = gain(P; P_1^*, \cdots, P_t^*) + gain(P; P_{t+1}^*, \cdots, P_j^* | P_1^*, \cdots, P_t^*).$$

Hence to prove inequality (2.19), it suffices to show the following inequality

$$(k-1)gain(P; P_2, \cdots, P_m | P_1) \geq gain(P; P_{t+1}^*, \cdots, P_j^* | P_1^*, \cdots, P_t^*). \quad (2.20)$$

Let $T_{k-smt}(P; P_2', \cdots, P_i' | P_1)$ denote a k-SMT given that P_1 is chosen. Then we have

$$(k-1)gain(P; P_2, \cdots, P_m | P_1) \geq gain(P; P_2', \cdots, P_i' | P_1)$$
$$\geq gain(P; P_{t+1}^*, \cdots, P_j^* | P_1), \quad (2.21)$$

where the first inequality follows from the inductive assumption and the second from the minimality of $T_{k-smt}(P; P'_2, \cdots, P'_m \mid P_1)$.

By Corollary 2.1, we have

$$gain\big(P; P^*_{t+1}, \cdots, P^*_j\big) = gain\Big(P; P^*_{t+1}, \cdots, P^*_j \mid \bigcup_{e \in T_{mst}(P_1)} P_e\Big)$$

$$\geq gain\big(P; P^*_{t+1}, \cdots, P^*_j \mid P^*_1, \cdots, P^*_t\big),$$

which, together with inequality (2.21), implies inequality (2.20). The proof is then finished.   □

From Theorem 2.3 along with definition (2.1), we can obtain, by dividing $l_{smt}(P)$ on both sides of (2.18), a upper bound on the approximation performance ratio α_{k_G} of algorithm 2.1.

Corollary 2.2   *For any $k \geq 3$, $\alpha_{k-G} \leq \frac{k-2}{k-1}\rho_2^{-1} + \frac{1}{k-1}\rho_k^{-1}$.*

It follows immediately from the above corollary that, if

$$\rho_k > \rho_2, \quad \text{for some } k \geq 3, \tag{2.22}$$

then the greedy algorithm 2.1 has approximation performance ratio $\alpha_{k-G} < \rho_2^{-1}$, that is, Algorithm 2.1 (with such ks) is a better approximation algorithm. In the following we will prove inequality (2.23) is true.

Theorem 2.4   *For any $k \geq 3$, $\alpha_{k-G} \leq 1 + 1/\kappa$, where $\kappa = \lfloor \log k \rfloor$.*

Proof.   The theorem is trivially true for $n \leq k$. The general case of $n > k$ is proved by mathematical induction on n. Note that it suffices to prove the theorem by considering a full SMT T_{smt}. T_{smt} can be turned into a rooted binary tree T_r by choosing an arbitrary edge and setting its middle point r as the root for T_r. The root is considered at the first level of T_r, and a vertex v is at the i-th level if the path from the root to the vertex contains i vertices (including both r and v). By Lemma 2.6, let $f(\cdot)$ be the desired one-to-one mapping. Then length of an edge is its length in metric space M.

Let I_l be the set of interval nodes at the l-th level of T_r. Let $U_d = \bigcup_{i=d(mod\,\kappa)} I_i$. Then U_d for $d = 1, 2, \cdots, \kappa$ are disjoint. Let $p(u)$ denote the path from an interval node u to $f(u)$. Let $l_d = \sum_{u \in U_d} l(p(u))$ for each d. Then by Lemma 2.6(ii), we have $l(T_r) \geq \sum_{d=1}^{\kappa} L_d$. Thus there exists an index d' such that $l_{d'} \leq l(T_r)/\kappa$. Decomposing T_r at nodes of $U_{d'}$ yields a collection of rooted binary trees each rooted at a node of $U_{d'}$ and having leaves either in $U_{d'}$ or in the leaf-set of T_r. Denote such a tree by $T_{r'}$ if it is rooted at node r'. Clearly, each $T_{r'}$ has at most k leaves.

Now for each leave u of $T_{r'}$ which is an internal node of T_r, connect u to $f(u)$ with an edge. Thus $T_{r'}$ is transformed to a Steiner tree $\overline{T}_{r'}$ for at most k terminals. Clearly, the union of T_r and $T_{r'}$ is connected for $r' \in U_{d'}$, hence it is a k-Steiner tree. Moreover, we have

$$l(T_{k-G}) \leq l(\overline{T}_r) + \sum_{r' \in U_{d'}} l(\overline{T}_{r'}) \leq l(T_r) + l_d \leq \frac{\kappa}{\kappa + 1} l(T_{smt}),$$

which proves the theorem. The proof is then finished. $\qquad\square$

Corollary 2.3 *For every metric space M with $l_{smt}(P) \neq l_{mst}(P)$ for all $P \in M$, the greedy algorithm 2.1 returns a better approximate solution for the Steiner tree problem in M for large enough k.*

Corollary 2.4 *For the Steiner tree problem in Euclidean plane, the greedy algorithm 2.1 with $k = 128$ has approximation performance ratio $\alpha_{k-G} < 2/\sqrt{3}$.*

For the case of $k = 3$, Zelikovsky [276] obtained the lower bound on the k-Steiner ratio in general graphs.

Theorem 2.5 *For any graph G, $\rho_3 > 3/5$.*

Proof. Let T_r be the tree defined in the proof of Theorem 2.4. Let T_u be a subtree of T_r rooted at node u and I_u denote the set of terminals of T_u. For each $v \in I_r$, let $c(v)$ denote a leaf closest to v, and let $p(v)$ denote the path from v to $c(v)$. Then $c(v')$ may be equal to $c(v)$ when v' is a child-node of v. Suppose that u' and u'' are the two child-nodes of u. Then $p(c(u), c(u'))$ is the unique simple path connecting $c(u')$ and $c(u)$ in T_r. Let $P_u = \bigcup_{v \in I_u} p(c(u), c(v))$. Then P_u is a tree interconnecting I_u. It can be easily verified that $l(P_u) \leq 2 l(T_r) - 2 l(p(c(r), c(r)))$. Let $\delta(u) = l(P_u) - l(T_u)$. Then

$$\delta(u) = \begin{cases} 0, & \text{if } u \text{ is a leaf;} \\ \delta(v') + p(v') + \delta(w') + p(w''), & \text{otherwise.} \end{cases}$$

From the above we obtain that $\delta(u) = \sum_{v \in I_u \setminus \{u\}} p(v)$.

Let $e = (u, u')$ and u' has two child-nodes x and y. Let $P_e = \{c(u''), c(x), c(y)\}$, and let T_e be an SMT on P_e. Since $p(u) \cap p(u') = p(u')$, we have

$$l(T_e) \leq l(p(c(u), c(u'))) + l(p(c(u), c(u''))) - p(u'). \qquad (2.23)$$

Let $T(P_{e_1}, \cdots, P_{e_t})$ denote a tree obtained from P_r by substituting T_{e_i} for $p(c(u), c(u_i))$ and $p(c(u), c(v_i))$, $i = 1, 2, \cdots, t$. Then

$$l(P_r(P_e)) \leq l(P_r) - l(p(u')). \tag{2.24}$$

Furthermore, if $\{P_{e_1}, \cdots, P_{e_t}\}$ is a cycle-free set, then

$$l(P_r(P_{e_1}, \cdots, P_{e_t})) \leq l(P(r)) - \sum_{i=1}^{t} l(p(v_i)). \tag{2.25}$$

It can be verified that $\{P_{e_1}, \cdots, P_{e_t}\}$ is a cycle-free set if and only if $e_i \cap e_j = \emptyset$ for all $1 \leq i \leq j \leq t$. Since the edge-set E of a rooted binary tree can be easily partitioned into three disjointed subsets such that edges in each subset do not intersect, and one subset, say E', must satisfy

$$\sum_{e_i \in E'} l(p(v_i)) \geq \frac{1}{3} \sum_{e_i \in E} l(p(v_i)).$$

Thus we obtain

$$l(P_r(\{P_{e_i} \mid e_i \in E'\})) \leq l(P_r) - \frac{1}{3} \sum_{e_i \in E} l(p(v_i)) \tag{2.26}$$

$$\leq l(P_r) - \frac{1}{3} \delta(r) \tag{2.27}$$

$$\leq l(P_r) - \frac{1}{3} \big(l(P_r) - l(T_r) \big) \tag{2.28}$$

$$= \frac{2}{3} l(P_r) + \frac{1}{3} l(T_r) \tag{2.29}$$

$$\leq \frac{3}{5} l(T_r) - \frac{3}{4} l(P_r). \tag{2.30}$$

From the above it immediately follows that

$$\rho_3 \geq \frac{l(T_r)}{P_r(\{P_{e_i} \mid e_i \in E'\})} > \frac{3}{5},$$

which proves the theorem. The proof is then finished. $\qquad \square$

Recall that the Steiner ratio in graphs is bounded by $\frac{n}{2(n-1)}$ (refer to [163]), i.e., $\rho_2 \leq \frac{n}{2(n-1)}$, where n is the number of vertices in given graph. Therefore, for $n > 12$, from Theorem 2.5 and Corollary 2.2, we immediately get the following theorem [276].

Corollary 2.5 *For the Steiner tree problem in graphs, the greedy algorithm 2.1 with $k = 3$ has approximation performance ratio $\alpha_{3-G} \leq 11/6$.*

For the rectilinear plane, Zelikovsky [275] proved the following theorem giving a better lower bound on ρ_3.

Theorem 2.6 *For the rectilinear plane, $\rho_3 \geq 4/5$.*

Proof. To prove the theorem, we again just need to consider a full SMT T_{smt}. By the structural theorem of SMTs in the rectilinear plane [142], we consider the following two cases.

Case 1. All Steiner points lie on one line, say, $x = 0$. Each terminal point is connected to the line of $x = 0$ by a horizontal edge which rotates to the left and to the right as shown in Fig.2.12(a). The first and the last such edges can have length of zero. Let the i-th terminal point t_i have the Cartesian coordinates (x_i, y_i), $i = 0, 1, \cdots, n - 1$, where $y_0 \leq y_1 \leq \cdots \leq y_{n-1}$. Then

$$l(T_{smt}) = (y_{n-1} - y_0) + \sum_{i=0}^{n-1} x_i. \tag{2.31}$$

Now let $P_i = \{t_{2i}, t_{2i+1}, t_{2i+2}\}$, for $i = 0, 1, \cdots, n'$ where $n' = \lceil n/2 \rceil - 2$. In addition, let T_i denote an SMT for P_i. Then

$$l(T_i) = y_{2i+2} - y_{2i} + x_{2i+1} + \max\{x_{2i}, x_{2i+2}\}. \tag{2.32}$$

It is also easily to verify that

$$l(T(P; P_0, \cdots, P_{n'})) = y_{n-1} - y_{n-2} + x_{n-1} + x_{n-2}.$$

Since $\{P_0, \cdots, P_{n'}\}$ is a cycle-free set, the following tree

$$T' \equiv T(P; P_0, \cdots, P_{n'}) \bigcup (T_0 \cup T_1 \cup \cdots \cup T_{n'}) \tag{2.33}$$

is a 3-Steiner tree. Let $n^* = \lfloor n/2 \rfloor - 1$ and $x_i = 0$ for $i \geq n$. Then by equality (2.33), we have

$$l(T') = y_{n-1} - y_0 + \sum_{i=0}^{n^*} x_{2i+1} + \sum_{i=0}^{n^*} \max\{x_{2i}, x_{2i+2}\}.$$

We now consider a second cycle-free set which is obtained from $\{P_0, \cdots, P_{n'}\}$ by substituting $P'_{i-1} \neq \{t_{2i-2}, t_{2i-1}, t_{2i+1}\}$ for P_{i-1} if $x_{2i-1} > x_{2i+1}$. Note that

$$l(T'_{i+1}) \leq y_{2i+1} - y_{2i-2} + x_{2i-1} + x_{2i-2}.$$

By further considering two more cycle-free sets obtained similarly as the second set, it could be shown that the four trees together use up at most five copies of each edge of T_{smt}. Therefore, one of the four trees, say T', has a length at most $\frac{5}{4} l(T_{smt})$. The theorem then follows from $l(T_{3-smt}) \leq l(T')$.

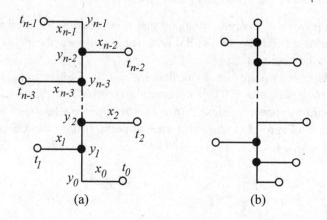

(a)   (b)

Fig. 2.12   (a) Case 1, and (b) Case 2.

Case 2. This case differs from Case 1 in that n must be even and the first Steiner point is connected to the second through a corner as shown in Fig.2.12(b). The theorem can be proved in a similar way.   □

As Hwang [142] proved that $\rho_2 = 2/3$ in the rectilinear plane, which is smaller than 8/11, the following corollary immediately follows from Theorem 2.6 and Corollary 2.2.

Corollary 2.6   *In the rectilinear plane, the greedy algorithm 2.1 with $k = 3$ has approximation performance ratio $\alpha_{3-G} \leq 11/8$.*

2.2.2   Variable Metric Method

Berman and Ramaiyer [29] employed a different idea to generalize Zelikovsky's result by introducing a *variable metric method*. It consists of two main phases, the *evaluation phase* and the *construction phase*.

In the evaluation phase, as the greedy algorithm 2.1, it initially computes an MST $T_{mst}(P)$ of terminal set P in the *distance graph* $G_D(P)$, then it computes every possible i-tree for $3 \leq i \leq k$ and determine those with positive gains. However, it doesn't decide if an i-tree should be used at this phase by the greedy strategy (as the greedy algorithm 2.1); Such a decision will be made in the next phase after all possible alternatives have been considered. In the construction phase, it maintains a Steiner tree $T_k(P)$ (initially it equals $T_{mst}(P)$ and at the end of this phase it gives the final solution). It considers the entries stored in the stacks in the reverse order in which they were pushed onto them in the evaluation phase (the algorithm terminates when the stack is empty).

Algorithm 2.2 *Variable Metric Algorithm*

Step 1 Process all *i*-trees, $3 \leq i \leq k$, sequentially as follows: for each *i*-tree T_i with positive saving in the current graph, put T_i in a stack and if two leaves u and v of T_i are connected by a path $p(u,v)$ in an MST without passing any other leaf of T_i, then put an edge between u and v with a weight equal to the length of the longest edge in $p(u,v)$ minus the savings of T_i.

Step 2 Repeatedly pop *i*-trees from the stack remodifying the original MST for all terminal points and keeping only *i*-trees with the current positive saving. Adding weighted edges to a point set would change the metric on the point set.

Let S_e be an arbitrary set of weighted edges such that adding them to the input metric space makes all *i*-trees for $3 \leq i \leq k$ have nonpositive saving in the resulting metric space M_{E_e}. Denote by $t_k(P)$ a supremum of the length of an MST for the terminal set P in metric space M_{E_e} over all such E_es. Then Berman and Ramaiyer proved the following theorem [29]

Theorem 2.7 *Algorithm 2.2 produces a k-Steiner tree with total length at most*

$$t_2(P) - \sum_{i=3}^{k} \frac{t_{i-1}(P) - t_i(P)}{i-1} = \frac{t_2(P)}{2} + \sum_{i=3}^{k-1} \frac{t_i(P)}{(i-1)i} + \frac{t_k(P)}{k-1}.$$

The following corollary follows immediately from the above theorem and the fact that $t_k(P) \leq \rho_k^{-1} l_{smt}(P)$.

Corollary 2.7 *The approximation performance ratio of variable metric algorithm 2.2 is at most*

$$\rho_2^{-1} - \sum_{i=3}^{k} \frac{\rho_{i-1}^{-1} - \rho_i^{-1}}{i-1}.$$

By *k*-Steiner ratio ρ_k (2.2), we know that the approximation ratio of the variable metric algorithm converges to 1.734 for sufficiently large *k*. Berman and Ramaiyer [29] also proved the following theorem giving a general result of Theorem 2.6.

Theorem 2.8   *For the rectilinear plane, $\rho_k \geq (2k-2)/(2k-1)$ for $k \geq 2$.*

Combining Theorems 2.3 and 2.8, they deduce that the performance ratio of Algorithm 2.2 with $k = 4$ is bounded by $97/72 < 3/2$, which is the performance ratio of MST-based algorithm in the rectilinear plane. In fact, it follows from Corollary 2.7 that the performance ratio of Algorithm 2.2 converges to a value close to 1.323 for some k.

Based on the above observation, we may have the following questions. Could we find another way to vary metric for a better bound? Could we discard the greedy idea and design a better approximation algorithm with only variable metric idea? Clearly, answering these questions requires deeper understanding of the variable metric method.

2.2.3   Relative Greedy Strategy

The *relative greedy algorithm* was proposed by Zelikovsky [277]. It adopts the same framework as the greedy algorithm 2.1. However, when choosing an i-SMT T_j for $i \leq k$, instead of maximizing the absolute gain it can achieve, it uses the following selection rule to maximize the *relative gain* of T_j defined as follows:

$$\frac{gain\left(P; P_1, P_2, \cdots, P_j, P_{j+1}\right) - gain\left(P; P_1, P_2, \cdots, P_j\right)}{l\left(T_{smt}(S_{j+1})\right)}$$

$$= \frac{l_{mst}\left(P; P_1, \cdots, P_j\right) - l_{mst}\left(P; P_1, \cdots, P_{j+1}\right) - l_{smt}\left(S_{j+1}\right)}{l_{smt}\left(S_{j+1}\right)}.$$

In other words, it relates the length of T_j to its benefit. Zelikovsky [277] proved the following theorem.

Theorem 2.9   *The relative greedy algorithm for any metric space has approximation performance ratio upper bounded by $(1+\ln \rho_2^{-1})$ for sufficiently large k.*

Recall that in graphs $\rho_2 = 1/2$ and in the Euclidean plane $\rho_2 = \sqrt{3}/2$. The following corollary immediately follows from the above theorem.

Corollary 2.8   *The relative greedy algorithm has approximation performance ratios $(1 + \ln 2)$ and $(1 + \ln \frac{2}{\sqrt{3}})$ in graphs and the Euclidean plane, respectively.*

2.2.4 *Preprocessing Technique*

Karpinski and Zelikovsky [165] observed that every time the above mentioned algorithms accept an *i*-tree, they also accept all its Steiner points. This may increase the cost of the minimum solution achievable at the current iteration. So they propose an approach to minimize this possible increase.

Let T_i be an *i*-tree and S_i be the set of its Steiner points. A subgraph F_i of T_i is called a *spanning forest* if for any $s \in S_i$, there is a path in F_i connecting s with terminal set P. The minimum spanning forest in T_i is called the *loss* of T_i, which is denoted by $L(T_i)$. (Refer to the next subsection for more detailed explanation.) For $\alpha \geq 0$, the α-*relative gain* of T_i is defined as follows:

$$gain(\alpha, T_i) \equiv gain(P; T_i) - \alpha \cdot l(L(T_i)).$$

Their algorithm [165] is similar to the algorithm of the variable metric algorithm [29]. But it uses the relative gain instead of the savings as a greedy function. Their method is to first use this procedure to choose some Steiner points and then run a better approximation algorithm on the union of the set of terminal points and the set of chosen Steiner points. Using this preprocessing technique, Karpinski and Zelikovsky [165] improved the variable metric algorithm and the relative greedy algorithm. Indeed, they proved the following theorem.

Theorem 2.10 *The preprocessing algorithm has performance ratios at most 1.644 and 19/15 for Steiner tree problems in graphs and in the rectilinear plane, respectively.*

Hougardy and Prömmel [136] realized that the idea behind the concept of loss is that we intend to choose only those Steiner points that an SMT contains. Although no approximation algorithm could achieve this goal, one attempts to avoid some bad choices by penalizing the choice of Steiner points that require long edges to connect them to a terminal point. They designed a sequence of algorithms each takes the returned solution of its predecessor as its input. When making greedy selections, all algorithms in the sequence use the weighted sum of the length and the loss of a Steiner tree, but with different weights in each round. By choosing the weights appropriately they obtained an approximation algorithm with performance ratio of 1.598, which beats the ratio of 1.644 due to Karpinski and Zelikovsky [165]. In fact, with the help of a computer program they estimate that the limit of performance ratio behaves roughly as $(1.588 + 0.114/k)$.

2.2.5   *Loss-Contracting Algorithm*

Robins and Zelikovsky [238] noticed that except the variable metric method, all above mentioned algorithms choose appropriate i-trees and then contract them in order to produce the final solution with them. Moreover, once we have decided to accept an i-tree, we will not exchange it with a better i-tree that has at least two common terminals with it (since accepting both will not make a cycle-free set).

They proposed the *loss-contracting algorithm*, which can be considered as a (relative) greedy algorithm. It does not contract the selected full components entirely (unlike the relative greedy algorithm), but only their losses. In fact, it tries to contract as little as possible while ensuring that

    (i) a selected full component may still be put into the solution but

    (ii) not many other full components would be rejected.

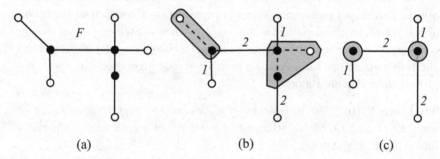

          (a)   (b)   (c)

Fig. 2.13   (a) A full component F, (b) the loss of F, and (c) the terminal-spanning tree.

Inspired by the above idea, Robins and Zelikovsky [238] introduced two concepts, terminal-spanning trees and relative length savings. A *terminal-spanning tree* is a Steiner tree that does not contain any Steiner points in current contracted graph (i.e., each full component of the tree is a single edge). The *relative cost savings* of a full component F is defined as the ratio of how much F saves the length of the current terminal-spanning tree over the cost connecting its Steiner points to terminals. Fig.2.13(a) shows a full component F, Fig.2.13(b) shows the loss $L(F)$ of F consisting of dashed edges and contracted components of $L(F)$. Fig.2.13(c) shows the corresponding terminal-spanning tree with the contracted $L(F)$.

The length savings of an arbitrary graph H with respect to a terminal-spanning tree T is the difference between the length of T and the length of Steiner tree obtained by using (possibly) some edges of H. More formally, let T_H^* be the graph of minimum length in $T \cup H$ that contains H and interconnects all terminals in P. The *gain* of H with respect to T is defined as

$$gain(T; H) \equiv l(T) - l(T_H^*). \tag{2.34}$$

Observe the above definition is a generalization of definition (2.13), where T is an MST of P and H a Steiner tree. In addition, we have $gain(T; H) \leq l(T) - l(T_{mst}(T \cup H))$ since the length of T_H^* is not less than that of the MST of $T \cup H$.

The basic idea of the loss-contracting algorithm is to contract the loss of full component F after F is accepted, that is, to collapse each connected component of loss $L(F)$ into a single node. In Fig.2.13(b), F has two connected components, one of them contains one terminal node and one Steiner node while the other contains one terminal node and two Steiner nodes. In Fig.2.13(c) they each are contracted to a single node. Note that a loss-contracted full component $C(F)$ is a terminal-spanning tree $T(F)$ for terminals in F and the length of any edge in $C(F)$ is the same as that of corresponding edge in F. Thus $l(F) - loss(F) = l(C(F))$.

Algorithm 2.3 *Loss-Contracting Algorithm*

$H := G_D(P)$ // the distance graph of terminal-set P)
$T := T_{mst}(P)$ // an MST of P in $G_D(P)$)
while $r > 0$ **do begin**
      find a full component F containing at most k terminals
            that achieves the maximum $r = gain(T; F)/loss(F)$
    $H := H \cup F$
    $T := T_{mst}(T \cup C(F))$
end-while
return $T_{k-LC} := T_{mst}(H)$

In the above, the algorithm iteratively modifies a terminal-spanning tree T, which initially set to be $T_{mst}(P)$, by putting into T loss-contracted full components $C(F)$. At each iteration, the full component F is chosen greedily by the *gain-over-loss ratio*. The reason that Robins and Zelikovsky [239] adopted this objective is as follows: The length of approximate solution T lies between the lengths of $T_{mst}(P)$ and T_{k-smt} of P. Upon accepting a component F, it increases the gap between $l(T_{mst}(P))$ and $l(T)$ by a gain of F, which is the profit obtained from choosing F. On the other hand, if F does not belong to T_{k-smt}, then accepting F pushes T away from T_{k-smt} since the cost $loss(F)$ for incorrectly connecting Steiner points occurs. It is an upper bound on the increase in the length gap between T_{k-smt} and the best achievable solution after F is accepted. In other words, $loss(F)$ is simply an estimate of connection expense. Maximizing the ratio of gain-over-loss aims at maximizing the profit per unit expense. Robins and Zelikovsky [239] proved the following theorem.

Theorem 2.11   *For Steiner tree problem in graphs with edge costs satisfying the triangle inequality, the approximation ratio of the loss-contracting algorithm 2.3 is at most $\rho_k^{-1}(1 + \frac{1}{2}\ln(4\rho_k - 1))$.*

It follows from (2.2) that $\rho_k^{-1} \leq 1 + (\lfloor \log_2 k \rfloor + 1)^{-1}$. Thus the approximation ratio of the loss-contracting algorithm converges to $(1 + \frac{1}{2}\ln 3) < 1.55$ when $k \to \infty$.

2.3   Discussions

Although the k-Steiner ratio ρ_k in graphs and the rectilinear plane have been completely determined for $k \geq 2$, the k-Steiner ratio in the Euclidean plane for $k \geq 3$ is still unknown. Du et al. [91] conjectured that the ratio is

$$\frac{(1 + \sqrt{3})\sqrt{2}}{1 + \sqrt{2} + \sqrt{3}}.$$

They also analyzed that k-Steiner ratio in the Euclidean plane might be determined in a similar way to the proof of Gilbert-Pollak conjecture (refer to Section 1.3). The difficulty arises only in the description of "critical structure".

In addition, although many "better approximation algorithms" have been proposed in past ten years, none of them has a performance ratio smaller than the inverse of 3-Steiner ratio, ρ_3^{-1}. It seems that this is the limit for the performance ratio of any polynomial time approximation algorithm for the Steiner tree problem in graphs to be able to reach.

Arora et al. [19] conjectured that their *backtrack greedy technique* gives a polynomial time approximation scheme to 3-size Steiner tree problem. If their conjecture is true, then their algorithms also yield approximate solutions for Steiner tree problem with performance ratios approaching to the inverse of the 3-Steiner ratio ρ_3^{-1}. This probably is the best possible performance ratio. Thus their conjecture is a very attractive problem worthy of further study.

A more accurate analysis [165; 275; 277] for the performance ratios of the variable metric algorithm and the preprocessing technique, which are discussed in Sections 2.2.2 and 2.2.4, respectively, requires bounds for t_k and a similar number $\bar{t}_k$. The techniques [39; 40] studied in Section 2.1 for determining the k-Steiner ratio seems very promising for establishing tight upper bounds for t_k and $\bar{t}_k$.

The lower bound for the achievable performance ratio of better approximation algorithms for Steiner tree problem in graphs is widely open. (The case of metric planes has been solved, which will be discussed in next chapter.) Bern and Plassmann [34] proved that the problem is *MAX SNP-hard* even for the graphs whose edges have weights either 1 or 2. This implies that there exists a lower bound larger than one for such a ratio unless $P = NP$. Now the largest known lower bound is 1.010 due to Thimm [255], which is far away from 1.550, the performance ratio of the loss-contracting algorithm discussed in Section 2.2.5.

As all instances resulting from known lower bound reductions are uniformly quasi-bipartite, it is interesting whether this special case can be approximated better than the general case. A graph $G(V, E)$ with terminal-set $P \subset V$ is called *uniformly quasi-bipartite*, if vertex-set $V \setminus P$ is stable and if, for each vertex in that set, all incident edges have the same length. It is called *quasi-bipartite graph* if vertex-set $V \setminus P$ is stable, but the edges incident with a vertex in that set may have different lengths. Rajagopalan and Vazirani [232] gave a $3/2 + \epsilon$ approximation algorithm based on the primal-dual method for quasi-bipartite case. Robins and Zelikovsky [238] showed that the popular 1-Steiner heuristic has a performance ratio of $3/2$ in this case. Moreover, they showed that the performance ratio of their loss contracting algorithm is 1.279 for quasi-bipartite instances. Gröepl et al. [121] proposed an approximation algorithm with performance ratio $73/60 < 1.217$ for the uniformly quasi-bipartite case. It uses a new method of analysis that combines ideas from the greedy algorithm for set cover problem with a matroid-style exchange argument to model the connectivity constraint.

Chapter 3

Geometric Partitions and Polynomial Time Approximation Schemes

In Chapter 1 we have determined the Steiner ratio ρ in the Euclidean plane proving that the minimum spanning tree algorithm is a ρ^{-1}-approximation algorithm for the Steiner tree problem in the Euclidean plane. In Chapter 2 we have determined the k-Steiner ratio in graphs and studied some *better approximation algorithms* whose performance ratios are smaller than ρ^{-1}. In this Chapter, we will present two best possible approximation algorithms for the Steiner tree problem in metric planes. They both use geometric partition techniques and could produce approximation solutions arbitrarily close to the optimal solutions.

A series of algorithms $\{A_\epsilon | \epsilon > 0\}$ is called a *Polynomial Time Approximation Scheme* (PTAS) for a minimization problem if given any instance I of the problem, it finds a solution for I with cost $c_{A_\epsilon}(I) \leq (1 + \epsilon)c_{opt}(I)$, and the running time of algorithm A_ϵ is bounded by a polynomial in $1/\epsilon$ and the input size of I.

In 1996, Arora [10] published a surprising result that many geometric optimization problems, including the Euclidean Steiner Minimum Tree (SMT), the rectilinear SMT, and the degree-restricted-SMT, have polynomial-time approximation schemes. Several weeks later, Mitchell [214] claimed that one of his earlier works [211] already contains an approach which is able to lead to the similar results. However, one year later, Arora [212] made another big progress that he improved running time from $n^{O(1/\epsilon)}$ to $n^3(\log n)^{O(1/\epsilon)}$. Soon later, Mitchell [212] claimed again that his approach can lead to a similar result.

In this section we will study these two approaches and describe how to use them to solve the Steiner tree problem in metric planes and some other related problems. We will also discuss some related open problems.

3.1 Guillotine Cut for Rectangular Partition

The approach that Mitchell [211; 214] proposed for the Euclidean Steiner
minimum tree problem was inspired by the technique that Du et al. [86]
introduced for studying *Minimum Rectangular Partition* (MRP) problem[1].
This paper initiated the idea of using guillotine cut to design approximation
algorithms [56].

The minimum rectangular partition problem was first studied by Lingas
et al. [191]. It can be stated as follows: Given a rectilinear polygon possibly
with some rectangular holes that are simple rectilinear polygons whose sides
are parallel or perpendicular to the sides of the rectilinear boundary and
located on the inside of a rectilinear boundary (but no hole is allowed in
a hole). A hole can be, possibly in part, degenerated into a line segment
or a point. A *rectangular partition* of the polygon is a set of line segments
lying within its boundary and not crossing any nondegenerate hole so that
when drawn into the figure the are not enclosed by holes is partitioned into
rectangles which do not contain degenerate holes. The partitioning line
segments are called *edges* of the partition. The goal of minimum rectangular
partition problem is to compute a rectangular partition with minimum total
length of edges in the partition.

Fig.3.1(a) shows such an instance of minimum rectangular partition
problem with four holes (three shaded areas and one vertical line segment
in the middle). A partition that consists of some (thin) edges divides the
given polygon and four holes into many rectangles.

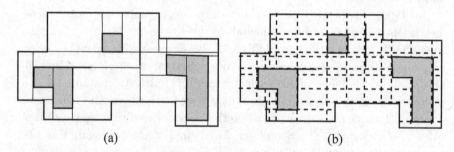

(a)   (b)

Fig. 3.1   (a) An instance of minimum rectangular partition problem and a partition,
(b) the basic grid.

---

[1]The problem is called the Minimum Edge-Length Rectangular Partition (MELRP)
problem in the literatures, in this book it is shortened to MRP problem for simplicity.

Lingas et al. [191] proved that the holes in the input make difference on the computational complexity. While the minimum rectangular partition problem in general is NP-hard, the problem for hole-free inputs can be solved in time of $O(n^4)$, where n is the number of vertices in the input rectilinear polygon. The polynomial algorithm is essentially a *dynamic programming* that is based on the following fact as shown in Fig.3.2(b): Through each vertex of the input rectilinear polygon, draw a vertical line and a horizontal line. Those lines will form a grid in the inside of the rectilinear polygon, which is called the *basic grid* for the rectilinear polygon. Lingas et al. [191] proved the following lemma.

Lemma 3.1   *There exists an optimal rectangular partition lying in the basic grid.*

Proof.   Suppose that there exists an optimal rectangular partition P_{opt} not lying in the basic grid (otherwise we are done). Then there is an edge e not lying in the basic grid. Consider the maximal straight segment in the partition that contains edge e. Say it is a vertical segment ab. Suppose that there are r horizontal segments touching the interior of ab from the right side of ab and l horizontal segments touching the interior of ab from the left side of ab. If $r \geq l$, then we can move ab to the right without increasing the total length of the rectangular partition. Otherwise, we can move ab to the left. We must be able to move ab into the basic grid because, otherwise, ab would be moved to overlapping with another vertical segment, so that the total length of the resulting rectangular partition is shorter than that of P_{opt}, that contradicts the optimality of P_{opt}. Therefore, we could be able to move every edge in P_{opt} that does not lie in the basic grid to the basic grid without increasing the total length, and then obtain the desired optimal rectangular partition.   □

A natural idea to design approximation algorithm for the minimum rectangular partition problem in general case is to use a *forest* interconnecting all holes to the boundary and then to solve the resulting hole free case in time of $O(n^4)$. Applying this idea, Lingas [192] gave the first constant-bounded approximation algorithm with performance ratio of 41. Later, Du and Zhang [89] improved the algorithm by significantly reducing the approximation performance ratio to 9.

Motivated from a work of Du et al. [84] on application of dynamic programming to optimal routing trees, Du et al. [86] initiated a novel idea, guillotine cut, which turns out to be important not only to the minimum rectangular partition problem, but also to many other geometric optimization problems including the Euclidean Steiner minimum tree problem.

A cut is called a *guillotine cut* if it breaks a connected area into at least two parts. A rectangular partition is called a *guillotine rectangular partition* if it can be performed by a sequence of guillotine cuts. Fig.3.2 shows a guillotine rectangular partition consisting of seven guillotine cuts (thin lines). Du et al. [86] noticed that there exists a minimum length guillotine rectangular partition lying in the basic grid, which can be computed by a *dynamic programming* in time of $O(n^5)$. Therefore, they suggested to use the minimum length guillotine rectangular partition as an approximate solution to the minimum rectangular partition problem. Unfortunately, they failed to get a constant approximation performance ratio for general case; Du et al. [78] proved the following theorem obtaining a constant ratio for a special case where the given polygon is a rectangle with some points inside (which remains NP-hard [117]).

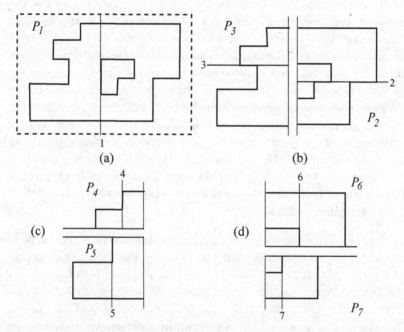

Fig. 3.2 (a) The first guillotine cut, (b) two separated parts are obtained after the first cut, (c) and (d) two parts in (b) are further divided by the second and third guillotine cuts, respectively.

Theorem 3.1   *The minimum length guillotine rectangular partition is a 2-approximation for the minimum rectangular partition problem with some points inside a rectangle.*

Proof.   Consider a rectangular partition P. Let $l_X(P)$ denote the total length of segments on a horizontal line covered by vertical projection of P. Partition P is said to be *covered* by a guillotine partition if each segment in P is covered by the guillotine cut. Let $l_G^*(P)$ denote the minimum total length of guillotine partition covering P, and let $l(P)$ denote the total length of P. We will prove the following inequality by mathematical induction on the number k of segments in P.

$$l_G^*(P) \leq 2\,l(P) - l_X(P). \tag{3.1}$$

For $k = 1$, we have $l_G^*(P) = l(P)$. If the segment is horizontal, then we have $l_X(P) = l(P)$, and thus we have $l_G^*(P) = 2\,l(P) - l_X(P)$. If the segment is vertical, then $l_X(P) = 0$, and hence we have $l_G^*(P) < 2\,l(P) - l_X(P)$.

Now we assume that inequality (3.1) is true for $(k-1)$ with $k \geq 2$ and study the case where P has k segments. Suppose that the input rectangle has each vertical edge of length a and each horizontal edge of length b. Consider the following two cases.

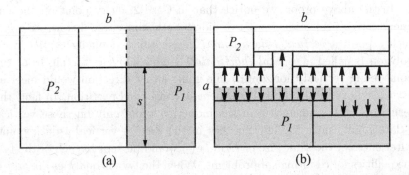

(a)   (b)

Fig. 3.3   (a) Case 1 and (b) Case 2 for the proof of Theorem 3.1.

Case 1. There exists a vertical segment s having length not less than $a/2$. In this case we apply a guillotine cut along this segment s. Then the remainder of P is divided into two partitions P_1 and P_2 which form a rectangular partition for two resulting small rectangles, respectively, as shown in Fig.3.3(a). By the induction hypothesis, we obtain

$$l_G^*(P_i) \leq 2\, l(P_i) - l_X(P_i), \quad \text{for } i = 1, 2. \tag{3.2}$$

Moreover, we have the following inequality and equalities

$$l_G^*(P) \leq l_G^*(P_1) + l_G^*(P_2) + a,$$
$$l(P) = l(P_1) + l(P_2) + l(s),$$
$$l_X(P) = l_X(P_1) + l_X(P_2),$$

from which we can deduce inequality (3.1).

Case 2. Every vertical segment in P has length less than $0.5a$. In this case we apply a horizontal guillotine cut to partition the input rectangle into two equal parts. Let P_1 and P_2 denote rectangular partitions of the two parts, obtained from P. Then inequality (3.2) also holds due to the induction hypothesis. Moreover, we have the following inequalities and equality

$$l_G^*(P) = l_G^*(P_1) + l_G^*(P_2) + b,$$
$$l(P) \geq l(P_1) + l(P_2),$$
$$l_X(P) = l_X(P_1) = l_X(P_2) = b,$$

from which we can also deduce inequality (3.1). The proof is finished.   □

In the above proof, we notice that in Case 2, every point on the cut line receives projection from two sides, both above and below. We call such a point a *vertical 1-dark point*. In general, a point inside the input polygon is called a *vertical (horizontal) 1-dark point* if starting from the point along vertical (horizontal) line going either direction would meet at least one horizontal (vertical) segment in considered partition. In fact, the term $l_X(\cdot)$ takes advantage in the induction proof only on those vertical 1-dark point, since the cut line lies in the area of vertical 1-dark point. After cutting, the same size of term $l_X(\cdot)$ would be kept in each of the two inequalities for resulting subproblems. When the two inequalities are added together, the size of term $l_X(\cdot)$ is doubled.

There is an alternative way to take the advantage of 1-dark points, which can be seen in the following alternative proof of Theorem 3.1.

Alternative Proof of Theorem 3.1 Consider a rectangular partition P.

Case 1. There exists a vertical segment s having length no less than $a/2$. Apply a guillotine cut along this segment s and charge 1 to the segment s.

Case 2. Every vertical segment in P has length less than $a/2$. Choose a horizontal guillotine cut which partitions the rectangle into two equal parts. Charge 1/2 to those horizontal segments, which directly face the cut. Note that every point on the cut is a vertical 1-dark point. Therefore, charged horizontal segments have a total length equal to exactly twice of the length of the cut.

Since each vertical segment in P is charged at most once and each horizontal segment is charged at most twice, the total length of added segments in guillotine cuts cannot exceed the total length of P. This completes the proof of Theorem 3.1.   □

3.1.1 *1-Guillotine Cut*

Finding a guillotine cut that consists of 1-dark points is a central part of the argument in [78; 86]. About ten years later, Mitchell [211; 213] made a significant progress in extending the idea of guillotine cut to the general case.

First, Mitchell [211] found a close relationship between 1-dark points and the guillotine cut by extending the guillotine cut to the 1-guillotine cut. A vertical (horizontal) cut is called *1-guillotine cut* if it consists of all vertical (horizontal) 1-dark points on the vertical (horizontal) line passing through the cut. This line will be called a *cut line*. An 1-guillotine cut can also be defined as a partition of a rectangle into two rectangles such that the cut line intersects considered rectangular partition with at most one segment. In fact, all cuts displayed in Fig.3.2 are 1-guillotine cuts.

Secondly, Mitchell [211] established a very important relationship between vertical 1-dark points and horizontal 1-dark points in the following lemma.

Lemma 3.2   (Mitchell's Lemma) *Let H and V be the sets of all horizontal and vertical 1-dark points, respectively. Then there exists either a horizontal line L_H such that $l(L_H \cap H) \le l(L_H \cap V)$, or a vertical line L_V such that $l(L_V \cap H) \ge l(L_V \cap V)$.*

Proof.   First, we assume that the area of H is not smaller than the area of V. Denote line $L_u = \{(x,y) \mid x = u\}$. Then areas of H and V can be represented by

$$\int_{-\infty}^{+\infty} l(L_u \cap H)du \quad \text{and} \quad \int_{-\infty}^{+\infty} l(L_u \cap V)du, \text{ respectively.}$$

Since

$$\int_{-\infty}^{+\infty} l(L_u \cap H) du \geq \int_{-\infty}^{+\infty} l(L_u \cap V) du,$$

there must exist a number u such that $l(L_u \cap H) \leq l(L_u \cap V)$. Similarly, if the area of H is smaller than the area of V, then there exists a horizontal line L_V such that $l(L_H \cap H) \geq l(L_H \cap V)$. The proof is then finished.   □

The above lemma actually means that there exists either a vertical 1-guillotine cut of length not exceeding the total length of segments consisting of all horizontal 1-dark points on the cut line, or a horizontal 1-guillotine cut of length not exceeding the total length of segments consisting of all vertical 1-dark points on the cut line. Namely, there always exists an 1-guillotine cut such that its length can be *symmetrically* charged to those segments parallel to the cut line, with value 1/2 to each side.

A rectangular partition is called an *1-guillotine rectangular partition* if it can be performed by a sequence of 1-guillotine cuts. Fig.3.2 shows such a cut. It can be showed that there exists a minimum 1-guillotine rectangular partition such that every maximal segment contains at least a vertex of the boundary.

Now, the question is whether the 1-guillotine cut also features the dynamic programming. The answer is yes. Refer to Fig.3.2. In fact, the 1-guillotine cut partitions a rectangular partition problem into two subproblems such that each maximal line-segment contains a vertex of the boundary, which is called the *boundary condition*, since after an 1-guillotine cut, two open segments may be created on the boundary. The boundary condition increases the number of subproblems in the dynamic programming. Since each subproblem is based on a rectangle with four sides, the condition on each side can be described by two possible open segments at the two ends. Hence each side has $O(n^2)$ possible conditions. So the total number of boundary conditions is $O(n^8)$, this implies that the total number of possible subproblems is $O(n^{12})$. For each problem, there are $O(n^3)$ possible 1-guillotine cuts. Therefore, the minimum 1-guillotine rectangular partition can be computed by a dynamic programming in time of $O(n^{15})$.

With 1-guillotine cuts, we can find a 2-approximation solution not only for the special case but also the general case as follows: Use a rectangle to cover the input rectangular polygon with holes. Cut the rectangle each time into two rectangles with an 1-guillotine cut, and then repeat this process to resulting rectangles until a rectangular partition is obtained. Mitchell [211] proved the following theorem.

Theorem 3.2 *The minimum 1-guillotine rectangular partition is a 2-approximation for the minimum rectangular partition problem.*

Proof. For any rectangular partition P, let P_G^* denote the minimum 1-guillotine rectangular partition covering P. Let $l_X(P)$ $(l_Y(P))$ denote the total lengths of segments on a horizontal (vertical) lines covered by vertical (horizontal) projection of partition P. To prove the theorem, it suffices to show that

$$l(P_G^*) \leq 2\,l(P) - l_X(P) - l_Y(P). \tag{3.3}$$

The argument is based on mathematical induction on the number k of segments in P, which is similar to that used in the proof of Theorem 3.1.

For $k = 1$, we have $l(P_G^*) = l(P)$. Assume, without loss of generality, that the segment is horizontal. Then we have $l_X(P) = l(P)$ and $l_Y(P) = 0$. Hence inequality (3.3) is satisfied. Next assume that inequality (3.3) is true for $k - 1$ with $k \geq 2$ and study the case where P has k segments. Consider the following two cases.

Case 1. There exists an 1-guillotine cut P_G. Without loss of generality, assume that P_G is vertical with length a. Suppose that the remainder of P is divided into two parts P_1 and P_2. Let P_i^* denote the minimum 1-guillotine rectangular partition covering P_i, $i = 1, 2$. Then, by induction hypothesis, we have

$$l(P_i^*) \leq 2\,l(P_i) - l_X(P_i) - l_Y(P_i), \quad \text{for } i = 1, 2. \tag{3.4}$$

It can be verified that the following inequalities and equalities hold true.

$$\begin{aligned}
l(P_G^*) &\leq l(P_1^*) + l(P_2^*) + a, \\
l(P) &= l(P_1) + l(P_2) + a, \\
l_X(P) &= l_X(P_1) + l_X(P_2), \\
l_Y(P) &\leq l_Y(P_1) + l_Y(P_2).
\end{aligned} \tag{3.5}$$

From the above inequalities and equalities, we can deduce inequality (3.3).

Case 2. There does not exist any 1-guillotine cut. In this case, we need to add a segment to partition P so that the resulting partition P' has an 1-guillotine cut and the length of added segment is at most $l_X(P_1) + l_X(P_2) - l_X(P)$ if the 1-guillotine cut is horizontal and $l_Y(P_1) + l_Y(P_2) - l_Y(P)$ if the 1-guillotine cut is vertical, where P_1 and P_2 are partitions obtained from P by the 1-guillotine cut.

By Lemma 3.2, we may assume, without loss of generality, that there exists a horizontal line L_H such that $l(L_H \cap H) \leq l(L_H \cap H)$, which implies $l(L_H \cap H) \leq l_X(P_1) + l_X(P_2) - l_X(P)$, where P_1 and P_2 are subpartitions obtained from P by the line which becomes an 1-guillotine cut after adding segment $L_H \cap H$ to the partition P. By induction hypothesis again, we obtain inequality (3.4). Moreover, inequality (3.5) also holds true in this case. Therefore, we obtain

$$l(P') = l(P_1^*) + l(G_2^*) + l(L_H \cap H)$$
$$\leq 2 \sum_{i=1}^{2} l(P_i) - \sum_{i=1}^{2} l_X(P_i) - \sum_{i=1}^{2} l_Y(P_i) + l(L_H \cap H)$$
$$\leq 2 l(P) - l_X(P) - l_Y(P)$$

As inequality (3.3) has been proved, the proof for the theorem is then finished.   □

In fact, Mitchell [211] used a different way to prove the above theorem. He symmetrically charged a half of the length of added segment to those parts of segments in P which face to 1-dark points. Since charging must be performed symmetrically, each point in P can be charged at most twice during the entire modification from a rectangular partition to an 1-guillotine rectangular partition. Therefore, the total length of added segments is at most $l(P)$ and hence the theorem is proved. Actually, this argument is equivalent to the above proof. In fact, only projections from both sides are considered (in Case 2), $l_X(P)$ or $l_Y(P)$ can contribute something against the length of the added segment.

3.1.2   *m-Guillotine Cut*

Mitchell [214] extended 1-guillotine cut to m-guillotine cut in the following way: A point p is a horizontal (vertical) m-*dark point* if the horizontal (vertical) line passing through p intersects at least $2m$ vertical (horizontal) segments of the considered rectangular partition P, among which at least m segments are on the left of p (above p) and at least m segments are on the right of p (below p).

A horizontal (vertical) cut is an m-*guillotine cut* if it consists of all horizontal (vertical) m-dark points on the cut line. In other words, let H_m (V_m) denote the set of all horizontal (vertical) m-dark points, then an m-guillotine cut is either a horizontal line L_H satisfying $L_H \cap H_m \subseteq L_H \cap P$ or a vertical line L_V satisfying $L_V \cap V_m \subseteq L_V \cap P$, where P is the considered

partition. A rectangular partition is called m-*guillotine* if it can be realized
by a sequence of m-guillotine cuts.

Fig.3.4(a) shows a vertical m-guillotine cut with $m = 3$, which results
in $2m$ open segments on each subproblem's boundary. The minimum m-
guillotine rectangular partition can also be computed by dynamic programming in time of $O(n^{10m+5})$. In fact, at each step, an m-guillotine cut has at
most $O(n^{2m+1})$ choices. There are $O(n^4)$ possible rectangles appearing in
the algorithm and each rectangle has $O(n^{8m})$ possible boundary conditions.
Therefore, the minimum m-guillotine rectangular partition can be used as
an approximation to the minimum rectangular partition for any natural
number $m \geq 1$.

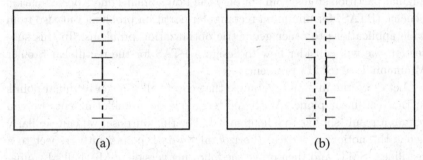

(a)   (b)

Fig. 3.4   (a) A vertical m-guillotine cut, and (b) $2m$ open segments on each subproblems's boundary.

Using arguments similar to the proofs of Lemma 3.2 and Theorem 3.2,
Mitchell [214] proved the following lemma and theorem, respectively.

Lemma 3.3   (Mitchell's Lemma) *Let H_m and V_m be the sets of all
horizontal and vertical m-dark points, respectively. Then there exists either
a horizontal line L_H such that $l(L_H \cap H_m) \leq l(L_H \cap V_m)$, or a vertical line
L_V such that $l(L_V \cap H_m) \geq l(L_V \cap V_m)$.*

Theorem 3.3   *The minimum m-guillotine rectangular partition is an $(1 + \frac{1}{m})$-approximation for the minimum rectangular partition problem.*

Proof.   Every time, if an m-guillotine cut already exists in P, then we use
it to divide considered rectangle into two. If there does not exist such an
m-guillotine cut, then by Lemma 3.3, there exists a vertical or horizontal
m-guillotine cut whose length can be symmetrically charged with value
$0.5/m$ to $2m$ vertical or horizontal segments in P; they face the horizontal
or vertical m-dark points on the cut line and each side of the cut line has m

layers. We can apply this m-guillotine cut to divide the considered rectangle into two. Since charging is performed symmetrically, no segment in P can be charged more than twice. Therefore, added segments have a total length not exceed one m-th of the total length of P. $\qquad\square$

Corollary 3.1 *There exists a polynomial time approximation scheme for the minimum rectangular partition problem with running time $O(n^{O(\log 1/\epsilon)})$.*

3.1.3 Guillotine Cut for Rectilinear Steiner Minimum Tree

The significance of techniques developed from the study of guillotine rectangular partition stems from not only the Polynomial Time Approximation Scheme (PTAS) for minimum rectangular partition problem, but also from wide applications to other geometric optimization problems. In this subsection, we will consider how to design a PTAS for the rectilinear Steiner Minimum Tree (SMT) problem.

Let Q be the minimal rectangle that covers all n given terminal points in the rectilinear plane. With the technique of m-guillotine cut, we can obtain a result similar to Theorem 3.3. For this purpose, we can similarly define the notion of vertical (horizontal) m-dark points with respect to a rectilinear SMT, and then prove the following version of Mitchell's Lemma.

Lemma 3.4 *Let H_m and V_m be the sets of all horizontal and vertical m-dark points, respectively, and T be a rectilinear Steiner tree. Then there exists either a horizontal cut line L_H that does not pass through any given point and satisfies $l(L_H \cap H_m) \leq l(L_H \cap V_m)$, or a vertical cut line L_V that does not pass through any given point and satisfies $l(L_V \cap H_m) \geq l(L_V \cap V_m)$.*

Proof. First, we assume that the area of H_m is not smaller than the area of V_m. Denote line $L_p = \{(x, y) \mid x = p\}$. Let the considered rectangle Q have four corner points with coordinates $(a, a'), (a, b'), (b, b')$, and (b, a'). Then areas of H_m and V_m can be represented by

$$\int_{-\infty}^{+\infty} l(L_u \cap H_m)du \quad \text{and} \quad \int_{-\infty}^{+\infty} l(L_u \cap V_m)du, \text{ respectively.}$$

Since

$$\int_{-\infty}^{+\infty} l(L_u \cap H_m)du \geq \int_{-\infty}^{+\infty} l(L_u \cap V_m)du,$$

and there are only finitely many u's such that L_u passes through a given point, there must exist a number $u \in (a, b)$ such that L_u does not pass through any given point and $l(L_u \cap H_m) \leq l(L_u \cap V_m)$. Similarly, if the area of H_m is smaller than the area of V_m, then there exists a desired horizontal line L_V. The proof is then finished. $\qquad\square$

By Hanan's classical result [130], there exists a rectilinear SMT T_{smt} lying in the *Hanan grid* (that consists of all horizontal and vertical lines each passing through a given terminal point). For such a T_{smt}, every line L between two adjacent lines of the Hanan grid has the same length of $L \cap H_m$ and the same length of $L \cap V_m$. Let S_L be the set of lines each lying at the middle between two adjacent lines of the Hanan grid. Fig.3.5(a) shows Hanan grid that consists of all those thin lines, together with all boundary lines, in the rectangle, and S_L that is composed of all dashed lines.

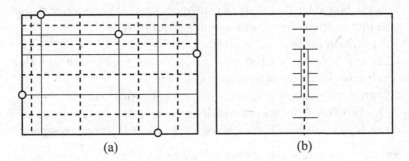

(a)   (b)

Fig. 3.5   (a) Hanan grid and cut lines, and (b) doubling an m-guillotine cut.

Theorem 3.4   *There exist an $(1 + 2/m)$-approximate rectilinear SMT T and an m-guillotine rectangular partition P of Q such that each cut of P intersects T at no more than $(2m - 1)$ points (called crosspoints) and all cut lines locate at $O(n)$ possible positions and all crosspoints also located at $O(n)$ possible positions.*

Proof.   In each iteration we always choose a cut line $L \in S_L$ and add an m-guillotine cut. Hence the number of possible locations of cut lines is $O(n)$. We may also double the m-guillotine cut and consider one of the two endpoints as a crosspoint as shown in Fig.3.5(b). Therefore, there are at most $(2m - 1)$ crosspoints all lying in the Hanan grid and thus the number of possible locations of crosspoints is also $O(n)$. Note that the cut line L is chosen in such a way that the twice length of added m-guillotine cut can be symmetrically charged with value $1/m$ to m layers of parallel line segments in each side. Therefore, the total length of added line segments is at most $\frac{2}{m}l(T_{smt})$. This completes the proof.   □

Corollary 3.2   *There exists a polynomial time $(1 + \epsilon)$-approximation for the rectilinear SMT problem with running time $O(n^{O(\log 1/\epsilon)})$.*

By Theorem 3.4, a *dynamic programming* can be employed to compute a shortest T described in the theorem. The running time can be estimated as follows: First, since each cut line has $O(n)$ possible positions and there are at most $(2m+1)$ crosspoints on each cut which yield $O(n^{2m-1})$ possibilities, each iteration runs in time of $O(n^{2m})$. Next, we study how many possible subproblems appearing in the computation of dynamic programming. Note that each subproblem can be specified in the following way:

(a) A rectangle R is given with a set S of at most $(8m - 4)$ crosspoints on the boundary (each edge has at most $(2m - 1)$ crosspoints), and all given terminal points lie in the interior of R.

(b) A partition of the set S is given.

(c) The goal is to find a forest with the minimum total length such that

(c.1) all crosspoints in each part are connected,

(c.2) two crosspoints in different parts are not connected,

(c.3) no two line segments cross each other, and

(c.4) each given terminal point is connected to at least one crosspoint.

There are $O(n^4)$ R's, and for each R there are $O(n^{8m-4})$ possible sets of crosspoints. Note that not every partition S can have a feasible solution (a rectilinear Steiner tree) satisfying conditions (c.1-3). We will prove in the following lemma that the number of partitions of S which has a feasible solution satisfying conditions (c.1-3) is $O(n^{O(m)})$. Therefore, the total number of subproblems generated in the dynamic programming is $O(n^4 \cdot n^{8m-4} \cdot n^{O(m)}) = n^{O(m)}$, and thus the dynamic programming runs in time of $n^{O(m)} \cdot O(n^{2m}) = n^{O(m)}$.

Lemma 3.5   *The number of partitions of S that leads to a feasible solution satisfying conditions (c.1-3) is $O(n^{O(m)})$.*

Proof.   Break the rectangle R at a point and put it into a straight line. Then the problem is reduced to count the number $n(k)$ of partitions of a set S of k ($k \le 8m - 4$) points on a horizontal line such that there exists a forest above the line satisfying conditions (c.1-3).

Denote by $p_1, p_2, \cdots, p_k$ the k points lying on the line from left to right.

When p_1 is connected to no one, the number of required partitions is $n(k-1)$. When p_1 is connected to p_i and p_i is the leftmost point other than p_1 among those points that are connected to p_1, the number of required partitions is $n(i-2)n(k-i+1)$, where $n(0) = 1$ for $i = 2$. Therefore,

$$n(k) = n(k-1) + \sum_{i=2}^{k} n(i-2)n(k-i+1)$$

$$= \sum_{j=0}^{k-1} n(k-2)n(k-1-j).$$

To determine $n(k)$, we define a generating function $f(x) = \sum_{k=0}^{\infty} n(k)x^k$. Then we have

$$f^2(x) = \sum_{k=0}^{\infty} \left(\sum_{j=0}^{k} n(k)n(k-j) \right) x^k = \sum_{k=0}^{\infty} n(k+1)x^k.$$

Hence, we obtain $xf^2(x) = f(x) - 1$. Thus $f(x) = (1 \pm \sqrt{1 - 4x})/(2x)$. Since

$$\lim_{x \to 0} f(x) = 1 \quad \text{and} \quad \lim_{x \to 0} \frac{1 + \sqrt{1 - 4x}}{2x} = \infty,$$

we have

$$f(x) = \frac{1 - \sqrt{1 - 4x}}{2x} = -\frac{1}{2x} \sum_{k=1}^{\infty} \binom{1/2}{k} (-4x)^k.$$

Therefore, we obtain

$$n(k) = -\frac{1}{2} \binom{1/2}{k+1} (-4)^{k+1} = 2^{O(k)}.$$

The proof is then finished. $\qquad\qquad\qquad\qquad\qquad\qquad\qquad\qquad\qquad\square$

3.2 Portals

The Polynomial Time Approximation Scheme (PTAS) proposed by Arora [10] is also based on a sequence of cuts on rectangles. To reduce the number of connections between two subproblems resulting from a cut, he introduced *portals* which enables him to equally divide the cut. In this section, we will study this technique and compare it with the technique of guillotine cuts.

3.2.1   Portals for Rectilinear Steiner Minimum Tree

In order to study the relationship between the technique of guillotine cut and Arora's seminal work [10], in this subsection we shall show how to design a PTAS for the rectilinear Steiner Minimum Tree (SMT) problem using his approach.

The basic idea is as follows: Initially, use a minimum square to enclose n given terminals. Then with a tree structure, partition this square into small rectangles each of which contains one given terminal. The tree structure is established as follows:

(i) Equally divide the initial square into a $(g \times g)$-grid with $g \geq n/\epsilon$.
    Move each terminal point to the center of the cell where it is located.
(ii) Choose cut line from grid lines in a range between $1/3$ and $2/3$ of a longer edge (or an edge for a square).

The following two lemmas explain how the above two techniques are used towards a PTAS for the rectilinear SMT problem.

Lemma 3.6 *Suppose $\epsilon < 1$ and $g \geq 4.5n/\epsilon$. Let P be the set of n terminal points and P' the set of n cell centers receiving n terminals. If there is a polynomial time $(1 + \epsilon)$-approximation for the rectilinear SMT on P', then there exists a polynomial time $(1 + 2\epsilon)$-approximation for the rectilinear SMT on P.*

Proof.   Let $T_\epsilon(P')$ be a polynomial time ϵ-approximation solution to the rectilinear SMT problem on P', and let $T_{smt}(P)$ and $T_{smt}(P')$ be the rectilinear SMTs on P and P', respectively. Then $l(T_\epsilon(P')) \leq (1+\epsilon)l(T_{smt}(P'))$. Note that $|l(T_{smt}(P)) - l(T_{smt}(P'))| \leq sn/g$, where s is the side length of the initial square. Since the square is minimal, s is not bigger than the length of the MST on P and $s \leq \frac{3}{2}l(T_{smt}(P))$.

Now construct a tree T interconnecting points in P from $T_\epsilon(P')$ by connecting each point in P' to its corresponding terminal point in P. As $g \geq 4.5n/\epsilon$, we have

$$l(T) \leq l(T_\epsilon(P')) + \frac{sn}{g}$$

$$\leq (1 + \epsilon)\, l(T_{smt}(P')) + \frac{sn}{g}$$

$$\leq (1 + \epsilon)\left(l(T_{smt}(P)) + \frac{sn}{g}\right) + \frac{sn}{g}$$

$$= (1 + \epsilon)\, l(T_{smt}(P)) + (2 + \epsilon)\frac{sn}{g}$$

$$\leq \left(1 + \epsilon + (2 + \epsilon)\frac{3n}{2g}\right) l\big(T_{smt}(P)\big)$$

$$\leq 2\epsilon \cdot l\big(T_{smt}(P)\big).$$

Note that for sufficiently large n, we have $(2+\epsilon/2)3/2n \leq \epsilon/2$, which implies $l(T) < (1 + \epsilon) l(T_{smt}(P))$. The proof is then finished. $\square$

By the above lemma, we will consider the rectilinear SMT on P' instead of the original terminal set P.

Lemma 3.7   *The binary tree structure of the partition obtained by technique* (ii) *has* $O(\log n)$ *levels.*

Proof.   With technique (ii), the rectangle at the i-th level has area of size at most $2^s(2/3)^{i-1}$. Since the ratio between the lengths of longer edge and shorter edge is at most three, the rectangle at the last level, say the l-th level, has area of size at least $\frac{1}{3}(s/n^2)^2$. Therefore, we obtain $2^s(2/3)^{l-1} \geq \frac{1}{3}(s/n^2)^2$, which leads to $l = O(\log n)$. The proof is then finished. $\square$

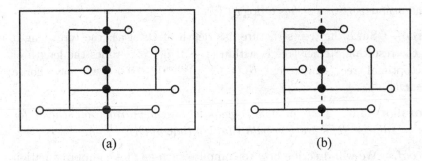

Fig. 3.6   (a) p-portals, and (b) p-portals vs m-guillotine cut.

To reduce the number of crosspoints at each cut line, Arora [10] introduced portals. A set of p points on a cut are called *p-portals* if they equally divide the guillotine cut into $(m + 1)$ segments. For rectilinear SMT (or Euclidean SMT, etc), crosspoints of the Steiner tree on a cut line can be moved to portals. This would reduce the number of crosspoints on the cut line. Fig.3.6(a) shows p-portals with $p = 5$. The following lemma shows that by properly choosing cut line, at each level of the tree structure moving crosspoints to portals would increase the length of the tree within a certain amount.

Lemma 3.8   *There exist a set of cut lines S_L such that at each level of the tree structure moving crosspoints to portals of S_L would increase the length of the tree within $3/p$ times the total length of Steiner tree.*

Proof.   Consider each rectangle R at a certain level. Suppose that its longer edge has length a and shorter edge has length b ($b \leq a$). Consider every possible cut in a range between $1/3$ and $2/3$ of a longer edge. Choose the cut line to intersect the Steiner tree with the smallest number of points, say c points. Then the length of the part of the Steiner tree lying in rectangle R is at least $c\,a/3$. Moving c crosspoints to portals requires to add some segments of total length at most

$$\frac{c\,b}{p+1} \leq \frac{c\,a}{p+1} \leq \frac{3/p}{c\,a/3},$$

which proves the lemma.   □

Theorem 3.5   *For any rectilinear SMT T_{smt}, there exists an $(1 + \epsilon)$-approximation tree T with a guillotine rectangular partition P of Q such that each guillotine cut intersects T with at most $m(= 1/\epsilon)$ crosspoints which are all located at p-portals where $p = O(\frac{\log n}{\epsilon})$. Moreover, the guillotine rectangular partition P has $O(\log n)$ levels.*

Proof.   Since the tree structure has depth of $O(\log n)$, the total length of the resulting Steiner tree is within $(1 + 3/p)^{O(\log n)}$ times the length of the optimal tree. Thus to ensure $(1 + 3/p)^{O(\log n)} \leq 1 + \epsilon$, we may choose $p = O(\frac{\log n}{\epsilon})$.   □

Corollary 3.3   *There exists a polynomial time approximation scheme for the rectilinear SMT problem with running time of $O(n^{O(1/\epsilon)})$.*

Proof.   We will describe how to compute such an $(1 + \epsilon)$-approximation specified in the above theorem. Again we employ the *dynamic programming* to find the shortest one among the trees with the same structure as the $(1 + \epsilon)$-approximation.

To estimate the running time of dynamic programming, we first note that each subproblem is characterized by a rectangle and conditions on the boundary. There are $O(n^8)$ possible rectangles each having four sides. One of them must contain p portals. However, each of other three may contain less than p portals resulting from previous cuts. Thus the number of positions for portals on each of these three sides is $O(n^4)$. Hence, the total number of portal positions on the boundary is $O(n^{20})$. For each fixed set of portal positions, we need also consider whether a portal is a crosspoint or not and how crosspoints are connected each other inside the rectangle. It yields $w^{O(p)}$ possibilities. Therefore, the total number of possible subproblems is $n^{O(1/\epsilon)}$.

Moreover, in each iteration of dynamic programming, the number of all possible cuts is $O(n^2)$. Therefore, the dynamic programming runs in time of $O(n^{O(1/\epsilon)})$. The proof is then finished.   □

3.2.2 *Portals versus Guillotine Cuts*

In this subsection we will compare those two techniques, m-guillotine cuts and portals, which were described in the previous two sections.

For those geometric optimization problems in three or higher dimensional spaces, cut lines should be replaced by cut planes or hyperplanes. The number of portals would be $O((\frac{\log n}{\epsilon})^2)$ or more. With so many possible crosspoints, the dynamic programming cannot be implemented in polynomial time. However, m-guillotine cuts have at most $2m$ crosspoints in each dimension and m is a constant with respect to n. Therefore, the polynomial time for the dynamic programming would be preserved in three or higher dimensional spaces.

In addition, the technique of portals cannot be applied to the minimum rectangular partition problem and the following three geometric optimization problems.

Problem 3.1 *Rectilinear Steiner Arborescence Problem* [196]

Instance	A set of n terminals in the first and the second quadrants of the rectilinear plane.
Solution	A directed tree T rooted at the origin that is connected to all terminals with only horizontal arcs (in either orientation) and vertical arcs oriented from bottom to top.
Objective	Minimizing the total length of arcs in T.

Problem 3.2 *Symmetric Rectilinear Steiner Arborescence Problem* [54]

Instance	A set of n terminals in the first quadrant of the rectilinear plane.
Solution	A directed tree T rooted at the origin that is connected to all terminals with only horizontal and vertical arcs oriented from left to right or bottom to top.
Objective	Minimizing the total length of arcs in T.

Problem 3.3 *Convex Partition Problem* [78]

Instance A polygon inside with some polygon-holes.
Solution A partition P of the polygon into convex areas.
Objective Minimizing the total length of partition in P.

In fact, for the above mentioned problems, moving crosspoints to portals is sometimes impossible. But the technique of m-guillotine cuts is applicable to them (refer to [196]).

On the other hand, the technique of m-guillotine cuts can not be applied to the following three problems while the technique of portals can.

Problem 3.4 *Euclidean k-Median Problem* [17]

Instance A set of n points in the Euclidean plane.
Solution Locations of k medians in the plane.
Objective Minimizing the sum of the distances from each given point to the nearest median.

Problem 3.5 *Euclidean Facility Location Problem* [17]

Instance A set of n points $p_1, p_2, \cdots, p_n$ in the Euclidean plane with a cost c_i for openning a facility at p_i for each $i = 1, 2, \cdots, n$.
Solution A subset S of $\{p_1, \cdots, p_n\}$.
Objective Minimizing $\sum_{p_i \in S} c_i + \sum_{i=1}^{n} \min\{d(p_i, p_j) | p_j \in S\}$.

Problem 3.6 *Euclidean Grade Steiner Tree Problem* [170]

Instance A sequence of terminal sets $P_1 \subset P_2 \subset \cdots \subset P_m$ in the Euclidean plane, and costs $c_1 > c_2 > \cdots > c_m$.
Solution A network $G(V, E)$ that contains a Steiner tree T_i for every P_i.
Objective Minimizing cost of G that equals $\sum_{e \in E} l(e) \cdot \max\{c_i | e \in T_i\}$.

In addition, the m-guillotine cut and partition require the number of locations for crosspoints is bounded by a polynomial of input size. For some problems (such as Euclidean SMT), without moving crosspoints, the number of possible locations may not be able to have a polynomial bound. Therefore, the m-guillotine alone may not work well. In this case, some other technique, such as banyan (see Section 3.4), must be used if the technique of portals is not used.

3.2.3   *Portals Integrated with Guillotine Cuts*

When both techniques of portals and guillotine cuts can be applied to a geometric optimization problem, such as the rectilinear SMT problem, it is natural to ask question: could two techniques be combined to yield a better result? Roughly speaking, this combination may reduce the running time for dynamic programming. In fact, one may first use the portal technique to reduce the number of possible locations for crosspoints to $O(\frac{\log n}{\epsilon})$ and then to choose $2m$ positions from them to form an m-guillotine cut with $m = 1/\epsilon$ as shown in Fig.3.6(b). Therefore, the dynamic programming for finding the best such partition runs in time of $n^c(\log n)^{O(1/\epsilon)}$, where c is a constant. However, when we implement this idea in detail, we may meet the following trouble: In these two techniques, two different principals are used for the choice of each cut. In the m-guillotine cut, the cut line satisfies the inequality in Mitchell's Lemma. But, when portals are used, the cut line is chosen to minimize the number of crosspoints. How can these two principles go together?

One method is to move our attention from a local optimal choice of each cut to the entire partition. To do so, let us focus on the rectilinear SMT problem again and define a family of partitions as follows.

(i) Construct a *quadtree partition* as shown in Fig.3.7(a): Initially, given a square Q. In each subsequent step, if a square contains more than one input terminal points, then partition it into four subsquares of equal size.

(ii) Divide the square Q into a $(2^q \times 2^q)$ grid, where $2^q = O(n/\epsilon)$, such that every input terminal point lies at the center of a cell.

We may assume that the initial square $Q = \{(x,y)|0 \le x \le 2^q, 0 \le y \le 2^q\}$, from (ii) we can easily prove the following lemma.

Lemma 3.9   *For any rectilinear Steiner tree T satisfying condition that every Steiner point lies at the center of a cell, the total number of crosspoints of T with vertical grid lines equals the total length of horizontal segments in T and the total number of crosspoints of T with horizontal grid lines equals the total length of vertical segments in T.*

Let T_{smt} be a rectilinear SMT lying in Hanan grid. Then every Steiner point is at a grid point of Hanan grid and hence is the center of a cell. So the above lemma can be applied to T_{smt}.

Now for each point (a,b) in the rectilinear plane with $0 \le a,b < 2^q$, we define a quadtree partition $P(a,b)$ as follows: Choose (a,b) as the center to draw an initial square Q with edge length twice that of Q. This square covers Q. From this initial square, construct a quadtree according to the definition. Let $\mathcal{P}$ be the family of $P(a,b)$ for (a,b) over all integer points in Q. Fig.3.7(b) shows such a quadtree partition $P(a,b)$ where Q is the shaded area. In the following we will first estimate the average of the total cost (the increased length) for moving crosspoints to p-portals.

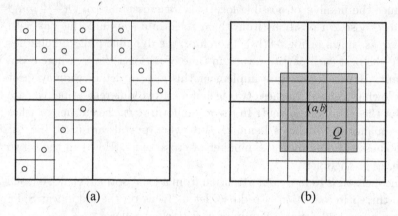

(a)   (b)

Fig. 3.7   (a) Quadtree partition, and (b) quadtree partition $P(a,b)$ covering Q.

Lemma 3.10   *Let $c_1(P,T)$ denote the total increased length for moving crosspoints of tree T to p-portals in partition P. Then*

$$\frac{1}{2^q} \sum_{0 \leq a < 2^q} c_1\big(P(a,a),T\big) \leq \frac{q+1}{2(p+1)} l(T). \qquad (3.6)$$

Proof.   The quadtree partition has a rooted tree structure with the initial square as the root. As usual, we call a vertex at the i-th level if the path from the root to it has length i. Hence, the root is at the 0-th level. A cut segment is said to be at the i-th level if it is one of four cut segments cutting a square at the i-th level into four squares at the $(i+1)$-th level. Thus, we may say that a grid line is in the i-th level if all cut segments on it are at the i-th level.

Consider a family of 2^q quadtree partitions $P(a,a)$ for $a = 0, 1, \cdots, 2^q - 1$. Let L_v be a vertical grid line. It is easy to see that L_v is at the 0-th level once, at the 1-st level once, at the 2-nd level twice, at the 3-rd level four times and so on; In general, it is at the i-th level 2^{i-1} times for $1 \leq i \leq q - 1$. Let $n(L_v, T)$ denote the number of crosspoints of T on L_v.

When L_v is at the i-th level, moving crosspoints on L_v to p-portals would increase the length of T within $n(L_v, T)2^{q-i}/(p+1)$. Therefore, the total increased length for moving crosspoints at the vertical cuts to portals when the partition is over all $P(a, a)$ for $a = 0, 1, \cdots, 2^q - 1$ is at most

$$\sum_{L_v \in S_{vgl}} n(L_v, T) \cdot \left(2^q + \sum_{i=1}^{q-1} 2^{i-1} \cdot 2^{1-i}\right) \frac{1}{p+1}$$

$$= l_{vs}(T) \cdot 2^{q-1}(q+1) \frac{1}{p+1}, \tag{3.7}$$

where S_{vgl} is the set of vertical grid lines and l_{vs} is the total length of vertical segments in T. Similarly, the total increased length for moving crosspoints at horizontal cuts to portals when the partition is over all $P(a, a)$ for $a = 0, 1, \cdots, 2^q - 1$ is at most

$$l_{vs}(T) \cdot 2^{q-1}(q+1) \frac{1}{p+1}, \tag{3.8}$$

where l_{hs} is the total length of vertical segments in T. Summarizing two terms of (3.7-8) will lead to inequality (3.6). The proof is then finished. $\square$

Next, we will show how to use global consideration to replace the local choice of each cut when the technique of m-guillotine cut applies.

The main purpose of using m-guillotine cuts is to reduce the number of crosspoints to $O(m)$ at each cut. The local choice of each m-guillotine cut is based on Mitchell's Lemma which guarantees that the symmetric charge can be done successfully. Such symmetric charge makes the total increased length is bounded by $l(T)/m$. Thus, we can choose $m \geq 1/\epsilon$ in order to ensure the bound at most $\epsilon l(T)$.

We now need to estimate the average of the total cost (the increased length) for reducing the number of crosspoints to no greater than m at each cut segment. In the following we will show first how to carry out the reduction and then how to make the analysis.

In fact, we would not consider symmetric charge any more. So, we may not need to make m-guillotine cut. A simple way to reduce the number of crosspoints at each cut is to add a guillotine cut segment. We may also double the added segment to make only one crosspoint left as shown in Fig.3.5(b). Such a process is sometimes called *patching*. An important property of m-guillotine cuts is that the total cost of patching may be bounded by $\epsilon \cdot l(T)$.

However, keeping the total increased length bounded by $\epsilon \cdot l(T)$ is not an easy job. It, in fact, requires to combine costs accumulated at different levels. The following procedure gives such a technique. Let $[x, y]$ denote the line segment with endpoints x and y. For each cut line segment $[x, y]$ at the i-th level, denote by $x(d)$ the point in $[x, y]$ with distance d from x, e.g., $x = x(0)$ and $y = x(2^{q-i})$.

Algorithm 3.1 *Patching Procedure*

---

for $k = 0$ **to** $q - i$ **do**
　　for $j = 0$ **to** $2^{q-i-k} - 1$ **do**
　　　　if $[x(j2^k), x((j+1)2^k)]$ has more than m crosspoints
　　　　then patch $[x(j2^k), x((j+1)2^k)]$
　　end-for$-j$
end-for$-k$

Lemma 3.11　Let $c_2(P, T)$ denote the total increased length for reducing the number of crosspoints to no more than m at each cut segment in partition P. Then

$$\frac{1}{2^q} \sum_{0 \le a < 2^q} c_2\big(P(a, a), T\big) \le \frac{2}{m} l(T). \tag{3.9}$$

Proof.　Let $n_s(k', L)$ denote the number of segments, or grid line L, patched when $k = k'$ in the patching procedure (Algorithm 3.1). Then, when L is at the i-th level, the total increased length for patching on grid line L is at most

$$\sum_{k=0}^{q-i} n_s(k, L) \cdot 2^{k+1},$$

where each cut segment is doubled for patching.

Now consider a family of w^q quadtree partitions $P(a, b)$ for $a = 0, 1, \cdots, 2^q - 1$ and an arbitrary vertical grid line L_v. Recall that L_v is at the 0-th level once and at i-th level 2^{i-1} times for $1 \le i \le q - 1$. Therefore, the total increased length for patching on L_v and the partition over all $P(a, a)$ for $a = 0, 1, \cdots, 2^q - 1$ is at most

$$\sum_{k=0}^{q} n_s(k, L_v) \cdot 2^{k+1} + \sum_{i=1}^{q} 2^{i-1} \sum_{k=0}^{q-i} n_s(k, L_v) \cdot 2^{k+1}$$

$$= \sum_{k=0}^{q} n_s(k, L_v)\Big(2^{k+1} + \sum_{i=1}^{q-k} 2^{k+i}\Big) = \sum_{k=0}^{q} n_s(k, L_v) \cdot 2^{q+1}.$$

Note that patching once would reduce at least m crosspoints. Thus we have

$$\sum_{k=0}^{q} n_s(k, L_v) \leq \frac{n_c(L_v, T)}{m},$$

where $n_c(L_v, T)$ is the number of crosspoints of T on L_v. It follows that the total increased length for patching on L_v and the partition over all $P(a, a)$ for $0 \leq a < 2^q$ is at most $2^{q+1} n_c(L_v, T)/m$. Therefore, we obtain

$$\frac{1}{2^q} \sum_{0 \leq a < 2^q} c_2(P(a, a), T) \leq \sum_{L_v \in S_{vgl}} \frac{2\, n_c(L_v, T)}{m} = \frac{2}{m} l(T),$$

where S_{vgl} is the set of vertical grid lines. The proof is then finished. □

Theorem 3.6 *For $p \geq 2q/\epsilon$ and $m \geq 8/\epsilon$, there are at least a half number of partitions $P(a, a)$ for $0 \leq a < 2^q$ having an ϵ-approximation T for the rectilinear SMT such that each cut intersects T with at most m crosspoints at p-portals.*

Proof. Let T_{smt} be a rectilinear SMT such that every maximal segment contains a terminal point and a quadtree partition $P(a, a)$. We first reduce the number of crosspoints of T_{smt} to at most m on each cut of $P(a, a)$ by patching procedure, and then move all crosspoints to p-portals. Let $c_3(P(a, a), T_{smt})$ denote the total increased length in the above process. Clearly,

$$c_3(P(a, a), T_{smt}) \leq c_1(P(a, a), T_{smt}) + c_2(P(a, a), T_{smt}).$$

Choose $p \geq 2(q + 1)/\epsilon - 1$ and $m \geq 8/\epsilon$. By Lemmas 3.10-11, we have

$$\frac{1}{2^q} \sum_{1 \leq a < 2^q} c_3(P(a, a), T_{smt}) \leq \frac{\epsilon}{2} l(T_{smt}).$$

Therefore, there are a half number of $P(a, a)$ for $1 \leq a < 2^q$ such that $c_3(P(a, a), T_{smt}) \leq \epsilon \cdot l(T_{smt})$. The proof is then finished. □

Theorem 3.7 *With probability at least $1/2$, an $(1 + \epsilon)$-approximation for the rectilinear SMT can be produced in time of $n((\log n)/\epsilon)^{O(1/\epsilon)}$.*

Proof. First, randomly choose a quadtree partition $P(a, a)$, and then compute the shortest Steiner tree T interconnecting all terminal points such that each cut of $P(a, a)$ intersects T with at most m crosspoints which are all located at p-portals. In fact, T can be computed with dynamic programming. Next choose $2^q = O(n/\epsilon)$, $p = O(q/\epsilon) = O(\log(n/\epsilon)/\epsilon)$, and

$m = O(1/\epsilon)$. In each iteration, all possible sets of crosspoints need to be considered. Since the number of possible sets of crosspoints is $O(p^m) = (n/\epsilon)^{O(1/\epsilon)}$, each iteration takes time of $(\log n/\epsilon)^{O(1/\epsilon)}$.

Note that in quadtree partition, there are at most $2n$ squares containing terminal points. Each subproblem can be specified by a particular square, a set of crosspoints on the boundary, and connected pattern of those crosspoints. By considering these three parameters, it is not hard to find out that there are totally $2n \cdot O(p^m)^4 \cdot 2^{O(m)} n((\log n)/\epsilon)^{O(1/\epsilon)}$ subproblems. Therefore, the running time of the dynamic programming is $n((\log n)/\epsilon)^{O(1/\epsilon)}$. The proof is then finished.   □

To derandomize, we need only to apply dynamic programming to every quadtree partition $P(a, a)$ for $0 \le a < 2^q$. Therefore, we have the following corollary.

Corollary 3.4   *There exists a PTAS for the rectilinear SMT problem with running time of $n^2((\log n)/\epsilon)^{O(1/\epsilon)}$.*

3.2.4   *Two-Stage Portals*

In this subsection, we will show how to improve the running time of the PTAS described in the previous subsection by using two-stage portals, which is motivated from the idea due to Mitchell [212].

Consider a cut $[x, y]$. Choose two point x', y' from p_1−portals of $[x, y]$. Then all p_2-portals of $[x', y']$ form a group of (p_1, p_2)-portals of $[x, y]$, which are called *two-stage portals*. There are $O(p_1^2)$ groups of (p_1, p_2)-portals.

Lemma 3.12   *Let $c_4(p(a, a), T)$ denote the total length of such $[x, y]$ over all cuts in $P(a, b)$ such that all crosspoints on the cut are located in $[x, y]$. Then*

$$\frac{1}{2^q} \sum_{0 \le a < 2^q} c_4(P(a, a), T) \le 4\,l(T). \tag{3.10}$$

Proof.   For each cut C in $P(a, a)$, we break it into two equal parts if all crosspoints keep on one part. Continue this process until further breaking would not keep all crosspoints in one part. Then let $[u(C), v(C)]$ denote the resulting segment in the above process.

If cut C contains only one crosspoint, then the length of $[u(C), v(C)]$ must be one. Thus, the total length of $[u(C), v(C)]$ for C over all cuts each containing one crosspoints is bounded by the number of crosspoints of T in

$P(a, b)$, that is, by Lemma 3.9 it is bounded by $l(T)$. For the cut C that contains at least two crosspoints, $[u(C), v(C)]$ must be patched for doing patching with $m = 1$. By Lemma 3.11, the total length of $[u(C), v(C)]$ for C over all cuts each with at least two crosspoints is bounded by $c_2(P(a, a), T)$ with $m = 1$. Now, moving $u(C)$ and $v(C)$ to p_1-portals would increase totally at most $2c_1(P(a, a), T)$. Therefore, we have

$$\frac{1}{2^q} \sum_{0 \leq a < 2^q} c_4\big(P(a, a), T\big) \leq l(T) + 2\, l(T) + \frac{q+1}{p_1+1}\, l(T) \leq 4\, l(T),$$

which proves the lemma. $\qquad\qquad\qquad\qquad\qquad\qquad\qquad\qquad\qquad\square$

Lemma 3.13 *For $m \leq 8/\epsilon$, $p_2 = 2m^2$ and $p_1 \geq q$, there are at least a half number partition $P(a, a)$ for $0 \leq a < 2^q$ having an ϵ-approximation T for the rectilinear SMT such that each cut intersects T with at most m crosspoints which are all located at a group of (p_1, p_2)-portals.*

Proof. Let T_{smt} be a rectilinear SMT lying in Hanan grid. For each quadtree partition $P(a, a)(0 \leq a < 2^q)$, we first reduce the number of crosspoints on each cut to at most m by patching procedure. By Lemma 3.11, the total increased length for patching is bounded by $c_1(p(a, a), T)$. Then for each cut we choose two p_1-portals x and y with minimum length of $[x, y]$ such that all crosspoints on the cut are located in $[x, y]$. Let $c_4(p(a, a), T)$ denote the total length of such $[x, y]$ over all cuts in $P(a, b)$. Moving all crosspoints to (p_1, p_2)-portals in the group corresponding to $[x, y]$, the total increased length is bounded by $(m/p_2)c_4(P(a, a), T)$. By Lemma 3.12, we have

$$\frac{1}{2^q} \sum_{0 \leq a < 2^q} c_4\big(P(a, a), T\big) + \frac{m}{p_2}c_4\big(P(a, b), T\big) \leq \Big(\frac{2}{m} + \frac{4m}{p_2}\Big)l(T) \leq \frac{\epsilon}{2}\, l(T),$$

which proves the lemma. $\qquad\qquad\qquad\qquad\qquad\qquad\qquad\qquad\qquad\square$

Theorem 3.8 *With probability at least $1/2$, an $(1+\epsilon)$-approximation for the rectilinear SMT can be produced in time of $n(\log n)^{10}(1/\epsilon)^{O(1/\epsilon)}$.*

Proof. To find the desired approximation, randomly choose a quadtree partition $P(a, a)$. We compute the shortest Steiner tree T interconnecting terminals and satisfying the property that each cut of $P(a, a)$ intersects T with at most m crosspoints which are all located at p-portals. By Theorem 1.7, with probability $1/2$, we have an $(1+\epsilon)$-approximation T for rectilinear SMT.

In fact, T can be computed using dynamic programming as follows: First, choose $2^q = O(n/\epsilon)$, $p_1 = O(q/\epsilon) = O(\log(n/\epsilon)/\epsilon)$, and $m = O(1/\epsilon)$. In each iteration, we need to consider all possible sets of crosspoints. Note that each set of crosspoints is chosen in two steps. In the first step, choose a group of two-stage portals. There are $O(p_1^2) = O((\log n/\epsilon)^2)$ groups of two-stage portals. In the second step, choose no more than m crosspoints from this group of (p_1, p_2)-portals. There are $O(p_2^m) = (1/\epsilon)^{O(1/\epsilon)}$ such sets. Thus, there are totally $(\log n)^2 (1/\epsilon)^{O(1/\epsilon)}$ possible sets of crosspoints on each cut. Hence, each iteration takes time $(\log n)^2 (1/\epsilon)^{O(1/\epsilon)}$. Note that in quadtree partition, there are at most $2n$ squares containing terminal points. Moreover, each subproblem can be determined by a particular square, a set of crosspoints on the boundary, and connected pattern of those crosspoints. By considering these three parameters, it is not hard to figure out that there are totally $2n \cdot O(p_1^2 p_2^m)^4 \cdot 2^{O(m)} n(\log n)^8 (1/\epsilon)^{O(1/\epsilon)}$ subproblems. Therefore, the running time of the dynamic programming is $n(\log n)^{10}(1/\epsilon)^{O(1/\epsilon)}$. The proof is then finished.   □

To derandomize, we need only to apply dynamic programming to every quadtree partition $P(a, a)$ for $0 \leq a < 2^q$. Therefore, we have the following corollary.

Corollary 3.5   *There exists a PTAS for the rectilinear SMT problem with running time of $n^2 (\log n)^{10}(1/\epsilon)^{O(1/\epsilon)}$.*

3.3   Banyan and Spanner

From the discussion in Section 3.2, we can see that portals are used extensively to reduce the running time of Polynomial Time Approximation Scheme (PTAS). However, to get further improvement, we have to give up portals and refer to some novel technique. In fact, the cost for moving to portals depends on the depth of partitions. This is really a disadvantage compared with patching. In this subsection we will discuss how to use *spanner* and banyan, instead of portals, to get a even faster PTAS. This approach was first proposed by Rao and Smith [234], and later it was extended to some problems in planar graphs by Arora et al. [13].

Patching with quadtree partition has the same power as the technique of guillotine cuts. To apply dynamic programming, the number of possible locations for crosspoints on each cut has to be limited. In the rectilinear SMT problem, Hanan grid does the job. A wild idea is to apply patching to Hanan grid instead of the rectilinear SMT lying in Hanan grid.

In fact, consider quadtree partition $P(a, a)$ which covers Hanan grid. We patch Hanan grid to make each cut of $P(a, a)$ contain at most m crosspoints. By the same argument as the proof of Lemma 3.11, it is easy to know that, when $0 \le a < 2^q$, this would increase the total length in average to at most $(1 + 2/m)h$ where h denotes the length of Hanan grid. Now, we compute a shortest rectilinear Steiner tree T from the patched Hanan grid with dynamic programming. Since each iteration takes time of $O(2^m)$ and there are at most $O(n \, 2^{O(m)})$ subproblems possibly appearing in the dynamic programming, the running time would be in $O(n \, 2^{O(m)})$. In addition, the length of tree T in average is at most $l(T_{smt}) + (2/m)h$. If $h \le O(1)l(T_{smt})$, then we may choose $m = O(1/\epsilon)$ to get an $(1 + \epsilon)$-approximation while the dynamic programming runs only in time of $n \, 2^{O(1/\epsilon)}$.

This idea looks so wonderful and very promising. However, there are two technical problems to solve. The first problem is that it is in general not true that $h \le O(1)l(T_{smt})$; The second problem is that the dynamic programming is applying to the patched Hanan grid. So computing patched Hanan grid is a part of algorithm for approximation. As a result, we also need to count the computation time for patching.

If we directly estimate the running time of the patching procedure, then the running time would be not what we expected. Indeed, for each quadtree partition $P(a, a)$, there are four cuts at the 0-th level, 4^2 cuts at the 1-st level, $\cdots$, r^{q+1} cuts at the $(q - 1)$-th level. For each cut at the i-th level, the patching procedure runs in time $O(2^{q-i})$. Thus, the total running time for patching is bounded by

$$O\Big(\sum_{i=0}^{q-1} 4^{i+2} 2^{q-i} \Big) = O(q \, 2^{2q}) = O(n^2 \log n).$$

How can we do patching more efficiently? In fact, the patching procedure wastes lots of time on some segments without any crosspoint. Given a quadtree partition $P(a, a)$, the following equivalent procedure can avoid such wastage.

Algorithm 3.2 *Modified Patching Procedure*

---

for each square S in $P(a, a)$ starting from the lowest level **do**
    for each $e \in S$ **do**
        if e contains more than m crosspoints **then** patch e.
    end-for-e
end-for-S

Now suppose that Hanan grid has v vertices. Then $P(a, a)$ has only $O(v)$ squares and hence the modified patching procedure would run in time of $O(vm)$. This means that if $v = O(n)$, then $O(vm) = (n/\epsilon)$, which is what we really desire. Unfortunately, in general, the number of vertices in Hanan grid is $O(n^2)$ instead of $O(n)$.

The above mentioned two technical problems motivate us to consider something else with more properties instead of Hanan grid, but still containing an optimal or near optimal solution.

An $(1 + c)$-*banyan* of a set S of points in the rectilinear plane is a graph with vertex set containing S such that for any subset S' of S, the shortest subgraph of the banyan interconnecting S' has the total length within a fact of $(1 + c)$ from the rectilinear SMT on S'. the following result shows that $(1 + c)$-banyan can really replace Hanan grid to make our wild idea work.

Lemma 3.14   *For any $c > 0$, there exists an $(1 + c)$-banyan of any set S of n points in the rectilinear plane such that*
  (i)   *the total length of the $(1 + c)$-banyan is at most a constant factor longer than the rectilinear SMT on S;*
  (ii) *the number of vertices in the $(1 + c)$-banyan is $O(n)$;*
  (iii) *the $(1 + c)$-banyan can be computed in time of $O(n \log n)$.*

The property (i) solves the above first problem while the property (ii) solves the second. In addition, the property (iii) is clearly also necessary. Now using $(1 + \epsilon/2)$-banyan instead of Hanan grid and choose m such that $2/m$ times the total length of the banyan is no more than $(\epsilon/2) l(T_{smt})$. Then from the above lemma we can easily deduce the following theorem.

Theorem 3.9   *With probability at least $1/2$, an $(1 + \epsilon)$-approximation for the rectilinear SMT can be computed in time of $O(n \log n)$ for fixed ϵ. The derandomization needs only time of $O(n/\epsilon)$.*

3.4   Discussions

Clearly, from the 1-guillotine cut to the m-guillotine cut, there is no any technical difficulty. However, why had Mitchell not did such a natural extension before Arora [10] published his remarkable results that we mentioned at the beginning of this section? The answer is that before Arora's breakthrough, nobody was thinking towards that goal. In fact, Arora was well known for his work showing that for many problems[2], no PTAS can exist unless $P = NP$. He was trying to prove the same result for the Euclidean travelling salesman problem. When he realized it would not work, he began to try to find a PTAS and he succeeded. In deed, the importance of Arora's work [10] is more on opening people's mind than proposing new techniques. We observed that Arora [11] used the quadtree partition and, by absorbing some nutrition from m-guillotine cuts, improved the patching procedure that proposed in his early work [10].

It is an open problem whether the problems of minimum edge-length rectangular partition, the rectilinear Steiner arborescence, the symmetric rectilinear Steiner arborescence, the Euclidean k-medians, and the Euclidean facility location each admits a PTAS with running time of $n^c(\log n)^{O(1/\epsilon)}$.

Bern and Eppstein [32] and Arora [12] give two nice surveys on approximation algorithms for many geometric problems including many Steiner tree related problems, some of them will be discussed in the following chapters of this book.

---

[2] Arora had his first work reported in *New York Times* in 1992 about the probablistic checkable proof system [16; 18], and his work on PTAS for many geometric optimization problems [10] was also reported in *New York Times*.

Chapter 4

Grade of Service Steiner Tree Problem

The *Grade of Service Steiner Tree* (GoSST) problem arises from two different applications: the general design of interconnection networks with different *grade of service* requests [71] and the multimedia distribution for users with different bitrate requests [205]. In the first application, all major universities are supposed to be connected to the Internet via a T3-line, while other universities and colleges are supposed to be connected to the Internet via a T1-line. It is clear that the Cost-Per-Unit-Length (CPUL) of a T3-line is more than the CPUL of a T1-line. In the second application, each node possesses a rate and the cost of a link is not constant but depends both on the cost per unit of transmission bandwidth and the maximum rate routed through the link.

The geometric version of minimum GoSST problem could be considered as a natural generalization of the Steiner tree problem in the Euclidean plane. Given a set P of n terminal points in the Euclidean plane, $P = \{t_i | i = 1, 2, \cdots, n\}$, each terminal t_i has a *service request of grade* $r(t_i) \in \{1, 2, \cdots, n\}$. Each edge e is assigned a specific grade of service $r(e)$, which is classified as a number in $\{1, 2, \cdots, n\}$. Let $0 < c(1) < c(2) < \cdots < c(n)$ be n given real numbers. The objective is to find the Steiner tree interconnecting all terminals in set P and some Steiner points each with a service request of grade zero such that

(1) between each pair of terminal points t_i and t_j there is a path whose minimum grade of service assigned is at least $\min\{r(t_i), r(t_j)\}$, and

(2) the cost of the tree is minimum among all Steiner trees satisfying (1), where the cost of an edge with service of grade r is the product of the Euclidean length of the edge $l(e)$ with $c(r)$.

The minimum cost Steiner tree interconnecting P and satisfying conditions (1-2) is called a minimum GoSST. The problem is more formally formulated as follows.

Problem 4.1   *The GoSST Problem in Euclidean Plane*

Instance   A set of n terminal in Euclidean plane $P = \{t_i \mid i = 1, 2, \cdots, n\}$ each with a service request of grade $r(t_i) \in \{1, 2, \cdots, n\}$, and a set of costs for n service grades $0 < c(1) < c(2) < \cdots < c(n)$. Grades of service for points not in P are zero.

Solution   A Steiner tree T interconnecting all points in P with an assigned grade of service to each edge $e = (p, q) \in T$ such that $r(e) \geq \min\{r(p), r(q)\}$ for any path between p and q including edge e.

Objective Minimizing the total cost of the grade of service of edges in T, $\sum_{e \in T} c(r(e)) \cdot l(e)$.

Fig.4.1 gives an instance of Problem 4.1. There are seven terminal points in P which form a hexagon of unit length with one terminal located at the center. The number in each node gives the grade of service of the associated terminal. Fig.4.1(a-b) show two GoSSTs, which are two Steiner Minimum Trees (SMTs) for P, the number on each edge gives the grade of service of that edge. It can be verified that the total cost of GoSST of Fig.4.1(a) is $5\sqrt{3}$ and that of GoSST of Fig.4.1(a) $14\sqrt{3}/3$, which, in fact, is the minimum GoSST, the optimal solution to Problem 4.1. In addition, Fig.4.1(c) shows the minimum grade of service spanning tree that has total cost of 9. It is easy to see that there exists a Minimum Spanning Tree (MST) whose total grade of service is 12.

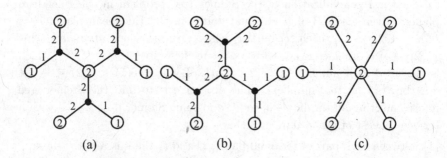

(a)   (b)   (c)

Fig. 4.1   An instance of Problem 4.1: (a) a GoSST, (b) the minimum GoSST, and (c) the minimum grade of service spanning tree.

The graph version of minimum GoSST problem, which is known as the *quality of service multicast tree problem* [49; 164], has a different formulation. Given a graph $G(V, E)$ with a specified source vertex $s \in V$, each edge $e \in E$ has a length $l(e)$ and each vertex $v \in V\}$ has a rate $r(v) \geq 0$ with $r(s) = 0$. Let r_i be the range of r and S_i be the set of vertices with rate r_i for $i = 1, 2, \cdots n$ with $r_0 = 0$. The objective is to find a minimum cost Steiner tree T of G spanning a given source vertex s and all vertices in $\cup_{i \geq 1} S_i$, all of which are referred to as *terminal vertices*. The cost of an edge $e \in T$ is the product of the length of the edge $l(e)$ with the rate of the edge $r(e)$, where $r(e)$ is the maximum rate in the component of $T \setminus \{e\}$ that does not contain the source vertex. Note that all vertices in S_0 with zero rate are not required to be connected to the source vertex s, but some of them may be included in T as Steiner vertices. The minimum cost Steiner tree interconnecting all terminal vertices in P is called a minimum GoSST. The problem is more formally formulated as follows.

Problem 4.1' *Minimum GoSST Problem in Graphs*

Instance	A graph $G(V, E)$ with a source vertex $s \in V$, each vertex $v \in V$ has a rate $r(v) \geq 0$ and each edge $e \in E$ has a length $l(e) > 0$.
Solution	A Steiner tree T interconnecting vertex s and all vertices having nonzero rates with a rate $r(e)$ on each edge $e \in T$ equal to the maximum rate in the component of $T \setminus \{e\}$ excluding source s.
Objective	Minimizing the total rate of all edges in T, $l(T) \equiv \sum_{e \in T} r(e) l(e)$.

Fig.4.2 gives an instance of Problem 4.1'. Fig.4.2(a) shows a graph of seven vertices, four of them have rates greater than 0 while the other all have zero rate and one of them is the source. Each edge has a length between one and seven. Fig.4.2(b) shows a Steiner tree T that has six edges, it has total cost of rates $c(T) = 4 \times 2 + 6 \times 4 + 6 \times 1 + 6 \times 1 + 6 \times 2 + 4 \times 2 = 64$ and total cost of lengths $l(T) = 2 + 4 + 1 + 1 + 2 + 2 = 12$. Fig.4.2(c) shows the minimum GoSST with $c(T) = 62$ and $l(T) = 11$.

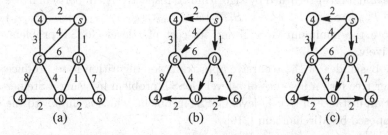

Fig. 4.2   An instance of Problem 4.1': (a) a graph, (b) a GoSST, and (c) the minimum GoSST.

Karpinski et al. [164] observed that the above two problems are equivalent. In fact, an instance of Problem 4.1' can be transformed into an instance of Problem 4.1 in graphs by assigning the maximum rate of all vertices to the service request of grade of the source vertex. The length of an edge will remain the same. Note that each edge e in a Steiner tree T will be on the path from the source to the vertex of the maximum rate in the component of $T \setminus \{e\}$ that does not contain the source. Conversely, an instance of Problem 4.1 in graphs can be transformed into an instance of Problem 4.1' by letting the vertex with the maximum service request of grade to be the source vertex.

In this chapter, we will first study some approximation algorithms for the GoSST problem in Euclidean plane in Section 4.1, and then for the problem in graphs in Section 4.2. In the end we will also discuss some related problems.

4.1   GoSST Problem in the Euclidean Plane

4.1.1   *Recursive Approximation Algorithm*

In this section we will focus on GoSST problem in the Euclidean plane with r different grades of service request, which will be called *r-level GoSST problem*. We will study the recursive approximation algorithm for this problem [272].

For an instance of r-level GoSST problem, we can partition the terminal set P into r subsets $S_1, S_2, \cdots, S_r$ with points in S_i all having a grade of service request r_i, and it is assumed that $1 \leq r_1 \leq r_2 \leq \cdots \leq r_r$. We will use GoSST$(r; S_1, r_1; S_2, r_2; \cdots; S_r, r_r)$ to denote an instance of r-level GoSST problem. In addition, we will use $T_A(r; S_1, r_1; S_2, r_2; \cdots; S_r, r_r)$ and $c_A(r; S_1, r_1; S_2, r_2; \cdots; S_r, r_r)$ to denote the tree and its cost returned by algorithm A, respectively. In particular, we will use $T_{opt}(r; S_1, r_1; S_2, r_2; \cdots; S_r, r_r)$ and $c_{opt}(r; S_1, r_1; S_2, r_2; \cdots; S_r, r_r)$ to denote the minimum GoSST and its cost of r-level GoSST problem, respectively.

The basic idea of the recursive approximation algorithm is to produce a Steiner tree for an instance of r-level GoSST problem by using a Steiner tree for an instance of $(r-1)$-level GoSST problem, which is similar to the one proposed by Mirchandani [210].

Algorithm 4.1 *Recursive Approximation Algorithm*

Step 1 When $r = 1$, produce an α_0-approximate SMT for terminal set P (since in this case the r-level GoSST problem is equivalent to the Steiner tree problem).

Step 2 Produce a Steiner tree T_1 for $(r - 1)$-level GoSST problem, $\text{GoSST}(r - 1; S_1, r_1; S_2, r_2; \cdots ; S_{r-2}, r_{r-2}; S_{r-1} \cup S_r, r_r)$.

Step 3 Produce a Steiner tree T' for $\text{GoSST}(1; S_r, r_r)$;
Produce a Steiner tree T'' for $\text{GoSST}(r - 1; S_1, r_1; S_2, r_2; \cdots ; S_{r-2}, r_{r-2}; S_{r-1} \cup S_r, r_{r-1})$;
Merge T' and T'' into a Steiner tree T_2 by removing the longest edge in any (possible) cycle with the lowest grade of service.

Step 4 Return T_1 if $c(T_1) \leq c(T_2)$; Otherwise return T_2.

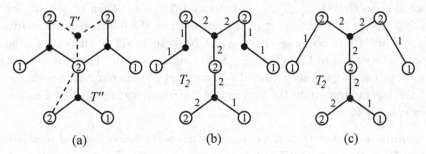

Fig. 4.3 Demonstration of the recursive algorithm: (a) Steiner trees T' and T'' and (b) Steiner tree T_2 obtained through the merging operation.

Fig.4.3 illustrates the recursive algorithm applied to an instance of GoSST problem 4.1 with seven terminals and $r = 2$. Fig.4.3(a) shows Steiner tree T' for spanning terminals of $r_2 = 2$ (which consists of dashed edges), and Steiner tree T'' for spanning all terminals (which consists of solid edges), four Steiner points are represented by black nodes. Fig.4.3(b) shows Steiner tree T_2 with grades of service on each edge in T_2 after merging T' and T''. Note that the cost of T_2 could be reduced through shortcut operation as shown in Fig.4.3(c).

The performance analysis for the above recursive algorithm for Problem 4.1 applies the same recursive idea as introduced by Mirchandani [210] for Problem 4.1'. However, the proof is a combinatorial one, rather than using a nonlinear programming approach. Mirchandani [210] proved the following three technical lemmas, which will be used in our discussions.

Lemma 4.1   *For any $\beta > 1$, the following function achieves its maximum $4/3$ at $\beta = 2$*

$$f(\beta) = \frac{\beta^2}{\beta^2 - \beta + 1}.$$

Lemma 4.2   *For any $\gamma > \beta > 1$, the following function achieves its maximum $3/2$ at $(\gamma, \beta) = (3, 2)$*

$$f(\gamma, \beta) = \frac{\gamma^2}{\gamma^2 - \gamma\beta + \beta^2 - \beta + 1}.$$

Lemma 4.3   *For any $\gamma > \beta > 1$, the following function achieves its maximum $(5 + 4\sqrt{2})/$ at $(\gamma, \beta) = ((4 + 3\sqrt{2})/2, 1 + \sqrt{2})$*

$$f(\gamma, \beta) = \frac{\gamma^2\beta - \gamma^2 + \gamma\beta}{\gamma^2\beta - \gamma^2 - \gamma\beta^2 + \gamma\beta + \gamma - \beta^2 + \beta^3}.$$

For the easy presentation, we will use c_i^* to denote the cost of the optimal solution to GoSST$(1; \cup_{k=i}^r S_k, 1)$ for $i = 1, 2, \cdots, r$. Then $c_{i+1}^* \le c_i^*$ for $i = 1, 2, \cdots, r$ and the cost of the Steiner tree T_1 returned by recursive algorithm 4.1 is no more than $\alpha_0 c_i^*$. Note that αc_i^* is the cost of the optimal solution to GoSST$(1; \cup_{k=i}^r S_k, \alpha)$ for any $\alpha > 0$. Thus we assume, without loss of generality, that $c_1^* = 1$ and $c(r_1) = 1$ in our analysis. In the following we first study the performance of recursive algorithm 4.1 in the case of $r = 2$.

Lemma 4.4   *Let α_2 be the approximation performance ratio of recursive algorithm 4.1 for r-level GoSST problem with $r = 2$. Then we have*

$$\alpha_2 \le \frac{\min\{c(r_2), c(r_2)c_2^* + 1\}}{(c(r_2) - 1)c_2^* + 1} \alpha_0. \qquad (4.1)$$

Proof.   Let T_{2-opt} be an optimal solution to 2-level GoSST$(2; S_1, r_1; S_2, r_2)$. Since the subtree of T_{2-opt} reduced by the terminals in S_2 is a tree with grade service r_2 which spans N_2, we then have

$$c_{opt}(2; S_1, r_1; S_2, r_2) \ge \big(c(r_2) - c(r_1)\big)c_2^* + c(r_1)c_1^*. \qquad (4.2)$$

As we assume that $c(r_1) = 1$ and $c_1^* = 1$, we thus have

$$c_{opt}(2; S_1, r_1; S_2, r_2) \ge \big(c(r_2) - 1\big)c_2^* + 1. \qquad (4.3)$$

Now let T_{RA} be the tree returned by recursive algorithm 4.1. If $T_{RA} = T_1$, which is the Steiner tree produced by applying the algorithm to GoSST$(1; S_1 \cup S_2, r_2)$ at Step 2, then we obtain $c(T) \le \alpha_0 c(r_2) c_1^* = \alpha_0 c(r_2)$. Otherwise, $T_{RA} = T_2$, which is the Steiner tree produced by applying the algorithm at Step 3, we obtain $c(T_{RA}) \leq c(r_2)\, c_2^* + c(r_1)\, c_1^*$. Therefore, for both cases, we have

$$c_{RA}(2; S_1, r_1; S_2, r_2) = c(T_{RA}) \leq \alpha_0 \min\left\{c(r_2), c(r_2)c_2^* + 1\right\}, \quad (4.4)$$

which, together with inequality (4.3), yields inequality (4.1). The proof is then finished. □

Theorem 4.1 *Recursive algorithm 4.1 for r-level GoSST problem with $r = 2$ has approximation ratio $\alpha_2 \leq \frac{4}{3}\alpha_0$.*

Proof. By Lemma 4.4, to prove the theorem it suffices to show that for any $\beta > 1$ and $c_2^* \in [0, 1]$

$$\min\left\{\frac{\beta}{(\beta - 1)c_2^* + 1}, \frac{\beta c_2^* + 1}{(\beta - 1)c_2^* + 1}\right\} \leq \frac{4}{3}. \quad (4.5)$$

Note that $\beta/((\beta - 1)c_2^* + 1)$ is monotonically decreasing with respect to c_2^* for $c_2^* \in [0, 1]$, and $(\beta c_2^* + 1)/((\beta - 1)\, c_2^* + 1)$ is monotonically increasing with respect to c_2^* for $c_2^* \in [0, 1]$. Therefore, the minimum of the two is achieved only when $\beta = \beta\, c_2^* + 1$. In this case by Lemma 4.1 we obtain

$$\frac{\beta}{(\beta - 1)c_2^* + 1} = \frac{\beta^2}{\beta^2 - \beta + 1} \leq \frac{4}{3},$$

which leads to inequality (4.5). The proof is then finished. □

Applying the same argument we can obtain the parallel results in case of $r = 3$.

Lemma 4.5 *Let α_3 be the approximation performance ratio of recursive algorithm 4.1 for the r-level GoSST problem with $r = 3$. Then we have*

$$\alpha_3 \leq \frac{\min\left\{c(r_3), c(r_3)c_2^* + 1, c(r_3)c_3^* + c(r_2), c(r_3)c_3^* + c(r_2)c_2^* + 1\right\}}{(c(r_3) - c(r_2))c_3^* + (c(r_2) - 1)c_2^* + 1} \alpha_0. \quad (4.6)$$

Proof. Let T_{3-opt} be an optimal solution to 3-level GoSST$(3; S_1, r_1; S_2, r_2; S_3, r_3)$. Since the subtree of T_{3-opt} reduced by the terminals in S_3 is a Steiner tree with grade service r_3 which spans N_3, and the subtree of T_{3-opt} reduced by the terminals in $S_3 \cup S_2$ is a Steiner tree with grade of service at least r_2 which spans $S_3 \cup S_2$, we then have

$$\begin{aligned}
c(T_{3-opt}) &= c_{3-opt}(2; S_1, r_1; S_2, r_2; S_3, r_3) \\
&\geq (c(r_3) - c(r_2))c_3^* + (c(r_2) - c(r_1))c_2^* + c(r_1)\, c_1^*.
\end{aligned}$$

from which, together with the assumption that $c(r_1) = 1$ and $c_1^* = 1$, we deduce

$$c(T_{3-opt}) \geq (c(r_3) - c(r_2))c_3^* + (c(r_2) - 1)c_2^* + 1. \qquad (4.7)$$

Now let T_{RA} be the tree returned by recursive algorithm 4.1, then we obtain

$$c(T_{RA}) = c_{RA}(3; S_1, r_1; S_2, r_2; S_3, r_3) \leq c_{RA}(2; S_1, r_1; S_2 \cup S_3, r_3), \text{ and}$$

$$c(T_{RA}) = c_{RA}(3; S_1, r_1; S_2, r_2; S_3, r_3) \leq c_{RA}(1; S_3, r_3) + c_{RA}(2; S_1, r_1; S_2 \cup S_3, r_2).$$

Following the proof of inequality (4.4), we deduce the following three inequalities

$$c_{RA}(2; S_1, r_1; S_2 \cup S_3, r_3) \leq \alpha_0 \min\{c(r_3), c(r_3)c_2^* + 1\}\rho,$$
$$c_{RA}(1, S_3, r_3) \leq c(r_3)\, c_3^*\, \alpha_0$$
$$c_{RA}(2; S_1, r_1; S_2 \cup S_3, r_2) \leq \alpha_0 \min\{c(r_2), c(r_2)c_2^* + 1\}.$$

Therefore we have

$$c(T_{RA}) \leq \min\left\{c(r_3), c(r_3)c_2^* + 1, c(r_3)c_3^* + c(r_2), c(r_3)c_3^* + c(r_2)c_2^* + 1\right\},$$

from which, along with inequality (4.7), we obtain inequality (4.6), the proof is then finished. □

Theorem 4.2 *Recursive algorithm 4.1 for the r-level GoSST problem with $r = 3$ has approximation ratio $\alpha_3 \leq \frac{5+4\sqrt{2}}{7}\alpha_0$.*

Proof. By Lemma 4.5, to prove the theorem it suffices to show that for any $\gamma > \beta > 1$ and $1 > c_2^* > c_3^*$

$$\frac{\min\left\{\gamma, \gamma c_2^* + 1, \gamma c_3^* + \beta, \gamma c_3^* + \beta c_2^* + 1\right\}}{(\gamma - \beta)c_3^* + (\beta - 1)c_2^* + 1} \leq \frac{5 + 4\sqrt{2}}{7}. \qquad (4.8)$$

In the following discussion, we consider four cases separately with respect to the value of c_2^*.

Case 1. $c_2^* \geq c_3^* \geq (\gamma - 1)/\gamma$. In this case we have

$$\frac{\min\left\{\gamma, \gamma c_2^* + 1, \gamma c_3^* + \beta, \gamma c_3^* + \beta c_2^* + 1\right\}}{(\gamma - \beta)c_3^* + (\beta - 1)c_2^* + 1}$$

$$\leq \frac{\gamma}{(\gamma - \beta)c_3^* + (\beta - 1)c_2^* + 1}$$

$$\leq \frac{\gamma}{(\gamma - \beta)c_3^* + (\beta - 1)c_3^* + 1}$$

$$\leq \frac{\gamma}{(\gamma - 1)c_3^* + 1}$$

$$\leq \frac{\gamma^2}{(\gamma^3 - \gamma) + 1} \leq \frac{4}{3} \qquad \text{(By Lemma 4.1).}$$

Case 2. $c_2^* \geq (\gamma - 1)/\gamma \geq c_3^*$. In this case we have

$$\frac{\min\left\{\gamma, \gamma c_2^* + 1, \gamma c_3^* + \beta, \gamma c_3^* + \beta c_2^* + 1\right\}}{(\gamma - \beta)c_3^* + (\beta - 1)c_2^* + 1}$$

$$\leq \frac{\min\left\{\gamma, \gamma c_3^* + \beta\right\}}{(\gamma - \beta)c_3^* + (\beta - 1)c_2^* + 1}$$

$$\leq \frac{\min\left\{\gamma, \gamma c_3^* + \beta\right\}}{(\gamma - \beta)c_3^* + (\beta - 1)(\gamma - 1)/\gamma + 1}. \qquad (4.9)$$

Since the function

$$\frac{\gamma}{(\gamma - \beta)c_3^* + (\beta - 1)(\gamma - 1)/\gamma + 1}$$

is monotonically decreasing in c_3^* and the function

$$\frac{\gamma c_3^* + \beta}{(\gamma - \beta)c_3^* + (\beta - 1)(\gamma - 1)/\gamma + 1}$$

is monotonically increasing in c_3^*, and $0 \leq c_3^* < c_2^* < 1$, the righthand term in inequality (4.9) achieves its maximum only when $c_3^* = (\gamma - \beta)/\gamma$, and the maximum value is

$$\frac{\gamma^2}{(\gamma - \beta)^2 + (\beta - 1)(\gamma - 1) + \gamma} \leq \frac{3}{2},$$

where the above inequality follows from Lemma 4.2.

Case 3. $(\gamma - 1)/\gamma \geq c_2^* \geq (\beta - 1)/\beta$. In this case we have

$$\frac{\min\left\{\gamma, \gamma c_2^* + 1, \gamma c_3^* + \beta, \gamma c_3^* + \beta c_2^* + 1\right\}}{(\gamma - \beta)c_3^* + (\beta - 1)c_2^* + 1} = \frac{\min\left\{\gamma c_2^* + 1, \gamma c_3^* + \beta\right\}}{(\gamma - \beta)c_3^* + (\beta - 1)c_2^* + 1}.$$

Case 3.1. $\gamma c_2^* + 1 \leq \gamma c_3^* + c_2^*$. It follows that $c_2^* \leq c_3^* + (\beta - 1)/\gamma$, and

$$\frac{\min\{\gamma c_2^* + 1, \gamma c_3^* + \beta\}}{(\gamma - \beta)c_3^* + (\beta - 1)c_2^* + 1} = \frac{\gamma c_2^* + 1}{(\gamma - \beta)c_3^* + (\beta - 1)c_2^* + 1}$$

$$\leq \frac{\gamma c_2^* + 1}{(\gamma - \beta)(c_2^* - (\beta - 1)/\gamma) + (\beta - 1)c_2^* + 1}$$

$$= \frac{\gamma c_2^* + 1}{(\gamma - 1)c_2^* + (1 - (\gamma - \beta)(\beta - 1)/c_3^*)}$$

The term in the last equality is a decreasing function with respect to c_2^* when $(\gamma - \beta)(\beta - 1) \geq 1$, and it is an increasing function with respect to c_2^* when $(\gamma - \beta_2)(\beta - 1) \leq 1$. In the former case, it achieves its maximum at $c_2^* = (\beta - 1)/\beta$, here we have $c_3^* = (\beta - 1)/\beta - (\beta - 1)/\gamma$, and the maximum value is

$$\frac{\gamma^2 \beta - \gamma^2 + \gamma\beta}{\gamma^2\beta - \gamma^2 - \gamma\beta^2 + \gamma\beta + \gamma - \beta^2 + \beta^3} \leq \frac{5 + 4\sqrt{2}}{7}.$$

In the latter case, it achieves its maximum at $c_2^* = (\gamma - 1)\gamma$, here we have $c_3^* = (\gamma - 1)/\gamma - (\beta - 1)/\gamma = (\gamma - \beta)/\gamma$, and the maximum value is

$$\frac{\gamma^2}{\gamma^2 - \gamma\beta + \beta^2 - \beta + 1} \leq \frac{3}{2}.$$

Case 3.2. $\gamma c_2^* + 1 \geq \gamma c_3^* + c_2^*$. It follows that $c_2^* \geq c_3^* + (\beta - 1)/\gamma$, and

$$\frac{\min\{\gamma c_2^* + 1, \gamma c_3^* + \beta\}}{(\gamma - \beta)c_3^* + (\beta - 1)c_2^* + 1} \leq \frac{\gamma c_3^* + \beta}{(\gamma - \beta)c_3^* + (\beta - 1)c_2^* + 1}.$$

Note that for any fixed c_3^*, the righthand term in the above inequality is a decreasing function with respect to c_2^*. Moreover, if $c_3^* \geq (\beta - 1)/\beta - (\beta - 1)/\gamma$, it achieves the maximum when $c_2^* = c_3^* + (\beta - 1)/\gamma$; And if $c_3^* \leq (\beta - 1)/\beta - (\beta - 1)/\gamma$, it achieves the maximum when $c_2^* = (\beta - 1)/\beta$. In the former case, we have

$$\frac{\gamma c_3^* + \beta}{(\gamma - \beta)c_3^* + (\beta - 1)c_2^* + 1} = \frac{\gamma c_3^* + \beta}{(\gamma - \beta)c_3^* + (\beta - 1)(c_3^* + (\beta - 1)/c_3^*) + 1}$$

$$= \frac{\gamma c_3^* + \beta}{(\gamma - 1)c_3^* + ((\beta - 1)^2/\gamma + 1)}.$$

Observe that the right term in the above equality is a decreasing function of c_3^* when $\beta \geq ((\beta - 1)^2 + \gamma)/\gamma - 1$ while an increasing function of c_3^* when

$\beta \le ((\beta - 1)^2 + \gamma)/\gamma - 1$, and it achieves the maximum respectively at

$$c_3^* = (\beta - 1)/\beta - (\beta - 1)/\gamma \quad \text{and}$$
$$c_3^* = (\gamma - 1)/\gamma - (\beta - 1)/\gamma = (\gamma - \beta)/\gamma,$$

and the maximum values are

$$\frac{\gamma^2\beta - \gamma^2 + \gamma\beta}{\gamma^2\beta - \gamma^2 - \gamma\beta^2 + \gamma\beta + \gamma - \beta^2 + \beta^3} \le \frac{5 + 4\sqrt{2}}{7} \quad \text{and}$$
$$\frac{\gamma^2}{\gamma^2 - \gamma\beta^2 + \beta^2 - \beta + 1} \le \frac{3}{2}, \quad \text{respectively.}$$

In the latter case, we have $c_3^* \le (\beta - 1)/\beta - (\beta - 1)/\gamma$ and

$$\frac{\gamma c_3^* + \beta}{(\gamma - \beta)c_3^* + (\beta - 1)c_2^* + 1} = \frac{\gamma c_3^* + \beta}{(\gamma - \beta)c_3^* + (\beta - 1)((\beta - 1)/\beta) + 1}$$
$$= \frac{\gamma c_3^* + \beta}{(\gamma - 1)c_3^* + ((\beta - 1)^2/\beta + 1)}.$$

Observe that the right term in the above equality is a decreasing function of c_3^* when $\gamma \ge \beta^3/(\beta - 1)$ while an increasing function of c_3^* when $\gamma \le \beta^3/(\beta - 1)$, and it achieves the maximum at $c_3^* = 0$ and $c_3^* = (\beta - 1)/\beta - (\beta - 1)/\gamma$, respectively, and the maximum values are

$$\frac{\beta^2}{\beta^2 - \beta + 1} \le \frac{4}{3} \quad \text{and}$$
$$\frac{\gamma^2\beta - \gamma^2 + \gamma\beta}{\gamma^2\beta - \gamma^2 - \gamma\beta^2 + \gamma\beta + \gamma - \beta^2 + \beta^3} \le \frac{5 + 4\sqrt{2}}{7}, \quad \text{respectively.}$$

Case 4. $0 \le c_3^* \le c_2^* \le (\beta - 1)/\beta$. In this case we have

$$\frac{\min\{\gamma, \gamma c_2^* + 1, \gamma c_3^* + \beta, \gamma c_3^* + \beta c_2^* + 1\}}{(\gamma - \beta)c_3^* + (\beta - 1)c_2^* + 1} = \frac{\min\{\gamma c_2^* + 1, \gamma c_3^* + \beta c_2^* + 1\}}{(\gamma - \beta)c_3^* + (\beta - 1)c_2^* + 1}.$$

Case 4.1. $\gamma c_2^* + 1 \le \gamma c_3^* + \beta c_2^* + 1$. It follows that $c_3^* \le c_2^* \le (\gamma/(\gamma - \beta))c_3^*$

and

$$\frac{\min\left\{\gamma c_2^* + 1, \gamma c_3^* + \beta c_2^* + 1\right\}}{(\gamma - \beta)c_3^* + (\beta - 1)c_2^* + 1} = \frac{\gamma c_2^* + 1}{(\gamma - \beta)c_3^* + (\beta - 1)c_2^* + 1}$$

$$\leq \frac{\gamma c_2^* + 1}{(\gamma - \beta)((\gamma - \beta)/\gamma)c_2^* + (\beta - 1)c_2^* + 1}$$

$$\leq \frac{\gamma((\beta - 1)/c_2^*) + 1}{((\gamma - \beta)^2/\gamma + \beta - 1)((\beta - 1)/c_2^*) + 1}$$

$$\leq \frac{\gamma^3\beta - \gamma^2 + \gamma\beta}{\gamma^3\beta - \gamma^2 - \gamma\beta^2 + \gamma\beta + \gamma - \beta^2 + \beta^3}$$

$$\leq \frac{5 + 4\sqrt{2}}{7}.$$

Case 4.2. $\gamma c_2^* + 1 \geq \gamma c_3^* + \beta c_2^* + 1$. It follows that $c_2^* \geq (\gamma/(\gamma - \beta))c_3^*$ and

$$\frac{\min\left\{\gamma c_2^* + 1, \gamma c_3^* + \beta c_2^* + 1\right\}}{(\gamma - \beta)c_3^* + (\beta - 1)c_2^* + 1} = \frac{\gamma c_3^* + \beta c_2^* + 1}{(\gamma - \beta)c_3^* + (\beta - 1)c_2^* + 1}$$

$$= \frac{1}{\gamma - \beta}\left(\gamma + \frac{(\gamma - \beta^2)c_2^* - \beta}{(\gamma - \beta)c_3^* + (\beta - 1)c_2^* + 1}\right).$$

Observe that the righthand term in the above equality is an increasing function with respect to c_3^* for any fixed c_2^*. Therefore, it achieves the maximum at $c_3^* = ((\gamma - \beta)/\gamma)c_2^*$, and we obtain

$$\frac{\min\left\{\gamma c_2^* + 1, \gamma c_3^* + \beta c_2^* + 1\right\}}{(\gamma - \beta)c_3^* + (\beta - 1)c_2^* + 1} \leq \frac{\gamma c_2^* + 1}{((\gamma - \beta)^2/c_3^* + (\beta - 1))c_2^* + 1}$$

$$\leq \frac{\gamma^2\beta - \gamma^2 + \gamma\beta}{\gamma^2\beta - \gamma^2 - \gamma\beta^2 + \gamma\beta + \gamma - \beta^2 + \gamma^3}$$

$$\leq \frac{5 + 4\sqrt{2}}{7}.$$

When $\gamma - \beta^2 \geq 0$, we have

$$\frac{\min\left\{\gamma c_2^* + 1, \gamma c_3^* + \beta c_2^* + 1\right\}}{(\gamma - \beta)c_3^* + (\beta - 1)c_2^* + 1} = \frac{\gamma c_3^* + \beta c_2^* + 1}{(\gamma - \beta)c_3^* + (\beta - 1)c_2^* + 1}$$

$$= \frac{1}{\beta - 1}\left(\beta + \frac{(\beta^2 - \gamma)c_3^* - 1}{(\gamma - \beta)c_3^* + (\beta - 1)c_2^* + 1}\right).$$

Observe that the righthand term in the above equality is an increasing function with respect to c_2^* for any fixed c_3^*. Therefore, it achieves the maximum at $c_2^* = (\beta - 1)/\beta$, and we obtain

$$\frac{\min\{\gamma c_2^* + 1, \gamma c_3^* + \beta c_2^* + 1\}}{(\gamma - \beta)c_3^* + (\beta - 1)c_2^* + 1} \leq \frac{\gamma c_3^* + \beta}{(\gamma - \beta)c_3^* + (\beta - 1)^2/\beta + 1}$$

$$= \frac{\gamma \beta c_3^* + \beta^2}{\beta(\gamma - \beta)c_3^* + (\beta^2 - \beta + 1)}.$$

Observe that the righthand term in the above equality is an increasing function of c_3^* when $\gamma \leq \beta^3/(\beta - 1)$ while a decreasing function of c_3^* when $\gamma \geq \beta^3/(\beta - 1)$, and it achieves the maximum at $c_3^* = 0$ and $c_3^* = ((\gamma - \beta)/\gamma)((\beta - 1)/\beta)$, respectively, and the maximum values are

$$\frac{\beta^2}{\beta^2 - \beta + 1} \leq \frac{4}{3} \text{ and}$$

$$\frac{\gamma^2\beta - \gamma^2 + \gamma\beta}{\gamma^2\beta - \gamma^2 - \gamma\beta^2 + \gamma\beta + \gamma - \beta^2 + \beta^3} \leq \frac{5 + 4\sqrt{2}}{7}, \text{ respectively.}$$

To summarize, we have shown that in any of the above four cases inequality (4.8) holds true. The proof is then finished. □

4.1.2 *Branch-and-Bound Algorithm*

Given an instance of Problem 4.1, a *tree topology* $T(P)$ for terminal set $P = \{t_i | i = 1, 2, \cdots, n\}$ is an undirected tree graph $T(P) = (V, E)$ where $V = \{v_1, v_2, \cdots, v_n, v_{n+1}, \cdots, v_{n+m}\}$ for some $m \leq n - 2$ is the vertex-set of $T(P)$ such that $\{v_i | i = 1, 2, \cdots, n\}$ is the terminal set P and $\{v_i | i = n + 1, n + 2, \cdots, n + m\}$ is the set of Steiner points and E is the edge-set $T(P)$. A *realization* $R(T)$ of a Steiner topology $T(P)$ is obtained by the following operations:

(1) Assign a grade of service $r(v_i, v_j)$ to each edge $(v_i, v_j) \in E$ such that between each pair of terminal points t_i and t_j the minimum grade of service on the path connecting t_i and t_j is at least as large as $\min\{r(t_i), r(t_j)\}$;

(2) Fix the terminal vertices at their corresponding terminal points;

(3) Fix the Steiner vertices at some locations $\{s_i | i = n+1, n+2, \cdots, n+m\}$.

The *cost of realization* $R(T)$ is defined by

$$c(R(T)) \equiv \sum_{(v_i, v_j) \in E} c(r(v_i, v_j)) \, l(v_i, v_j),$$

and the *cost of Steiner topology* T is given by $c(T) = \min\{c(R(T)) | R$ is a realization of T}. The realization $R(T)$ is called the *minimum cost Steiner tree under topology* T if $c(R(T)) = c(T)$. Given a Steiner topology T with $m = n - 2$, for any realization $R(T)$, the degree of every terminal vertex in $R(T)$ is 1, and the degree of every Steiner vertex in $R(T)$ is 3, so such a topology T is called a *full Steiner topology*.

Let T be Steiner tree topology. An edge $e = (s, t) \in T$ is called a *Steiner-terminal edge* if t is a terminal vertex and s a Steiner vertex. For a Steiner-terminal edge $e = (s, t) \in T$, we can *shrink* edge e by deleting e and replacing vertices s and t with a new terminal vertex t', which corresponds to the terminal vertex t, and any vertex other than s and t is adjacent to t' if and only if it was adjacent to s or t before the shrinking operation. Notice that tree topology T is changed to another tree topology T' after a Steiner-terminal edge is shrunk as shown in Fig.4.4. A tree topology T' is called a *degeneracy* of T if T' can be obtained from T using zero or more number of shrink operations. We will use $S(T)$ to denote the set of tree topologies which are degeneracies of T.

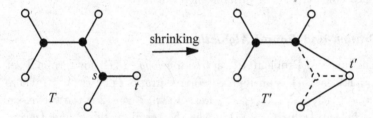

Fig. 4.4   Shrinking operation.

Lemma 4.6   *Given any full Steiner topology T for n terminals, the grades of service for all edges that will appear in the minimum cost network under T can be computed in time of $O(n)$.*

Proof.   Assume, without loss of generality, that t_1 has the largest grade of service request. Then consider T as a tree rooted at v_1, which corresponds to t_1 and make an assignment as follows: For each terminal vertex v_i, set $r(v_i) = r(t_i)$. For each Steiner vertex v_j, $j = n + 1, n + 2, \cdots, n + m$, initially set $r(v_j) = 0$. As long as there is a leaf edge which is not incident to v_1, say (v_i, v_j) where a degree-one vertex v_i is a child of v_j with $v_j \neq v_1$, set $r(v_i, v_j) = r(t_i)$ and $r(v_j) = \max\{r(v_j), r(v_i)\}$. It is clear that such a method assigns the minimum possible grade of service to each edge satisfying the service requirement in time of $O(n)$. The proof is then finished.   □

Lemma 4.7 *Given an instance of the GoSST problem in Euclidean plane, there exists a full Steiner topology T such that the minimum cost network under T has the same cost as that of the optimal solution to the problem.*

Proof. Let T_{opt} be an optimal solution to Problem 4.1. We can modify it into a minimum cost network under a full Steiner topology as follows: If there is a terminal vertex t in T_{opt} that is adjacent to vertices $u_1, u_2, \cdots, u_k$ for $k \geq 2$, then we can introduce a Steiner point s located at the terminal point t and then connect s and t with a zero-length edge while replacing each edge (t, u_i) with a new edge (t', u_i). If there is a Steiner vertex s whose degree $deg(s)$ is greater than three, then we can split s into $deg(s) - 3$ degree-3 Steiner points. It is easy to see that such a modification yields a desired topology. □

Theorem 4.3 *Given an instance of the GoSST problem with n terminals in Euclidean plane and any given full Steiner tree topology T, an $(1 + \epsilon)$-approximation to the minimum cost network under T can be computed in time of $O(n^{1.5}(\log n + \log(1/\epsilon)))$.*

Proof. By Lemma 4.6, we know that the optimal grades of service of edges in the minimum cost network under a given full Steiner tree topology can be computed in time of $O(n)$ without knowing the optimal locations of the Steiner points. In addition, computing the minimum cost network under a full Steiner tree topology is a special case of the problem of *minimizing the sum of Euclidean norms* [6; 7; 74]. This general optimization is to minimize $f(\vec{x}) = \sum_{i=1}^{m} ||r_i(\vec{x})||_2$, where $r_i(\vec{x}) = A_i^t \vec{x} - \vec{b}_i$ with $\vec{x} \in R^n$, $\vec{b}_i \in R^m$, and $A_i \in R^{n \times m}$.

Xue and Ye [274] presented a primal-dual interior point algorithm for computing an ϵ-optimal solution to the problem of minimizing a sum of Euclidean norms and proved that their algorithm requires $O(n^{1.5}(\log n + \log(c/\epsilon)))$ arithmetic operations if the problem has a tree structure, where c is a constant dependent on the input. Using a similar approach introduced in [272] that transforms the terminal points so that the weighted center is at the origin), we can show that this algorithm is also a polynomial time approximation scheme for computing the minimum cost network under a given tree topology that computes an $(1 + \epsilon)$-approximation in time of $O(n^{1.5}(\log n + \log(1/\epsilon)))$. The proof is then finished. □

A brute-force algorithm for solving the GoSST problem is to compute the minimum cost network under a full Steiner topology for every full Steiner topology interconnecting the n terminal points. Note, however, that the number of full Steiner topologies for n terminal points, denoted by $f(n)$, satisfies $f(n+1) = (2n-3)f(n)$ for $n \geq 2$, which yields $f(n) = (2n-4)!/((n-2)!2^{n-2})$ (refer to [144]) or $f(n) = 1 \cdot 3 \cdot \cdots \cdot (2n-7) \cdot (2n-5)$. Thus Lemma 4.7 and Theorem 4.3 could not guarantee an efficient approximation algorithm for the GoSST problem. Smith [247] proved the following lemma. It leads to a branch-and-bound algorithm for the GoSST problem that partially enumerates the Steiner tree topologies.

Lemma 4.8   *There is an one-to-one correspondence between full Steiner topologies on n terminal points and integral vectors $\vec{x} = (x_1, x_2, \cdots, x_{n-3})$ with $1 \leq x_i \leq 2i+1$.*

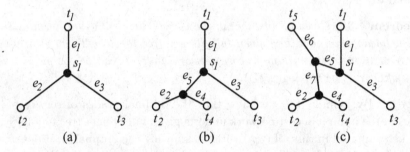

Fig. 4.5   Correspondence between full Steiner tree topologies and $(n-3)$-element vectors.

Fig.4.5 illustrates the one-to-one correspondence for the case of $n = 5$ and $\vec{t} = (2, 5)$. Fig.4.5(a) shows the topology for three terminal points with three edges each incident to a terminal and a common Steiner point s_1, which is the only moving point. To add the fourth terminal point t_4 into the network, we connect it to an interior point, which is a Steiner point s_2, on edge e_2 since $x_{4-3} = 2$ in the corresponding vector. s_2 becomes the second moving point. Observe that edge e_2 is broken into two parts with one part still labeled e_2 while the other labeled e_5 (since $5 = 2 \times 4 - 3$). The edge interconnecting t_4 and s_2 is labeled e_4 (since $4 = 2 \times 4 - 4$). After the above process, we obtain the topology for the first four terminal points as shown in Fig.4.5(b). Similarly, Fig.4.5(c) shows the topology for the first five terminal points, which is obtained by breaking the edge labeled e_5 (since $5 = x_{5-3}$).

Using Lemma 4.8, we can use a backtrack method to enumerate only part of the full topologies and then find the topology that leads to a minimum cost network. Before we present the branch-and-bound algorithm, we need the following lemma that can be used to prune hopeless branches, which generalizes the result obtained in [244].

Lemma 4.9 *For any $k = 3, 4, \cdots, n-1$ and any $x_{k-2} \in \{1, 2, \cdots, 2k - 3\}$, the minimum cost network under a full Steiner topology for terminal set $\{t_1, t_2, \cdots, t_k\}$ with the topology vector $(x_1, x_2, \cdots, v_{k-3})$ has cost no greater than that of the minimum cost network under the full Steiner topology for terminal set $\{t_1, t_2, \cdots, t_{k+1}\}$ with the topology vector $(x_1, x_2, \cdots, x_{k-3}, x_{k-2})$.*

Proof. Let T_{k+1}^* be a minimum cost network under the topology for $\{t_1, t_2, \cdots, t_{k+1}\}$. Then deleting the leaf vertex v_{k+1}, which corresponds to t_{k+1}, along with the edge incident to v_{k+1} forms a new tree T'. Clearly, the optimal grade of service for any edge in the topology of T' is no greater than the optimal grade of service for the (corresponding) edge in the topology of T_{k+1}^*. Moreover, the Steiner vertex originally adjacent with v_{k+1} by the deleted edge can now be removed since it is now a degree-2 interior vertex, which could further shorten the cost of T' by the triangle inequality. Optimizing the resulting network will yield a desired network T_k^*, which may have even less cost. The proof is then finished. $\square$

Algorithm 4.2 *Branch-and-Bound Algorithm*

Step 1 Compute an upper bound U_B on the cost of optimal solution to Problem 4.1 or simply set $U_B := \infty$,
$k := 4$ and $x_1 := 3$.

Step 2 Compute the minimum cost network under the topology with topological vector $(x_1, x_2, \cdots, x_{k-3})$.
Set c^* be the cost of current network.

Step 3 if $c^* \geq U_B$ **then go to** Step 4
                 else go to Step 5.

Step 4 if $x_{k-3} > 1$ **then** $x_{k-3} := x_{k-3} - 1$; **go to** Step 2.
                 else if $k = 4$ **then return** the topological vector
$$(bx_1, bx_2, \cdots, bx_{n-3})$$
(U_B is the minimum cost)
                 else $k := k - 1$; **go to** Step 4.

Step 5 if $k = n$ **then** $U_B := c^*$;
                 $(bx_1, \cdots, bx_{n-3}) :=$ the current topological vector;
                 go to Step 4;
        else $k := k + 1$; $x_{k-3} := 2k + 1$;
                 go to Step 2.

It follows from Lemma 1.9 that the above algorithm correctly computes the optimal solution to any given instance of Problem 4.1. However, in order to make the algorithm to be practically efficient, Xue et al. [272] proposed some methods that produce a good initial upper bound and a good ordering of the terminal points so that the algorithm will generate only a small portion of the whole tree.

4.2   Minimum GoSST Problem in Graphs

In this section we will present the β-convex Steiner tree approximation algorithms proposed by Karpinski et al. [164] for minimum GoSST problem in graphs (Problem 4.1'). The performance analysis is based on more detailed analysis of the k-Steiner ratio ρ_k of graphs in Section 2.1. They proved the following lemmas.

Lemma 4.10   *Let T be a full Steiner tree. Then $l_{k-smt} \leq \rho_k(l(T)- l_{max}(T)) +l_{max}(T)$, where $l_{max}(T)$ is the length of the longest path in T.*

Proof.   For full Steiner tree T, there exits k-restricted Steiner trees T_i, $i = 1, 2, \cdots, r2^r + s$ such that $l(T_i) = l(T) + l_i$, where l_i is the sum of the lengths of the paths $p(u)$ from intermediate leaves u in T_i to tree leaves. By the argument of Theorem 2.2, we know that

$$l_1 + l_2 + \cdots + l_{s+r2^r} \leq 2^r \big(l(T) - l_{max}(T) \big). \tag{4.10}$$

From inequality (4.10) we deduce that there exists an index i such that

$$l_i \leq l(T) + \frac{2^r}{s + r2^r} \Big(l(T) - l_{max}(T) \Big).$$

Therefore, we obtain

$$l_{k-smt} \leq l(T_i) \leq l(T) + \frac{2^r}{s + r2^r} \Big(l(T) - l_{max}(T) \Big)$$
$$= \Big(1 + \frac{2^r}{s + r2^r} \Big) \big(l(T) - l_{max}(T) \big) + l_{max}(T)$$

which, together with Theorems 2.1-2, proves the lemma.   □

Lemma 4.11   *Suppose that Steiner tree T for terminal set P is partitioned into edge-disjoint full components C_i. Let $l_{max}(T)$ be the length of the longest path in T. Then*

$$l_{k-smt}(P) \leq \sum_i \Big(\rho_k \big(l(C_i) - l_{max}(C_i) \big) + l_{max}(C_i) \Big). \qquad (4.11)$$

Proof. Let $l_{k-smt}(C_i)$ be the length of the optimal k-Steiner tree for the full component C_i. Then we have $l_{k-smt}(P) \leq \sum_i l_{k-smt}(C_i)$, which, together with Lemma 4.10, implies inequality (4.11). □

4.2.1 β-Convex α-Approximation Steiner Tree Algorithms

Karpinski et al. [164] introduced the concept of β-convex α-approximation Steiner tree algorithms, which is a uniform treatment of many approximation algorithms for Steiner tree problems. They obtained tighter upper bounds on the performance of algorithms by applying this technique to the minimum GoSST problem in graphs.

An algorithm $A_{\alpha\beta}$ is called a β-convex α-approximation Steiner tree algorithm if there exist an integer m and real numbers $\lambda_i \geq 0$, $i = 1, 2, \cdots, m$, with $\beta = \sum_{i=2}^m \lambda_i$ and $\alpha = \sum_{i=2}^m \lambda_i \rho_i$ such that the length $l(T_{\alpha\beta})$ of the tree T_A computed by $A_{\alpha\beta}$, is at most $l(T_{\alpha\beta}) \leq \sum_{i=2}^m \lambda_i l_{i-smt}$, where l_{i-smt} is the length of the optimal i-Steiner tree.

By the above definition, the minimum spanning tree algorithms [179; 253] are 1-convex 2-approximation Steiner tree algorithms since the returned solutions are the optimal 2-Steiner trees of length l_{2-smt}. Every k-restricted approximation algorithm in [29] is an 1-convex (the sum of coefficients in the approximation ratio always equals to 1), e.g., for $k = 3$, it is an 1-convex $\frac{11}{6}$-approximation algorithm since the returned solution has length upper bounded by $\frac{1}{2}l_{2-smt} + \frac{1}{2}l_{3-smt}$. In addition, the returned solution by the polynomial time approximation scheme in [227] converges to the optimal 3-Steiner tree and has length $(1 + \epsilon)l_{2-smt}$, thus it is an $(1 + \epsilon)\frac{5}{3}$-approximation algorithm. However, the $(1 + \frac{\sqrt{3}}{2})$-approximation algorithm in [238] is not known to be β-convex for any value of β.

Let $A_{\alpha\beta}$ be a β-convex α-approximation algorithm. Then it follows from Lemma 4.10 that

$$l(T_{\alpha\beta}) \leq \sum_i \lambda_i l_{i-smt} \leq \sum_i \lambda_i \rho_i \big(l(T_{smt}) - l_{max}(T_{smt}) \big) + \beta \, l_{max}(T_{smt})$$

$$= \alpha \big(l(T_{smt}) - l_{max}(T_{smt}) \big) + \beta \, l_{max}(T_{smt}) \qquad (4.12)$$

Let $c(T_{opt})$ be the cost of optimal GoSST tree T_{opt}, and let w_i^* be the total length of rate r_i edges in T_{opt}. Then

$$c(T_{opt}) = \sum_{i=1}^{n} r_i \, w_i^*.$$

Let T_k^* be the subtree of T_{opt} induced by edges of rates r_i, $i \geq k$. Then T_k^* interconnects the source s and all nodes of rate r_k. Thus an optimal Steiner tree interconnecting s and nodes of rate r_k cannot have length longer than

$$l(T_k^*) = \sum_{i=k}^{n} w_i^*.$$

The main idea of the proposed algorithms by Karpinski et al. [164] for the minimum GoSST problem in graphs (Problem 4.1') is to reuse connections for the higher rate nodes when interconnecting lower rate nodes. When interconnecting nodes of rate r_k, they collapse nodes of rate strictly higher than r_k into the source s, thus allowing to reuse higher rate connections for free. (Recall that it is assumed that $r_i < r_{i+1}$ for each $1 \leq i \leq n-1$.)

Let $T(s; r_k)$ be an α-approximation of Steiner tree interconnecting s and all nodes of rate r_k after collapsing all nodes of rate strictly higher than r_k into the source s and treating all nodes of rate lower than r_k as Steiner points. Then we have

$$l\big(T(s; r_k)\big) \leq \alpha \cdot l\big(T_k^*\big) = \alpha \, w_k^* + \alpha \, w_{k+1}^* + \cdots + \alpha \, w_n^*.$$

Karpinski et al. [164] proved the following lemma, which shows that one can obtain a tighter upper bound on the length of $T(s; r_k)$ by using the β-convex α-approximation Steiner tree algorithm.

Theorem 4.4   *Given an instance of the minimum GoSST problem in graphs, the β-convex α-approximation Steiner tree algorithm computes a tree $T(s; r_k)$ that has cost at most $c\big(T(s; r_k)\big) \leq \alpha \, r_k \, w_k^* + \beta(r_k \, w_{k+1}^* + r_k \, w_{k+2}^* + \cdots + r_k \, w_n^*).$*

Proof.   Let T_{opt} be the optimal GoSST and T_k^* be the subtree of T_{opt} induced by edges of rate r_i, for $i \geq k$. By duplicating nodes and introducing edges of zero length as shown in Fig.2.8, we can modify T_k^* into a complete binary tree with the set of leaves consisting of the source s and all nodes of rate at least r_k. The edges of rate r_{k-1} form subtrees attached to the tree T_k^* connecting nodes of rate r_{k-1} to T_k^*. See Fig.4.6(a) where edges of rate greater than r_k form a Steiner tree for $\{s\} \cup S_{k+1} \cup \cdots \cup S_n$ (white nodes) with some Steiner (black) nodes while edges of rate r_k are attached to it.

Now we partition the binary tree T_k^* into edge-disjoint paths as follows: Each internal node v including the degree-2 root is split into two nodes v' and v'' such that v' becomes a leaf incident to one of the downstream edges and v'' becomes a degree-2 node (or a leaf if v is the root) incident to an edge connecting v to its parent (if v is not the root) and another downstream edge. Note since each node is incident to a downstream edge, each resulted connected component will be a path containing exactly one leaf of T_k^*, which is adjacent with an internal node of T_k^*.

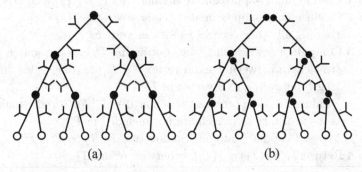

Fig. 4.6 (a) T_k^* is a complete binary tree, and (b) $T_k^*(r_i)$ contains a single terminal of rate r_i, $i > k$.

Denote by $T_k^*(r_i)$ the connected component of T_k^* that contains a single terminal of rate r_i, for each $i > k$, and denote by $P_k(r_i)$ the path that contains all edges of rate r_i, $i > k$. Thus T_k^* is decomposed into edge-disjoint connected components $T_k^*(r_i)$, where each component consists of a path $P_k(r_i)$ and attached Steiner trees with edges of rate r_k as shown in Fig.4.6(b). Moreover, the total length of paths $P_k(r_i)$ is

$$l\big(T_{k+1}^*\big) = w_{k+1}^* + w_{k+2}^* + \cdots + w_n^*.$$

In the end, we decompose the tree $T(s; r_k)$ along these full components T_k^* and, by Lemma 4.11 and the definition of β-convex α-approximation Steiner tree algorithm, we obtain

$$l\big(T(s; r_k)\big) \le \sum_i \Big(\alpha\big(l\big(T_k^*(r_i)\big) - l\big(P_k(r_i)\big)\big) + \beta \cdot l\big(P_k(r_i)\big) \Big)$$

$$= \alpha \cdot w_k^* + \beta \left(w_{k+1}^* + w_{k+2}^* \cdots + w_n^* \right).$$

Multiplying the above inequality by r_k yields the theorem.   □

4.2.2   *Algorithm for Two Non-Zero Rates*

Karpinski et al. [164] proposed a generic approximation algorithm for the minimum GoSST problem in graphs with two non-zero rates, which belongs to the class of β-convex α-approximation Steiner tree algorithms described in the previous subsection. It first computes two GoSSTs, and then returns the better solution.

Algorithm 4.3 *β-Convex α-Approximation Algorithm for Two Rates*

Step 1 Produce an α_1-approximate Steiner tree T_1 for $\{s\} \cup S_1 \cup S_2$.

Step 2 Produce an α_1-approximate Steiner tree T' for $\{s\} \cup S_2$
treating all other vertices as Steiner vertices.

Step 3 Compute a new graph G' by contracting T' into the source s
(the length between s and a vertex $v \notin T'$ of rate r_1 is the shortest
length between v to a vertex in T').
Produce an α_2-approximate Steiner tree T'' for s and remaining
vertices of rate r_1 (which are not included in T');
Merge T' and T'' into one Steiner tree T_2, $T_2 := T' \cup T''$.

Step 4 Return T_1 if $c(T_1) \leq c(T_2)$, otherwise return T_2.

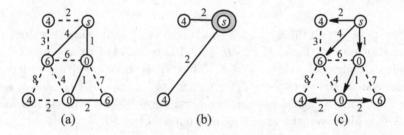

Fig. 4.7   (a) T' is produced, (b) T' is contracted into s, and (c) T'' is produced.

Fig.4.7 demonstrates the Steps 2-3 of the above algorithm when it is applied to the instance of the problem as shown in Fig.4.2. Fig.4.7(a) shows the Steiner tree T' for s and vertices of rate $r_2 = 6$, which consists of four (solid) edges with total length of 8. Fig.4.7(b) shows the resultant graph G' after T' is contracted into the source s, and T'' consists of two edges with total length of 4. Fig.4.7(c) shows the Steiner tree T_2 of total length 12 while the cost of T_2 is 64 (the minimum GoSST has cost 62 as shown in Fig.4.2(c)).

Theorem 4.5 *The Algorithm 4.3 for the minimum GoSST problem with two rates has an approximation ratio at most*

$$\max\left\{\alpha_2, \max\left\{\frac{\alpha_1(\alpha_1 - \alpha_2 r + \beta r)}{\alpha_1 - \alpha_2 r + \beta r^2}\, l_{smt} \,\middle|\, r\right\}\right\}. \tag{4.13}$$

Proof. In the Step 1, the algorithm produces a solution T_1 with cost $c(T_1) \leq \alpha_1 r_2 (w_1^* + w_2^*)$. In the Steps 2-3, the algorithm produces a Steiner tree T' with cost $c(T') \leq \alpha_1 r_2 w_2^*$ and T'' with cost $c(T'') \leq \alpha_2 r_1 w_1^* + \beta r_1 w_2^*$ (due to Theorem 4.4). Thus the algorithm produces a solution T_2 with cost $c(T_2) \leq \alpha_1 r_2 w_2^* + \alpha_2 r_1 w_1^* + \beta r_1 w_2^*$. To obtain the upper bound of (4.12), we consider the following two cases.

Case 1. $\beta r_1 \leq (\alpha_2 - \alpha_1)r_2$. In this case we have

$$\begin{aligned} c(T_2) &\leq \alpha_1 r_2 w_2^* + \alpha_2 r_1 w_1^* + (\alpha_2 - \alpha_1)r_2 w_2^* \\ &\leq \alpha_2 (r_2 w_2^* + r_1 w_1^*) = \alpha_2 l_{smt} \end{aligned}$$

Case 2. $\beta r_1 > (\alpha_2 - \alpha_1)r_2$. In this case, let $x_2 = r_2 - r_1$ and

$$x_1 = \frac{r_1}{\alpha_1 r_2}(\beta r_1 - (\alpha_2 - \alpha_1)r_2).$$

Then both x_1 and x_2 are positive. In the following we will bound the linear combination of $c(T_1)$ and $c(T_2)$

$$\begin{aligned} &x_1 c(T_1) + x_2 c(T_2) \\ &= \frac{r_1}{\alpha_1 r_2}(\beta r_1 - (\alpha_2 - \alpha_1)r_2)c(T_1) + (r_2 - r_1)c(T_2) \\ &\leq r_1(\beta r_1 - (\alpha_2 - \alpha_1)r_2)(w_1^* + w_2^*) + (r_2 - r_1)(\alpha_1 r_2 w_2^* + \alpha_2 r_1 w_1^* + \beta r_1 w_2^*) \\ &= ((\beta - \alpha_2)r_1^2 + \alpha_1 r_1 r_2)w_1^* + ((\beta - \alpha_2)r_1 r_2 + \alpha_1 r_2^2)w_2^* \\ &= ((\beta - \alpha_2)r_1 + \alpha_1 r_2)(r_1 w_1^* + w_2^*) \\ &\leq (\beta r_1 + \alpha_1 r_2 - \alpha_2 r_1)l_{smt} \tag{4.14} \end{aligned}$$

Now let $c(T_{\alpha\beta})$ be the cost of the solution returned by Algorithm 4.3. Then it follows from inequality (4.13) that

$$c(T_{\alpha\beta}) = \min\{c(T_1), c(T_2)\} = \frac{x_1 \min\{c(T_1), c(T_2)\} + x_2 \min\{c(T_1), c(T_2)\}}{x_1 + x_2}$$

$$\leq \frac{x_1 c(T_1) + x_2 c(T_2)}{x_1 + x_2}$$

$$\leq \frac{\beta r_1 + \alpha_1 r_2 - \alpha_2 r_1}{\frac{r_1}{\alpha_1 r_2}\left(\beta r_1 - (\alpha_2 - \alpha_1)r_2\right) + r_2 - r_1} l_{smt}$$

$$\leq \alpha_1 \frac{\beta r_1 r_2 + \alpha_1 r_2^2 - \alpha_2 r_1 r_2}{\beta r_1^2 - (\alpha_2 - \alpha_1)r_1 r_2 + \alpha_1 r_2^2 - \alpha_1 r_1 r_2} l_{smt}$$

$$\leq \alpha_1 \frac{\alpha_1 - \alpha_2 r + \beta r}{\alpha_1 - \alpha_2 r + \beta r^2} l_{smt}, \quad \text{where } r = \frac{r_1}{r_2}.$$

By the analysis of the two cases, we deduce that $c(T_{\alpha\beta})$ is at most the maximum of the following two values

$$\alpha_1 l_{smt} \text{ and } \alpha_1 \frac{\alpha_1 - \alpha_2 r + \beta r}{\alpha_1 - \alpha_2 r + \beta r^2} l_{smt},$$

which proves the theorem.   □

Karpinski et al. [164] applied Theorem 4.5 to obtain a few bounds on the approximation ratios of Algorithm 4.3 by combining some known algorithms for Steiner tree problem in graphs (refer to Section 2.2). They obtained the following corollaries.

Corollary 4.1   *If Algorithm 4.3 uses the algorithm from [238] to generate the α_1-approximate Steiner tree and the algorithm from [227] to generate the α_2-approximate Steiner tree, then it has approximation ratio of $1.960 + \epsilon$.*

Corollary 4.2   *If Algorithm 4.3 uses the algorithm from [227] to generate both α_1-approximate and α_2-approximate Steiner trees, then it has approximation ratio of $2.059 + \epsilon$.*

Corollary 4.3   *If Algorithm 4.3 uses the algorithm from [29] to generate both α_1-approximate and α_2-approximate Steiner trees, then it has approximation ratio of 2.237.*

Corollary 4.4   *If Algorithm 4.3 uses the algorithm from [60; 179; 253] to generate both α_1-approximate and α_2-approximate Steiner trees, then it has approximation ratio of 2.414.*

4.2.3 Algorithm for Arbitrary Number of Rates

For the general case of the minimum GoSST problem with arbitrarily many rates in graphs, Karpinski et al. [164] proposed a randomized algorithm that is a modification of the algorithm designed by Charikar et al. [49].

Both algorithms round up node rates to the closest power of some number a starting with a^p, where p is picked uniformly at random between 0 and 1, that is, node rates are rounded up to numbers in the set $\{a^p, a^{p+1}, \cdots\}$. The only difference is that Karpinski et al. [164] contract each approximate Steiner tree T_k for nodes of rounded rate a^{p+k}, instead of simply taking their union as Charikar et al. [49] do. Such a modification enables them to reuse contracted edges at zero cost by Steiner trees when interconnecting lower rate nodes. This idea is used when they design Algorithm 4.3 for the special case of two rates and enables them to reduce the approximation ratio from 4.211 to 3.802 for the general case of unbounded number of node rates.

Algorithm 4.4 *Randomized Algorithm for Multiple Rates*

Step 1 Choose a real number $a > 1$ and randomly pick $p \in (0, 1)$.
　　　Round up each rate to the closest number in $\{a^p, a^{p+1}, \cdots\}$.
　　　Order new rates $r'_1 < r'_2 < \cdots r'_m$, where $m \leq n$.
　　　Compute S'_i that contains terminals whose rates rounded to r'_i.
Step 2 Set $T := \emptyset$ and $i := 1$.
　　while $i \leq m$ **do**
　　　　Produce a β-convex α-approximate Steiner tree T_i of $\{s\} \cup S'_i$.
　　　　Update $T := T \cup T_i$ and $i := i + 1$.
　　　　Contract T_i into source s.
　　end-while
Step 3 Return $T_R := T$.

To obtain the approximation ratio of Algorithm 4.4, Karpinski et al. [164] proved the following lemma, which applies the same idea from the proof of Lemma 4 in [49].

Lemma 4.12 *Let T_{opt} be the optimal GoSST and w_i^* be the total length of the edges of T_{opt} with rates rounded to a^{p+i}. Then the expected cost $c(T'_{opt})$ of T'_{opt} after the rounding node rates*

$$c(T'_{opt}) = \sum_{i=0}^{m} w_i^* a^{p+i} \leq \frac{a-1}{\ln(a)} c(T_{opt}). \tag{4.15}$$

Proof. Note that an edge e used at rate r in T_{opt} will be used at rate a^{p+j}, where j is the smallest integer i such that $a^{p+j} \geq r$. Now let $r = a^{q+j}$. If $q \leq p$, then the cost of e caused by rounding up is a^{p-q} times the original cost, i.e., $c'(e) = a^{p-q}c(e)$; otherwise, $c'(e) = a^{p+1-q}c(e)$. Hence the expected factor by which the increasing cost of edge e is

$$\int_0^q a^{p+1-q}dp + \int_q^1 a^{p-q}dp = \frac{a-1}{\ln(a)}.$$

By linearity of expectation, we can obtain the expected cost of T'_{opt} in inequality (4.14). The proof is then finished. $\square$

A *randomized algorithm A* for a minimization problem is said to have an α-approximation ratio if, given any instance of the problem, the expected cost of the solution returned by A is at most α times that of the optimal solution. Karpinski et al. [164] proved the following theorem, which gives a upper bound on the approximation ratio of Algorithm 4.4.

Theorem 4.6   *Randomized algorithm 4.4 has approximation ratio at most*

$$\min_a \left\{ (\alpha - \beta)\frac{a-1}{\ln a} + \beta\frac{a}{\ln a} \right\}. \tag{4.16}$$

Proof.   Let $c(T_R)$ be the expected cost of the solution returned by Algorithm 4.4, and $c(T_R(r_i))$ be the cost the tree T_i for $\{s\} \cup S'_i$ produced by the algorithm. Then by Theorem 4.4, we have

$$c(T_R(r_i)) \leq \alpha \, a^{p+i}w_i^* + \beta\big(a^{p+i+1}w_{i+1}^* + a^{p+i+2}w_{i+2}^* + \cdots + a^{p+m}w_m^*\big).$$

Thus we obtain an upper bound on the total cost of the returned solution

$$
\begin{aligned}
c(T_R) \leq \ & \alpha w_1^* a^p &+\beta w_2^* a^p &&+\beta w_3^* a^p &&+\cdots &&+\beta w_{m-1}^* a^p &&+\beta w_m^* a^p \\
& &+\alpha w_2^* a^{p+1} &&+\beta w_3^* a^{p+1} &&+\cdots &&+\beta w_{m-1}^* a^{p+1} &&+\beta w_m^* a^{p+1} \\
& & &&\ddots && &&\ddots \\
& & && && &&+\alpha w_{m-1}^* a^{p+m-1} &&+\beta w_m^* a^{p+m-1} \\
& & && && && &&+\alpha w_m^* a^{p+m}
\end{aligned}
$$

$$
= (\alpha-\beta)c(T'_{opt})+\beta\cdot
\begin{pmatrix}
w_1^* a^p & +w_2^* a^p & +w_3^* a^p & \cdots & +w_{m-1}^* a^p & +w_m^* a^p \\
& +w_2^* a^{p+1} & +w_3^* a^{p+1} & \cdots & +w_{m-1}^* a^{p+1} & +w_m^* a^{p+1} \\
& & \ddots & & \ddots \\
& & & & +w_{m-1}^* a^{p+m-1} & +w_m^* a^{p+m-1} \\
& & & & & +w_m^* a^{p+m}
\end{pmatrix}
$$

$$\leq (\alpha - \beta)c(T'_{opt}) + \beta \cdot \begin{pmatrix} \vdots \\ w_1^* a^{p-m+1} & \vdots \\ w_1^* a^{p-m+2} & +w_2^* a^{p-m+2} & \vdots \\ \vdots & \vdots & \ddots \\ w_1^* a^{p-1} & +w_2^* a^{p-1} & +\cdots & +w_{m-1}^* a^{p-1} & \vdots \\ w_1^* a^{p} & +w_2^* a^{p} & +\cdots & +w_{m-1}^* a^{p} & +w_m^* a^{p} \\ & +w_2^* a^{p+1} & +\cdots & +w_{m-1}^* a^{p+1} & +w_m^* a^{p+1} \\ \ddots & & +\cdots & \ddots & \vdots \\ & & & +w_{m-1}^* a^{p+m-1} & +w_m^* a^{p+m-1} \\ & & & & +w_m^* a^{p+m} \end{pmatrix}$$

Therefore, from Lemma 4.12 we deduce the following inequalities

$$c(T_A) \leq (\alpha - \beta)c(T'_{opt}) + \beta\Big(1 + \frac{1}{a} + \frac{1}{a^2} + \cdots\Big)c(T'_{opt})$$
$$\leq (\alpha - \beta)\frac{a-1}{\ln a}c(T_{opt}) + \beta\frac{a}{\ln a}c(T_{opt})$$

which yields the upper bound of (4.14). The proof is then finished. □

Immediately from the above theorem, Karpinski et al. [164] obtained the approximation ratio of Algorithm 4.4 is at most 3.802 and 4.059, respectively, when β-convex α-approximation Steiner tree algorithm at Step 2 applies the algorithms in [227; 29], and 4.311 if the algorithm from [60; 179; 253] is used.

4.3 Discussions

To find a better approximation algorithm for the Grade of Service Steiner Tree (GoSST) problem in the Euclidean plane, Kim et al. [170] observed that the optimal solution to the problem is determined by the concurrent combinations of two factors: (1) the service of grade for every edge, and (2) the choice of Steiner points. That is, (1) could not be determined before (2) is done, and vice versa. Therefore, to design a Polynomial Time Approximation Scheme (PTAS) for the problem applying rectangular partition technique with portals [10; 11] and using dynamic programming, they made some modifications to facilitate the running of the dynamic programming. In particular, they not only move every terminal point to the nearest grid point, but also assume that all Steiner points lie only at the grid points. It turns out that the GoSST problem after modification fits well to the PTAS so they could obtain an $(1 + \epsilon)$-approximation.

Khuller et al. [168] studied the *quality of service multicast routing problem* under a bicriteria optimization model. In order to guarantee that audio or video signals can be effectively used in interactive multimedia communications, they consider not only the network cost of a multicast routing tree, but also the delay of transmitting data over the tree from a source node to any destination node. They assumed that the cost of the edges reflects both the cost to install the link and the time for a signal to traverse the link once the link is installed. Notice that multicast routing of minimal network cost is equivalent to the Steiner tree problem in graphs (Problem 2.1), while the *Shortest Path Tree* (SPT), which consists of the shortest paths from the source to every destination, has the shortest delay to any destination. Sometimes an SPT has a high network cost and the Steiner Minimum Tree (SMT) causes long delay. They proved that a single (Steiner) tree can approximately achieve both goals. More precisely, for $\alpha \geq 1$ and $\beta \geq 1$, there exists such a spanning tree T of given graph G, called an (α, β)-LAST (*Light Approximate Shortest-path Tree*), that satisfies the following two requirements[1]:

(1) For every vertex v, the distance between r and v in T is at most α times the shortest distance from r to v in G.

(2) The cost of T is at most β times the cost of an SMT of G.

An (α, β)-LAST can be found as follows [168]: Traverse an (approximation of) SMT, and then check each vertex when it is encountered to ensure that the distance requirement for that vertex is met in the current tree. If not, the edges of the shortest path between the vertex and the root are added into the current tree while other edges are discarded so that a tree structure is maintained.

More recently, Charikar et al. [49] consider multicast routing problem that require various levels of quality-of-service (QoS). The objective is to compute a low-cost multicast/Steiner tree from a source that would provide k QoS levels requested by r receivers. They assume that QoS level required on a link is the maximum among the QoS levels of the receivers that are connected to the source through the link, and the cost of a link to be a function of the QoS level that it provides. They study a couple of variants of this problem all are proved to be NP-hard. They propose an $O(\min\{r, k\})$-

---

[1]The concept was originally defined for spanning trees, which, clearly, could be naturally extended to Steiner trees.

approximation algorithm for the problem and extend it the case of many multicast groups.

Most recently, Wang et al. [266] studied how to provide the QoS with a certain performance guarantee, such a mechanism is called *Differentiated service* (DiffServ). They assume that each link will not provide the service to receivers unless it receives a payment large enough to compensate its relay cost. Under such an assumption, each link is first asked to report its relay cost and then a payment to this link is computed based on mechanisms. When a link is paid whatever it asked, clearly to its best interest it may not report its cost truthfully. Consequently, instead of paying inks whatever they requested a paying scheme should be designed to make every link give its real cost out of its own interest, that is called *strategy proof*. After the cost of each link is determined, a payment sharing scheme should be designed to ensure that payment is fairly shared among the receivers. In addition to computing multicost/Steiner tree of minimum cost, they studied how to design a strategy proof payment scheme and a fair payment sharing scheme. Some positive and negative results are obtained.

Chapter 5

Steiner Tree Problem for Minimal Steiner Points

The Steiner tree problem for minimal Steiner points has an important application in Wavelength Division Multiplexing (WDM) optical network design [233]. Suppose that we need to connect n sites located at $t_1, t_2, \cdots, t_n$ with WDM optical network. Due to the limit in transmission power, signals can only travel a limited distance (say R) otherwise correct transmission may not be guaranteed. If some of the inter-site distances are greater than R, we need to provide it into shorter pieces. This problem also finds some applications in VLSI design [48] and the evolutionary/phylogenetic tree constructions in computational biology [144]. The problem is more formally stated as follows.

Problem 5.1 *Steiner Tree Problem for Minimal Steiner Points* [50]

Instance A set of n terminals in Euclidean plane $\Re^2$ and a constant $R > 0$.
Solution A Steiner tree T interconnecting all terminal points such that each edge in T has length no more than R.
Objective Minimizing the number of Steiner points in T

In this chapter, the *optimal Steiner tree* means the optimal Steiner tree for above problem unless specified otherwise. We will see that optimal Steiner trees have some structural properties different from those of Steiner minimum trees. These properties demand new approaches different from those used for classic Steiner tree problem. We shall study some of them in this chapter. In Sections 5.1 and 5.2 we will study the Steiner tree problem for minimal Steiner points in the Euclidean and rectilinear planes, respectively. In Section 5.3 will turn to the case of general metric space. In the end we will discuss some related problems.

5.1 In the Euclidean Plane

In the classical Euclidean Steiner tree problem (Problem 1.1), all Steiner points of Steiner minimum tree have degree 3. In the Steiner tree problem for minimal Steiner points, however, the optimal Steiner tree may have degree-2 Steiner points. For example, when $n = 2$ and the distance between t_1 and t_2, denoted by $\|t_1 - t_2\|$ is larger than R, then the optimal Steiner tree is a path containing $\lceil \|t_1 t_2\|/r \rceil - 1$ Steiner points, each has degree 2. In addition, Steiner points can also have degree larger than 3. Consider five points $t_1, t_2, \cdots, t_5$ that make a regular pentagon whose sides have length greater than R and the distance from the center to each vertex is R. The optimal Steiner tree contains only one Steiner point at the center. See Fig.5.1(a).

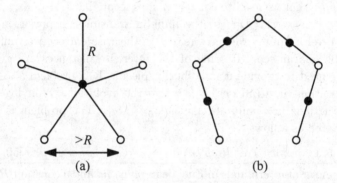

Fig. 5.1   (a) The Optimal Steiner tree, and (b) Steinerized minimum spanning tree.

The following two lemmas describe some properties of the optimal Steiner trees with minimal lengths.

Lemma 5.1   *Every shortest optimal Steiner tree T_{opt} must have the following properties.*
(i) *No two edges in T_{opt} cross each other.*
(ii) *Two edges meeting at a vertex in T_{opt} form an angle of at least $60°$.*
(iii) *If two edges in T_{opt} form an angle of exactly $60°$, then they have the same length.*

Proof.   (i) Suppose, by contradiction, that two edges ac and bd in T_{opt} cross at point e as shown in Fig.5.2(a). Note that quadrangle $abcd$ must have an inner angle of at least $90°$. Without loss of generality, assume $\angle abc \geq 90°$. Then $\angle bca < 90°$ and $\angle cab < 90°$. Hence $\|ab\| < \|ac\|$ and

$\|bc\| < \|ac\|$. When edge ac is removed from T_{opt}, T_{opt} would be broken into two parts containing vertices a and c, respectively. Observe that the part that contains a contains vertex b. Adding edge bc results in a shorter optimal tree. This contradicts that T_{opt} is the shortest one.

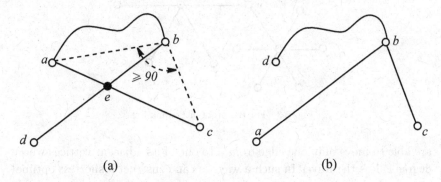

Fig. 5.2 For the proof of Lemma 5.1: (a) for (i), and (b) for (ii).

(ii) Suppose, by contradiction again, that two edges ab and bc in T_{opt} meet at vertex b with $\angle abc < 60°$ as shown in Fig.5.2(b). Then either $\angle cab > 60°$ or $\angle bca > 60°$. This implies that either $\|bc\| > \|ac\|$ or $\|ab\| > \|ac\|$. Now replacing either bc or ab with ac we can get a shorter optimal tree. This leads the same contradiction.

(iii) It can be proved by a similar argument for (ii). □

Lemma 5.2 *There exists a shortest optimal Steiner tree T_{opt} such that every vertex in T_{opt} has degree at most five.*

Proof. It follows immediately from Lemma 5.1(ii) that every vertex in a shortest optimal Steiner tree T_{opt} has degree at most six. If vertex u is of degree six in T, then by Lemma 5.1(ii) each angle at u equals 60° and by Lemma 5.1(iii) all edges incident to u have the same length.

Next consider any vertex v of degree d in T. Assume that v is adjacent to k degree-six vertices, where $k \geq 1$. Suppose that v is adjacent to u of degree six. Then u is incident to two edges uy and ux with $\angle zyv = \angle vux = 60°$ and $\|uv\| = \|uy\| = \|ux\|$. See Fig.5.3. Replacing uy and ux with vy and vx produces another shortest optimal tree. Vertex v, however, has degree increased by two. For all degree-six vertices adjacent to v, perform the same operation. We can get a shortest optimal tree such that v has degree $d + 2k \leq 6$ due to Lemma 5.1(ii). Now, for each degree-six vertex u we

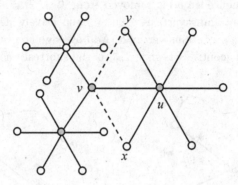

Fig. 5.3   For the proof of Lemma 5.2.

are able to move only one edge from u to one of its adjacent vertices whose degree is less than five. In such a way we can construct a shortest optimal tree such that every vertex has degree at most five.   □

5.1.1   *Complexity Study*

We now prove the decision version of Problem 5.1 is NP-complete. The proof [188] is based on a polynomial time reduction from the decision version of *discrete Euclidean Steiner tree problem*.

Given two points u and v in the Euclidean plane, the *discrete length* of the edge joining them is defined as $d'(u, v) \equiv \lceil d(u, v) \rceil$, where $d(u, v)$ is the Euclidean distance between u and v, and $\lceil \alpha \rceil$ is the least integer not less than α. For a Steiner tree T, the *discrete length* of T, denoted by $l'(T)$, is defined as the sum of the discrete Euclidean lengths of all edges in T.

Problem 5.2 *Discrete Euclidean Steiner Tree Problem*

Instance   A set Q of n terminals all have integral coordinators in $\Re^2$.

Solution   A Steiner tree T for Q whose Steiner points all have integral coordinators in the plane.

Objective Minimizing the discrete length of Steiner tree T.

Lemma 5.3   *There is a polynomial time reduction from the decision of discrete Euclidean Steiner tree problem to the decision version of Steiner tree problem for minimal Steiner points.*

Proof.   The decision version of discrete Euclidean Steiner tree problem is as follows: Given a set Q of points and a positive integer L, decide if there exists a Steiner tree T for Q such that all Steiner points in T have integral coordinators and T has discrete length $l'(T) \leq L$. This problem is known to be NP-complete [109].

The decision version of Steiner tree problem for minimal Steiner points is as follows: Given a set P of points, a positive constant R and a positive integer B, decide if there exists a Steiner tree T for P such that all edges in T have lengths no greater than R and the number of Steiner points is at most B.

Let I be an instance of the decision version of discrete Euclidean Steiner tree problem. We construct an instance I' of the decision version of Steiner tree problem for minimal Steiner points by letting $P := Q$, $R := 1$ and $B := L - (|Q| - 1)$. Clearly, the construction could be done in polynomial time.

First, suppose that T' is a solution to I'. Then the discrete length of each edge in T' is no more than 1 since the Euclidean length of each each in T' is no more than 1. Let P' be the set of all points in T', then we have $l'(T') \leq |P'| - 1$ since there are $(|P'| - 1)$ edges in T'. However,

$$|P'| - 1 = |P' \setminus P| + |P| - 1 \leq L - (|P| - 1) + |P| - 1 = L.$$

Therefore, T' is also a solution to I.

Next, suppose that T is a solution to I. Let Q' be the set of all points in T, then T has $(|Q'| - 1)$ edges. For each edge e in T, we equally place $(l'(e) - 1)$ degree-2 Steiner points on the edge dividing e into $l'(e)$ edges so that all have lengths at most 1. As a result, we obtain a tree T' spanning $P' \supseteq P$ such that each edge in T' has length no more than 1. Note that the number of added Steiner points is

$$|P'| - |Q'| = \sum_{e \in T} (l'(e) - 1) = \sum_{e \in T} l'(e) - |E(T)| \leq L - (|Q'| - 1).$$

Therefore, the number of Steiner points in T' is

$$\begin{aligned}
|P'| - |Q| &= |P'| - |Q'| + (|Q'| - |Q|) \\
&\leq L - (|Q'| - 1) + (|Q'| - |Q|) \\
&= L - (|Q| - 1) = B.
\end{aligned}$$

Therefore, T' is a solution to I'.

We have proved that the answer to I is "Yes" if and only if the answer to I' is "Yes". The lemma is then proved. $\qquad\square$

Theorem 5.1   *The decision version of Steiner tree problem for minimal Steiner points is NP-complete.*

Proof.   Given an instance of decision version of Steiner tree problem for minimal Steiner points and the topology of a Steiner tree, which specifies the edges between points in the tree. Because a *bottleneck tree* under the given topology and edge-bound constraint can be computed in polynomial time [245], the decision version of Steiner tree problem for minimal Steiner points belongs to the complexity class NP. This, together with Lemma 5.3, proves the theorem. $\square$

5.1.2   *Steinerized Minimum Spanning Tree Algorithm*

Observe that a (minimum) spanning tree T may not be a feasible solution for Steiner tree problem for minimal Steiner points since some edges in T may have length longer than R. To make it feasible, we can add some Steiner points in each long edge of T breaking them into small pieces each having lengths at most R. The resulting tree is called a *Steinerized spanning tree*. Recall that the Minimum Spanning Tree (MST) can be constructed in polynomial time, thus we can develop a simple approximation algorithm for Steiner tree problem for minimal Steiner points as follows.

Algorithm 5.1 *Steinerized Minimum Spanning Tree Algorithm*

Step 1 Construct an MST T_{mst} of P and set $T := T_{mst}$.
Step 2 for all edges e in T whose length $l(e) > R$ **do begin**
        Cut long edge e into $\lceil l(e)/R \rceil$ shorter ones of equal length
        by placing $(\lceil l(e)/R \rceil - 1)$ Steiner points on e.
        Replace edge e with $\lceil l(e)/R \rceil$ shorter edges.
   end-for
Output Steiner tree $T_{smst} := T$.

Lemma 5.4   *Every Steinerized minimum spanning tree has the minimal Steiner points among Steinerized spanning trees.*

Proof.   Every MST can be obtained from a spanning tree by a sequence of operations that each replaces an edge by a shorter edge. Since the shorter edge needs Steiner points no more than the longer edge needs when we Steinerize edges in a spanning tree. $\square$

Let T be a Steiner tree for terminal-set P without Steiner points of degree more than two. Removing Steiner points of degree two in T will make T a spanning tree. Thus from Lemma 5.4 we deduce that the number

of Steiner points in T_{smst} is no bigger than the number of Steiner points in
T. Using this fact, we can prove the following theorem [188].

Theorem 5.2   *The performance ratio of Steinerized minimum spanning
tree algorithm is at most 5.*

Proof.   By Lemma 5.2, let T_{opt} be an optimal solution to Steiner tree
problem for minimal Steiner points whose Steiner points have degree at
most five. Partition T_{opt} into *full subtrees*, $T_1, T_2, \cdots, T_k$, such that every
terminal point in P is a leaf of some subtree. For each subtree T_i, duplicating all edges in T_i will produce an Eulerian tour for all points in t_i (dashed
edges as shown in Fig.5.4(a)). Note that every Steiner point appears at
most five times in the tour while every Steiner terminal point in P appears
exactly once. Removing one (or more) edge in the tour will produce a tree
T_i', which, in fact, is a path (solid thin edges as shown in Fig.5.4(b)).

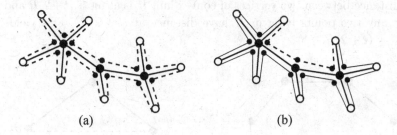

(a)   (b)

Fig. 5.4   (a) Produce an Eulerian tour t_i, and (b) obtain a tree T_i'.

Clearly, the number of Steiner points in T_i' is no more than five times
that Steiner points in T_i. Pasting all T_i's together, we obtain a spanning
tree T_{opt}' for P whose Steiner points have degree at most two. Therefore,
the number of Steiner points in T_{smst} is less than or equal to the number
of Steiner points in T_{opt}', which is at most five times that of Steiner points
in T_{opt}. The proof is then finished.   □

In the following we will make a more sophisticated analysis that proves
the performance ratio of Steinerized minimum spanning tree algorithm is
exactly 4. After some modifications, it can be used to prove that there exists a better approximation algorithm for Steiner tree problem for minimal
Steiner points, which will be discussed in the next subsection. In addition,
a much simple (but elegant) analysis [201] could prove the same result,
which is included in Section 5.3 since it is applicable to the Steiner tree
problem for minimal Steiner points in all metric spaces.

Before we make the performance analysis of the algorithm, we need two more lemmas on the geometrical properties of the shortest optimal Steiner tree. A path $q_1 q_2 \cdots q_m$ in T is called a *convex path* if for every $i = 1, 2, \cdots, m - 3$, $q_i q_{i+2}$ intersects $q_{i+1} q_{i+3}$. Angles bigger than 120° will play an important role in the following analysis, they, for simplicity, are called *big angles*.

Lemma 5.5   *Let $q_1 q_2 \cdots q_m$ be a convex path and $m \geq 2$. Suppose that there are t big angles among $(m - 2)$ angles $\angle q_1 q_2 q_3$, $\angle q_2 q_3 q_4$, $\cdots$, $\angle q_{m-2} q_{m-1} q_m$. Then $|q_1 q_m| \leq (t + 2)R$.*

Proof.   We prove the lemma by induction on m. For $m \leq 3$, it is true since $|q_1 q_3| \leq |q_1 q_2| + |q_2 q_3| \leq 2R \leq (t+2)R$. Now suppose $m \geq 4$. Consider the convex hull H formed by points $q_1, q_2, \cdots, q_m$. If at least one of q_1 and q_2 does not lie on the boundary of H, then by the induction hypothesis, any distance between two vertices of convex hull H is at most $(t+2)R$ and hence any two points lying in H have distance at $(t + 2)R$. This yields $|q_1 q_m| \leq (t + 2)R$.

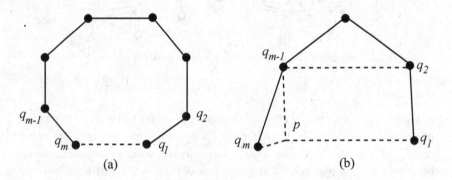

Fig. 5.5   For the proof of Theorem 5.3.

Next, we can assume that both q_1 and q_m lie on the boundary of H. It follows immediately that whole path $q_1 q_2 \cdots q_m$ lies on the boundary of H as shown in Fig.5.5(a). If $\angle q_1 q_m q_{m-1} \geq 90°$, then $|q_1 q_m| \leq |q_1 q_{m-1}|$ and by the induction hypothesis, $|q_1 q_{m-1}| \leq (t+2)R$. Hence $|q_1 q_m| \leq (t+2)R$. Similarly, $\angle q_2 q_1 q_m \geq 90°$ can also produce $|q_1 q_{m-1}| \leq (t+2)R$. Therefore, we may assume $\angle q_1 q_m q_{m-1} < 90°$ and $\angle q_2 q_1 q_m < 90°$. If follows that

$$(m - 2) \cdot 180° \leq 2 \cdot 90° + (m - t - 2) \cdot 120° + t \cdot 180°,$$

which implies $m - t - 2 < 3$. This means that path $q_1 q_2 \cdots q_m$ has at most two angles of degrees not more than $120°$.

If $\angle q_{m-2} q_{m-1} q_m$ is a big angle, then by the induction hypothesis, $|q_1 q_{m-1}| \le ((t-1) + 2)R$. Thus $|q_1 q_m| \le |q_1 q_{m-1}| + |q_{m-1} q_m| \le (t+2)R$. Similarly, if $\angle q_1 q_2 q_3$ is a big angle, then $|q_1 q_m| \le (t+2)R$. Therefore, we can assume $\angle q_{m-2} q_{m-1} q_m \le 120°$ and $\angle q_1 q_2 q_3 \le 120°$. They are the only two angles not big on the path $q_1 q_2 \cdots q_m$. Now draw a parallelogram $q_1 q_2 q_{m-1} p$ as shown in Fig.5.5(b). Since $\angle q_1 q_2 q_{m-1} \le \angle q_1 q_2 q_3 \le 120°$, we have $\angle q_2 q_{m-1} p \ge 60°$. Moreover, $\angle q_2 q_{m-1} q_m \le \angle q_{m-2} q_{m-1} q_m \le 120°$. Thus, $\angle p q_{m-1} q_m \le 60°$. If follows that

$$|p q_m| \le \max \{ |p q_{m-1}|, |q_{m-1} q_m| \} = \max \{ |q_1 q_2|, |q_{m-1} q_m| \} \le R.$$

Therefore, we obtain

$$|q_1 q_m| \le |q_1 p| + |p q_m| \le |q_2 q_{m-1}| + |p q_m| \le (t+1)R + R = (t+2)R.$$

That proves the lemma. $\qquad\square$

Lemma 5.6 *In a shortest optimal Steiner tree T_{opt}, there are at most two big angles at a vertex with degree three, there is at most one big angle at a vertex with degree four, and there is no big angle with degree five.*

Proof. Suppose that $\alpha_1, \alpha_2, \cdots, \alpha_d$ are all angles at a vertex with degree d and $k > 0$ of them are big angles. Since each angle is of at least $60°$, we have

$$360° = \alpha_1 + \alpha_2 + \cdots + \alpha_d > (d - k) \cdot 60° + k \cdot 120°,$$

which implies $6 < (d-k) + 2k = d + k$. Then the lemma follows immediately from this inequality. $\qquad\square$

Note that every leaf in a Steiner tree is a terminal point p_i for some i. Recall that a Steiner tree is *full* if every terminal point is a leaf. In the case that a Steiner tree is not full, that is, there exists at least one terminal point whose degree is greater than one. We can break the tree at these points and obtain several small full Steiner trees, that are called *full components* of the Steiner tree.

Lemma 5.7 *Let T_{opt} be a shortest optimal Steiner tree. Suppose that T_{opt} is a full Steiner tree. Let s_i denote the number of Steiner points with degree i in T_{opt}. Then $3s_5 + 2s_4 + s_3 = n - 2$.*

Proof. Since T_{opt} has $(s_5 + s_4 + s_3 + s_2 + n) - 1$ edges in total, we have $5s_5 + 4s_4 + 3s_3 + 2s_2 + n = 2(s_5 + s_4 + s_3 + s_2 + n - 1)$. This proves the lemma.   □

Theorem 5.3   *The Steinerized minimum spanning tree algorithm produces a Steiner tree T_{smst} whose number of Steiner points is at most four times that of the optimal Steiner tree T_{opt}.*

Proof.   By Lemma 5.2, there exists a shortest optimal Steiner tree T_{opt} in which every Steiner point has degree at most five. We consider two cases.

Case 1. T_{opt} is a full Steiner tree. By Lemma 5.6, $n = 3s_5 + 2s_4 + s_3 + 2$. Consider a spanning tree T consisting of $(n-1)$ edges each connecting two terminals at endpoints of a convex path in T_{opt}. See Fig.5.6. By Lemma 5.4, each edge ab in T has length upper bounded by $(t+2)R$ where t is the number of big angles on the convex path connecting a and b. Hence, we need at most $(t+1)$ Steiner points to Steinerize edge ab. By Lemma 5.5, any Steinerized minimum spanning tree contains at most $s_4 + 2s_3 + 2s_2 + n - 1$ Steiner points. Clearly,

$$s_4 + 2s_3 + 2s_2 + n - 1 = 3s_5 + 3s_4 + 3s_3 + 2s_2 + 1$$
$$\leq 3(s_5 + s_4 + s_3 + s_2) + 1.$$

If $s_5 + s_4 + s_3 + s_2 > 0$, then $s_4 + 2s_3 + 2s_2 + n - 1 \leq 4(s_5 + s_4 + s_3 + s_2)$. If $s_5 + s_4 + s_3 + s_2 = 0$, then $T = T_{opt}$. Therefore, in either case, every Steinerized minimum spanning tree contains at most $4(s_5 + s_4 + s_3 + s_2)$ Steiner points.

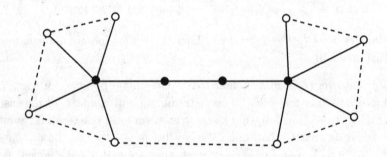

Fig. 5.6   For the proof of Theorem 5.3.

Case 2. T_{opt} is not a full Steiner tree. Then T_{opt} can be decomposed into several full components $T_{opt}^1, T_{opt}^2, \cdots, T_{opt}^k$. For each full component T_{opt}^i, by the above argument for Case 1, we know that the Steinerized minimum spanning tree T^i_{smst} on terminals in T^i_{opt} contains at four times Steiner points of T^i_{opt}. Note that the union of Steinerized minimum spanning trees $\bigcup_i T^i_{smst}$, each for terminals in a full component, is a Steinerized spanning tree T_{smst} for all terminals. Therefore, the theorem follows from Lemma 5.3. $\qquad\square$

5.1.3 *Greedy Algorithm*

Recall that in the proof of Theorem 5.1 we have shown that for any full component T' of an optimal Steiner tree with degrees at Steiner points upper bounded by five, the Steinerized minimum spanning tree on terminal points in T' contains at most $(3 \cdot c(T') + 1)$ Steiner points, where, for a Steiner tree T, $c(T)$ denotes the number of Steiner points in T. Now in the following we will improve this bound.

Lemma 5.8 *Let T_{opt} be an optimal Steiner tree whose Steiner points have degrees at most five. Suppose that T_{opt} consists of k full components, $T_1, T_2, \cdots, T_k$. Then for each i,*

(i) *The steinerized minimum spanning tree on terminal points in T_i contains at most $(3 \cdot c(T_i) + 1)$ Steiner points.*

(ii) *If T_i contains a Steiner point whose degree is at most four, then the Steinerized minimum spanning tree on terminal points in T_i contains at most $3 \cdot c(T_i)$ Steiner points.*

(iii) *If the Steinerized minimum spanning tree on terminal points in T_i contains an edge between two terminal points, then it contains at most $3 \cdot c(T_i)$ Steiner points.*

Proof. Since (i) and (iii) follow immediately from the proof of Theorem 5.3, we just need to prove (ii). Let s_i be the number of Steiner points with degree i in T_j, and let n_j be the number of terminals in T_j. Note that there are exactly n_j convex paths in T_j. So we can choose any $(n_j - 1)$ of them and connect two endpoints of each path. We will obtain a spanning tree and denote its Steinerization by T_s. Now, assume that vertex v is the Steiner point with degree at most four. If there is a big angle at v, then we choose $(n_j - 1)$ convex paths not containing the big angle. If there is no big angle at v, then we can choose any $(n_j - 1)$ convex paths. Choosing in such a way, we would have

$$c(T_s) \le s_4 + 2s_3 + 2s_2 - 1 + (n_j - 1) \le 3(s_4 + s_3 + s_2) = 3 \cdot c(T_j),$$

which proves (ii), and the proof is then finished. $\qquad\square$

To get a better approximation, we need to find a method for determining whether three or four terminal points could be connected to a common Steiner point. Note that an angle of less than 90° is *acute* and an angle of more than 90° is *obtuse*. A triangle is *acute* if its three angles are all acute, a triangle is *obtuse* if it has one obtuse angle, and a triangle is *right* if it has one right angle.

Lemma 5.9   *If triangle △abc is acute or right, then the disk of minimal radius that covers △abc is the one bounded by the circle circumscribing a, b, and c. If △abc is obtuse or right, then the disk of minimal radius that covers △abc is the one whose diameter is the longest edge of triangle.*

Proof.   Suppose that *ab* is the longest edge of △*abc*. When a disk covers △*abc*, we can always arrange the boundary of the disk passing through *a* and *b*. If *ab* is not a diameter of the disk and *c* is not on its boundary, then we are able to shrink the disk while it still covers △*abc*.   □

Lemma 5.10   *Four terminal points a, b, c, and d can be covered by a disk of radius R if and only if each of four triangles △abc, △bcd, △acd, and △abd can be covered by a disk of radius R.*

Proof.   The 'Only-if' part is easy to be verified. it suffices to show the 'If' part. Assume, without loss of generality, that those four points *a*, *b*, *c*, and *d* form a convex quadrilateral since if not, one of them must lie in the triangle of other three, which can be covered by a disk of radius *R*.

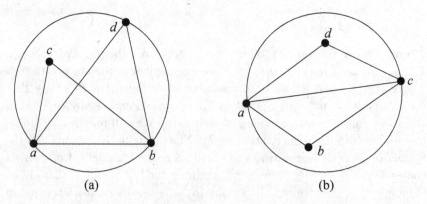

(a)   (b)

Fig. 5.7   For the proof of Lemma 5.10.

Consider the longest edge of complete quadrilateral *abcd* (that has six edges). Suppose that it is not a diagonal edge, say *ab*. We compare ∠*acb* with $\angle adb$. Assume, without loss of generality, that $\angle acb \leq \angle adb$. Then the minimum disk covering $\triangle abc$ also covers point d as shown in Fig.5.7(a). Now suppose that the longest edge of complete quadrilateral $abcd$ is a diagonal edge, say ac. Then we consider following two cases.

Case 1. Triangles $\triangle abc$ and $\triangle acd$ are obtuse or right. In this case, $\angle abc$ and $\angle cda$ are obtuse or right. Therefore the disk with diameter ac can cover a, b, c, and d as shown in Fig.5.7(b).

Case 2. Either triangle $\triangle abc$ or $\triangle acd$ is acute, say $\triangle abc$. Suppose that $\triangle abd$ is also acute. We compare $\angle acb$ with $\angle adb$. Assume, without loss of generality, that $\angle acb \leq \angle adb$. Then the minimum disk that covers $\triangle abc$ also covers point d as shown in Fig.5.7(a). Similar argument may apply to the subcases that $\triangle bcd$ is acute and that $\angle bdc \geq 90°$ or $\angle adb \geq 90°$. Note than $\angle cbd \leq \angle cba < 90°$ and $\angle bda \leq \angle cba < 90°$. Therefore, the remainder is the case of $\angle bad \geq 90°$ and $\angle dcb \geq 90°$. In this subcase, the disk with diameter bd can cover a, b, c, and d. □

Observe that the proof of Lemma 5.10 is constructive. Thus it enables us to find the Steiner point to connect four terminal points when it exists. We now are ready to present the improved approximation algorithm, which is called the *greedy algorithm* since it tries to use the shortest edges between terminal points.

Algorithm 5.2 *Greedy Algorithm*

Step 0 Given a set P of n terminals, sort all $n(n-1)/2$ possible edges between n points in length increasing order $e_1, e_2, \cdots, e_{n(n-1)/2}$. Initially, set $T_g := (P, \emptyset)$ and $i := 1$.

Step 1 while $|e_i| \leq R$ do begin
  if e_i connects two different connected components of T
    then $T_g := T_g \cup \{e_i\}$ and $i := i + 1$.
  end-while

Step 2 for each subset of four terminal points a, b, c and d respectively in four connected components of T_G do
  if there exists a point s within distance R from a, b, c and d
    then $T_G := T_G \cup \{sa, sb, sc, sd\}$
  end-for

Step 3 while $i \leq n(n-1)/2$ do begin
  if e_i connects two different connected components of T_G
    then $T_g := T_g \cup \{e_i\}$ (e_i is cut into $\lceil |e_i|/R \rceil$ segments)
      and $i := i + 1$
  end-while

return T_g

Theorem 5.4   *Given a set P of n terminal points and a constant R, the greedy algorithm returns a Steiner tree T_g whose number of Steiner points is at most three times that of the optimal Steiner tree T_{opt}.*

Proof.   Denote by T_i the T_g at the beginning of the i-th step for $i = 1, 2, 3$. Suppose that $T_3 - T_2$ contains k 4-stars consisting of $\{sa, sb, sc, sd\}$. Then $c(T_g) \leq C(T_s) - 2k$, where T_s is a Steinerized minimum spanning tree on all n terminal points. Let T_{opt} be a shortest optimal Steiner tree with Steiner points of degrees at most five. Suppose that T_{opt} has m full components $T_{opt}^1, T_{opt}^2, \cdots, T_{opt}^m$. Then we construct a Steinerized spanning tree T as follows: Initially, put T_2 into T. For each full component T_{opt}^i, add to T the steinerized minimum spanning tree H_i for terminal points in T_{opt}^i. If T has a cycle, then destroy the cycle by deleting some edges and Steiner points of H_i. An important observation is that if H_j does not contain an edge between two terminal points, then a Steiner point must be deleted for destroying a cycle in $H_i \cup T_2$. From this observation and Lemma 5.6, we have

$$c(T_s) \leq 3 \cdot c(T_{opt}) + h,$$

where h is the number of full components T_{opt}^i's with properties that every Steiner point in T_{opt}^i has degree five and $T_{opt}^i \cup T_2$ has no cycle. Hence we obtain

$$c(T_g) \leq 3 \cdot c(T_{opt}) + h - 2k.$$

Clearly, to prove the theorem it suffices to show $h \leq 2k$. Suppose that T_2 has q connected components. Then T_3 has $(q - 3k)$ connected components $C_1, C_2, \cdots, C_{q-3k}$. We now construct a graph H with vertex set P and edge set which is defined as follows:

(1) Put all edges of T_2 into H. Then consider each of those full component T_{opt}^i that every Steiner point in T_{opt}^i has degree five and $T_{opt}^i \cup T_2$ has no cycle. If T_i has only one Steiner point, then this Steiner point connects to five terminal points which must lie in at most three C_i's. Hence, among them there are two pairs of terminal points each lying in the same C_i.

(2) Connect the two pairs with two edges and put them into H. If T_{opt}^i has at least two Steiner points, then there must exist at least two Steiner points each connecting to four points. We can also find two pairs of terminal points among them such that each pair lies in the same C_i.

(3) Connect these two pairs with two edges and put the two edges into H.

Clearly, H has at most $(q - 2h)$ connected components. Since every connected component of H is contained by some C_i, we have $q - 3k \leq q - 2h$. Therefore, we have $h \leq 2k$, and the proof is finished. $\square$

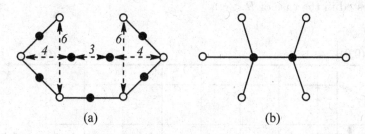

(a)   (b)

Fig. 5.8 The performance ratio of the greedy algorithm is at least 2.5.

Note that the exact value of the performance ratio of the greedy algorithm is between 2.5 and 3. Consider the example shown in Fig.5.8 with $R = 4$. The greedy algorithm returns a Steiner tree with five Steiner points as shown in Fig.5.8(a) while the optimal Steiner tree has two Steiner points as shown in Fig.5.8(b). Unfortunately, the exact value is not known.

5.1.4 *Polynomial Time Approximation Scheme*

In this subsection, we consider a variation of Steiner tree problem for minimal Steiner points. Instead of minimizing the number of Steiner points in a Steiner tree, we now aim at minimizing the number of *total points* in the Steiner tree, that is, both Steiner points and terminal points are counted. The new version is more formally defined as follows.

Problem 5.3 *Steiner Tree Problem for Minimal Total Points*

Instance	A set of n points $P = \{p_1, \cdots, p_n\}$ in the Euclidean plane $\Re^2$ and a positive constant R.
Solution	A Steiner tree T interconnecting all points in P such that each edge in T has length no more than R.
Objective	Minimizing the number of Steiner points in T plus n.

It is clear that the decision version of the above problem is the same as that of Steiner tree problem for minimal Steiner points, thus the above problem is also NP-hard. In the following, we will construct a Polynomial

Time Approximation Scheme (PTAS) when terminal points satisfy certain conditions.

A set P of terminal points is called *c-local* if in the Minimum Spanning Tree (MST) of P the length of the longest edge is at most c times of the length of the shortest edge. We assume, without loss of generality, that the distance between any pair of points in P is at least 1 and $c \geq 1$. We are interested in the case of $R < c$.

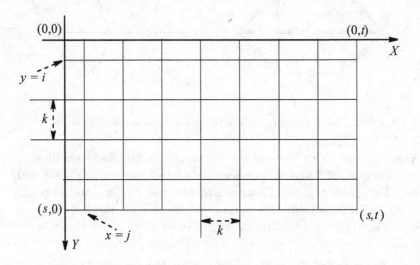

Fig. 5.9   Partition the whole area into many rectangles.

The basic idea of the scheme is to combine the shifting technique in [135] with a local optimization method. We will design a set of partitions, each of them partitions the whole area enclosing all terminal points in P into many rectangular *cells* of some constant size as shown in Fig.5.9. Each cell is further divided into *interior* and *boundary* areas as shown in Fig.5.10.

Then, with respect to each partition, we first organize the terminals contained in the interior area of each cell into several groups such that the distance between any two groups is greater than c, and then construct an optimal solution (a *local Steiner tree*) for each group. The collection of all the local Steiner trees in a cell constitute a *local Steiner forest* for the cell. After that, we connect all the local Steiner forests and the terminal points in the boundary areas using the spanning tree approach. At the end, we select a partition which yields an optimal global solution among all the partitions.

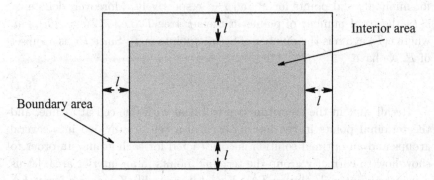

Fig. 5.10 Each cell is divided into interior and boundary areas.

5.1.4.1 *Partition Strategy*

We first study how to partition the whole area into cells (mostly squares). Assume, without loss of generality, that all terminal points in P are contained in a rectangle Rec with four corners at $(0,0),(s,0),(0,t)$, and (s,t) as shown in Fig.5.9. For any positive integer k, a *partition* of size k is a grid in which adjacent horizontal/vertical lines are separated by a distance of k. Clearly, there are k^2 different partitions of size k, depending on the locations of the top horizontal line and the leftmost vertical line. Denote $P_{i,j}$, where $0 \le i,j < k$, the partition in which the top horizontal line and the leftmost vertical line are $y = i$ and $x = j$, respectively as shown in Fig.5.9. Observe that the grid partitions the rectangle Rec into many cells, most of which are squares of size $k \times k$. Thus each cell contains at most k^2 points in P since the distance between any two points is assumed to be at least one. Moreover, each cell is divided into an *interior area* and a *boundary area*, with a boundary of width $l = (2 + 3 \log k)c$.

5.1.4.2 *Approximation Scheme*

Given a set P of points and a partition X, let $P_X \subseteq P$ be the set of points in the interior areas. Let T be a Steiner tree of P, an edge in T is a crossing edge if it is not completely contained in any interior area of a cell, a path in T is called a *stem* if every vertex in the path is Steiner points of degree-2 except that the two vertices at the ends are terminal points. A stem is called a *crossing stem* if at least one of the terminal points is in the boundary area.

Denote by T_{opt} and T_{opt}^X the optimal solution to the Steiner tree problem for minimal total points for P and P_X, respectively. Moreover, denote by $\bar{c}(T)$ the total number of points in Steiner tree T, i.e., $\bar{c}(T) \doteq c(T) + n$, where $c(T)$ again is the number of Steiner points in T. Since P_X is a subset of P, we have

$$\bar{c}(T_{opt}^X) \leq \bar{c}(T_{opt}) + n. \qquad (5.1)$$

Recall that in the algorithm, we will deal with one cell at a time, and the terminal points in the interior area of a cell are divided into several groups and an optimal solution is constructed for each group. In order to show how to correctly group the terminal points in an interior area, let us consider an optimal solution T_{opt}^X. We need to modify T_{opt}^X into a forest F^X such that each tree in F^X is completely included in the interior area of a cell for X. Note that each interior area of a cell may contain more than one tree in F_X. Define the *distance* between two trees to be the shortest distance between any pair of terminal points in the two trees. The following lemma shows that we may require some more properties.

Lemma 5.11 *For any partition X, T_{opt}^X can be modified into a forest F^X such that each tree in F^X is completely in an interior area of a cell for X and the distance between any pair of trees in F_X is greater than c. Moreover, the total cost $\bar{c}(F^X)$, which is the sum of the costs of all trees in F_X, is at most $\bar{c}(T_{opt}^X)$, and $\bar{c}(F_X) \leq \bar{c}(F_X) \leq \bar{c}(T_{opt})$.*

Proof.   First, we eliminate those stems from F_X that have lengths greater than c. Note that the distance between any pair of resulting trees is greater than c since T^X is optimal. For each tree T' in the forest obtained above, we reconstruct an optimal tree connecting terminal points in T'. Assume, without loss of generality, that each stem in the reconstructed trees has length at most c (otherwise we can repeat the procedure and further decompose the forest).

Secondly, we prove that each tree in the forest obtained above is completely in an interior area of a cell. For this purpose, it suffices to show that there is no Steiner point in the boundary area. Suppose, by contradiction, that there are some Steiner points in the boundary area. We call a Steiner point a *real Steiner point* if it has degree greater than 2. Recall that the distance between two cells is $(2 + 3 \log k)c$. It is easy to see that two terminal points in distinct cells can not be connected in the above resultant forest; otherwise, there must be a Steiner point s which is at least $\frac{3c}{2} \log k$ away from any boundary line. To reach any boundary line, s has to create at least $2^{1.5 \log k} = k^{1.5}$ real Steiner points and $(\lceil c/R \rceil - 1)k^{1.5}$ Steiner points of degree-2. Now we remove all those $\lceil c/R \rceil k^{1.5}$ Steiner points in the boundary areas and use them to connect the disconnected subtrees with distance less than c in the corresponding 4 neighboring cells as shown in Fig.5.11. At most $k\lceil 1/R \rceil$ Steiner points are required to be added in each of the eight boundary segments of the four cells.

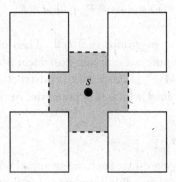

Fig. 5.11   Connect the disconnected subtrees in the corresponding four neighbor cells.

Since no two cells are connected, we can move the Steiner points in the boundary areas back to the interior areas. In such a way, all Steiner points in the boundary area can be eliminated. The proof is then finished. $\quad\square$

Unfortunately, it is difficult to compute the forest F_X since T_{opt}^X is unknown. Nevertheless, we can construct a forest which is similar to F_X. Consider the terminal points in the interior area of some fixed cell. By Lemma 5.9, if the distance between two terminal points is at most c, then they must belong to the same tree of F_X. Thus we can group the terminal points by forming an MST of these points and then deleting the edges longer than c. As a result, we obtain a set of spanning trees $\{T_1, T_2, \cdots, T_m\}$ each consisting of degrees of length at most c. We call these trees the *c-spanning trees*. Let S_i be the set of terminal points contained in the c-spanning trees T_i for $i = 1, 2, \cdots, m$. Clearly, the terminal points in the same group S_i belong to the same tree of the forest F_X. The converse, however, is not necessarily true. Namely, terminal points in different group may also belong to the same tree of F_X. Thus to find the best way of grouping the terminal points, we have to consider all possible ways for merging the groups $S_1, S_2, \cdots, S_m$. After each such possible merging process, we obtain a local Steiner forest by constructing an optimal solution for every *new* group. We are interested in a local Steiner forest with the minimum cost among all possible merging processes for each cell.

Lemma 5.12 *Let forest F'_X denote the collection of the minimum-cost local Steiner forests, one for each cell. Then*
(i) *Each tree in F'_X is completely contained in the interior area of a cell;*
(ii) *The distance between any tree pair T' and T'' in F'_X is greater than c;*
(iii) *The total cost of F'_X is at most $\bar{c}(F_X)$, and $\bar{c}(F'_X) \leq \bar{c}(F_X) \leq \bar{c}(T_{opt})$.*

Suppose that there are m groups in a cell. Then using the method in [263], we are able to compute a minimum-cost local Steiner forest in time of $O(2^m M(|S|))$, where $M(S)$ is the time to construct an optimal solution for the set of terminal points, which is exponential in the size of S.

5.1.4.3   *Exact Algorithm*

Assume, without loss of generality, that the terminal points in S are leaves in the tree. The number of possible topologies for S is at most $|S|!$. Now consider a fixed topology t for S. If the number of candidate points for each internal vertex in t is at most m, then a modification of a standard dynamic programming algorithm finds an optimal solution for the fixed topology in time of $O(m|S|)$ [143].

Lemma 5.13 *The number of candidate points for each internal vertex is at most $(|S|\sqrt{2}k/R)^{3^{|S|}-1}$ if terminal points in S are in a square of size k by k.*

Proof.   Let T^t_{opt} be an optimal solution for the fixed topology t. Consider an internal vertex v at the bottom whose children are leaves in t. Let v_1 and v_2 are two children of v. Without increasing the number of Steiner points, we are able to move the point adjacent with v such that the distance between v and v_i is Rh_i for $i = 1, 2$, where h_i's are integers. Thus, the number of candidate points for v is at most $|S|^2(\sqrt{2}k/R)^2$. The height of t is at most $(|S| - 1)$. For a vertex of height i, denote by $f(i)$ the number of candidate points. Then $f(i) \leq f(i-1)^2|S|^2(\sqrt{2}k/R)^2 \leq f(i-1)^3$. Therefore, for any internal vertex, the number of candidate points is at most $(|S|\sqrt{2}k/R)^{3^{|S|}-1}$. The proof is then finished. $\square$

From the above discussion, we can easily deduce that $M(S) = O\left(|S|!(|S|\sqrt{2}k/R)^{2^{|S|}}\right)$.

5.1.4.4 *Connecting Local Forests and Boundary Points*

We can now construct a Steiner tree for terminal-set P from the forest F'_X as follows. (1) Fix an MST T_{mst} for P and add degree-2 Steiner points to ensure that the length of each edge is at most R. Note that each stem in Steinerized T_{mst} has length at most c since P is $c - local$. Let E_X denote the set of crossing edges in T_{mst}. (2) Construct a graph G_X by adding all the crossing edges in E_X to F'_X and adding degree-2 Steiner points to ensure that the length of each edge is at most R.

Note that in the above G_X is connected. Now we are ready to present the whole algorithm.

Algorithm 5.3 *Polynomial Time Approximation Scheme*

Step 0 Construct an MST T_{mst} for P.
Step 1 for each possible partition $X_{i,j}$ **do begin**
        Find the set of crossing edges $E_{X_{i,j}}$.
        for each cell
            Compute a minimum-cost local Steiner forest.
        end-for
        $F'_{X_{i,j}} := \{$ all local Steiner forests $\}$
        Construct the graph $G_{X_{i,j}}$ consisting of $E_{X_{i,j}} \cup F'_{X_{i,j}}$
        Add degree-2 Steiner points if necessary
    end-for
Step 2 Select a $G_{X_{i,j}}$ with the smallest cost among all partitions.
Step 3 Prune $G_{X_{i,j}}$ into a tree T
Return $T_{ptas} := T$.

Theorem 5.5 *The performance ratio of approximation scheme 5.3 is* $1 + \left(16 + (4 + 3\log k)c/k\right)$.

Proof. Consider the stems in MST T_{mst} for P. Since the boundary area of each cell consists of at most $4(2 + 3\log k)ck$ terminal points, each terminal point of a crossing stem can be inside a boundary area at most $4(2 + 3\log k)ck$ times under the k^2 partitions. Since the length of a stem is at most c, a stem can be a crossing stem at most $4(2 + 3\log k)ck$ times. Therefore, the total cost of $k^2 G_{X_{i,j}}$'s is bounded by:

$$\sum_{i=0}^{k-1}\sum_{j=0}^{k-1} \bar{c}(G_{X_{i,j}}) \leq k^2 \bar{c}(T_{opt}) + \sum_{i=0}^{k-1}\sum_{j=0}^{k-1} \bar{c}(E_{X_{i,j}})$$

$$\leq k^2 \bar{c}(T_{opt}) + 4(4 + 3\log k)\,c\,k \cdot \bar{c}(T_{mst}).$$

From Theorem 5.1, we know that at least one partition yields a solution with cost at most $1 + \left(16 + (4 + 3\log k)c/k\right)$ times of the optimal solution. The proof is then finished. □

Corollary 5.1   *There exists a polynomial time approximation scheme for the Steiner tree problem for minimal total points when the set of terminal points is $c - local$.*

Corollary 5.2   *Let T_{ptas} be the Steiner tree produced by approximation scheme 5.3 and T_{opt} be an optimal Steiner tree for Steiner tree problem for minimal total points. Then*

$$\frac{c(T_{ptas})}{c(T_{opt})} \leq 1 + \frac{16(4 + 3\log k)c}{k}\left(1 + \frac{4n}{c(T_{smst})}\right),$$

where T_{smst} is a Steinerized minimum spanning tree for the same set of terminal points.

The above implies that there exists a PTAS for Steiner tree problem for minimal total points when the set P of n terminal points is $c - local$ and the MST on P has length at least $(1 + \alpha)nR$ for some positive constant α.

5.2   In the Rectilinear Plane

In this section, we will study the Steiner tree problem for minimal Steiner points in the rectilinear plane [195]. The analysis and discussion are similar to those for the case of Euclidean plane (i.e., Problem 5.1), but different techniques are used.

The Steiner Minimum Tree (SMT) for the rectilinear Steiner tree problem and the optimal Steiner tree for the rectilinear Steiner tree problem for minimal Steiner points may have different structures. Recall that in a rectilinear SMT, every vertex has degree less than five and Steiner points have degree either three or four. In an optimal Steiner tree, however, Steiner points may have degree two. For example, when V contains two terminals v_1 and v_2 with $|v_1 v_2| > R$, the optimal Steiner tree is a path containing $\lceil |v_1 v_2|/R - 1 \rceil$ Steiner points whose degrees are two; Moreover, some vertices may have degree as large as eight. Consider nine terminals (white nodes) in Fig.5.12(a). The SMT for the rectilinear Steiner tree problem in Fig.5.12(b) includes two Steiner points (black nodes) and has length of $6R$. Two optimal Steiner trees for minimal Steiner points in Fig.5.12(c,d) both include no Steiner point and have length of $8R$; Note in the first tree every terminal has degree less than three while in the second tree one terminal has degree eight. However, we can show the following lemma.

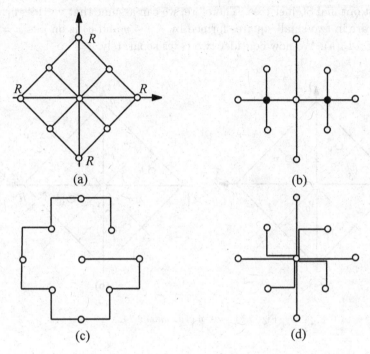

Fig. 5.12 (a) nine terminals in the rectilinear plane, (b) the SMT of the rectilinear Steiner tree problem, (c) and (d) two optimal Steiner trees for minimal Steiner points.

Lemma 5.14 *There exists a shortest optimal Steiner tree for the rectilinear Steiner tree problem for minimal Steiner points such that every vertex in the tree has degree at most four.*

Proof.   Suppose, by contradiction, that every shortest optimal Steiner tree has a vertex whose degree is more than four. Then let T_{opt} be such a tree with minimal number of vertices that have degrees more than four, and let vertex v_0 in T be adjacent to five vertices v_i, $i = 1, 2, \cdots, 5$. We assume, without loss of generality, that $v_0 = (0, 0)$. Now consider v_0's neighborhood of radius R, which is a square. Note that lines $x = y$ and $x = -y$ partition the square into four small squares. See Fig.5.13.

Suppose that vertices v_1 and v_2 are in same small square, say, the one formed by lines $-x \leq y$ and $x \leq y$. If neither v_1 nor v_2 is on one of these two lines, then it can be verified that $|v_1v_2| < \max\{|v_1v_0|, |v_2v_0|\}$. Thus replacing the longer edge of v_1v_0 and v_2v_0 with edge v_1v_2 will produce a Steiner tree whose length is shorter than T, contradicting that T_{opt} is a shortest optimal Steiner tree. Therefore, we can assume that vertices v_1, v_2 and v_3 are in two small squares formed by $-x \leq y$ and v_2 is on line $x = y$. See Fig.5.13(a). We now consider two cases separately.

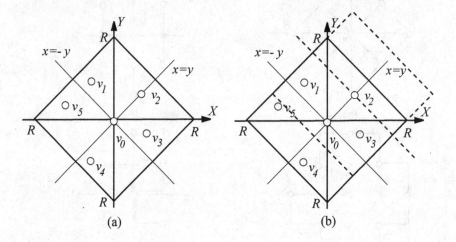

Fig. 5.13   Proof of Lemma 5.14.

Case 1. v_2 has degree less than four. Then substituting edge v_1v_0 with edge v_1v_2 will produce a shortest Steiner tree that has less number of vertices that have degree at least five than T_{opt}, contradicting that T_{opt} has minimal number of such vertices.

Case 2. v_2 has degree four (or bigger). Then v_2 is adjacent to three vertices u_1, u_2, and u_3 besides vertex v_0. If there exists a vertex u_i that is in the two small squares of v_2's neighborhood formed by the line in direction of $x = -y$ and passing vertex v_2, then either $|v_1u_i| < |v_1v_0|$ or $|v_3u_i| < |v_3v_0|$. In either case a shorter Steiner tree can be constructed by replacing edge v_1v_0 with edge v_1u_i or edge v_3v_0 with edge v_3u_i, contradicting that T is a shortest one. Hence we can assume that u_i for $1 \leq i \leq 3$ are all in the other two small squares while u_2 in line $x = y$. See Fig.5.13(b).

By repeating the above argument at vertex u_2 and so on, in the end we will reach Case 1 and find a contraction. The proof is then finished. □

Lemma 5.15   *There exists a shortest optimal Steiner tree for the rectilin-*

ear Steiner tree problem for minimal Steiner points such that no two edges cross each other.

Proof. Suppose that T_{opt} is a shortest optimal Steiner tree that includes two edges ac and bd crossing at point p, and ac and bd may have corners at points b' and c', respectively. See Fig.5.14.

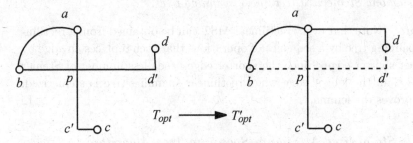

Fig. 5.14 Proof of Lemma 5.15.

We can assume that $|ap| \le |pc'|$, $|bp| \le |pd'|$ and $|ap| \le |bp|$. Then we have

$$|ad| = |ap| + |pd'| \pm |d'd| \le |bp| + |pd'| + |d'd| = |bd|,$$

$$|ab| = |ap| + |pb| \le |bp| + |pd'| + |d'd| = |bd|.$$

Therefore, replacing edge bd with edge ab if there is a path in T_{opt} between vertices b and a or c, and with edge ad if there is a path in T_{opt} between vertices d and a or c, can make a shortest Steiner tree T'. Moreover, the number of crossings in T'_{opt} is less than that in T_{opt}. By repeating this operation we can obtain a shortest optimal Steiner tree without any crossing. The proof is then finished. $\square$

In fact, by applying the combination of the arguments in the proofs of Lemmas 5.14-15, we are able to prove that there exists a shortest optimal Steiner tree for the rectilinear Steiner tree problem for minimal Steiner points such that every vertex in the tree has degree at most four and there is no two edges crossing each other. Hence at the rest of the section, we only consider such a kind of shortest optimal Steiner trees for the rectilinear Steiner tree problem for minimal Steiner points.

Note that as in the Euclidean plane, we can Steinerize a rectilinear Minimum Spanning Tree (MST) if it is a feasible solution for the rectilinear

Steiner tree problem for minimal Steiner points by breaking any edge longer than R into several small segments with lengths at most R through adding some Steiner points in the edge. The Steinerized rectilinear MST has the following interesting property.

Lemma 5.16   *Every Steinerized rectilinear MST has the minimal Steiner points among Steinerized rectilinear spanning trees.*

Proof.   Note that every rectilinear MST can be obtained from a rectilinear spanning tree by a sequence of operations that each replaces an edge by another shorter edge. Since the shorter edge needs less number of Steiner points than the longer edge when rectilinear spanning tree is steinerized. This proves the lemma.   □

5.2.1 *Steinerized Minimum Spanning Tree Algorithm*

To obtain the lower bound of performance ratio for the Steinerized spanning tree algorithm for the rectilinear case, consider a simple example shown in Fig.5.15. There are four terminals (white nodes) such that every pair of them has length of $2L$, where $L = (1 - \varepsilon)R$ and ε is a very small positive real number. Fig.5.15(a) shows the optimal Steiner tree for minimal Steiner points, which has only one Steiner point (black node). Fig.5.15(b) shows the Steinerized rectilinear MST that has three Steiner points. This implies that the performance ratio of Steinerized minimum spanning tree algorithm is at least three.

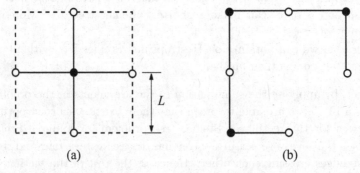

(a)   (b)

Fig. 5.15 The lower bound: (a) the optimal Steiner tree for minimal Steiner points, and (b) the Steinerized rectilinear MST.

To obtain the upper bound of performance ratio, we need to study the properties of rectilinear convex path in the shortest optimal Steiner tree for minimal Steiner points. A path $q_1 q_2 ... q_m$ is called a *rectilinear convex path* if one places a coordinate axis at any point on the path then at least one of the quadrants does not contain any point in the path. Note that angles can be defined along the convex path. The angles of 180° and 270° will play an important role in the the following analysis. For simplicity, such angles are called *big angles*.

Lemma 5.17 *Let $q_1 q_2 ... q_m$ be a rectilinear convex path and $m \geq 2$. Suppose that there are t_1 angles of 180° and t_2 angles of 270° among $m - 2$ angles $\angle q_1 q_2 q_3, \angle q_2 q_3 q_4, \cdots, \angle q_{m-2} q_{m-1} q_m$. Then $|q_1 q_m| \leq (t_1 + 2t_2 + 2)R$.*

Proof. First, we define a staircase to be a continuous path of alternating vertical lines and horizontal lines such that their projections on the vertical and horizontal axes have no overlapping intervals.

Then we prove the lemma by induction on m. For $m \leq 3$, it is true, since $|q_1, q_3| = 2R \leq (t_1 + 2t_2 + 2)R$. Now, suppose $m \geq 4$. Consider the rectilinear convex hull H of points $q_1, q_2, \cdots, q_m$. If at least one of q_1 and q_m does not lie on the boundary of H, then by the induction hypothesis, any rectilinear distance between two vertices of the convex hull H is at least $(t_1 + 2t_2 + 2)R$. Therefore, we have $|q_1 q_m| \leq (t_1 + 2t_2)R$.

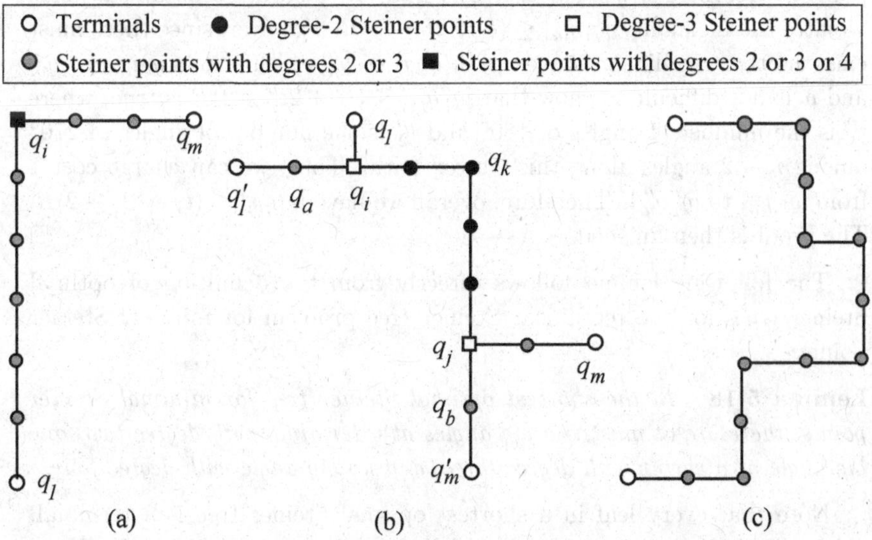

Fig. 5.16   Proof of Lemma 5.17.

Next, we may assume that both q_1 and q_m lie on the boundary of H. Then the whole path $q_1 q_2 \cdots q_m$ lies on the boundary of H as shown in Fig.5.16(c). Since there are overlapping intervals if a rectilinear convex path is not a staircase, we can always obtain $|q_1 q_m| \leq (t_1 + 2t_2 + 2)R$ by induction hypothesis. So it suffices to show that this inequality is true for staircase as shown in Fig.5.16(a,b). We consider the following cases:

Case 1. When the rectilinear convex path is straight. Clearly, we have $|q_1 q_m| = (t_1 + 1)R < (t_1 + 2t_2 + 2)R$.

Case 2. When the rectilinear convex path is a L-shaped path. As shown in Fig.5.16(a), the turning point of the L-shaped path may be a Steiner point with degrees 2 or 3 or 4. It can also just be a corner. For the latter we have $|q_1 q_m| = (t_1 + 1)R$ from Case 1. If the turning point is a Steiner point, say q_i, then we have $|q_1 q_m| = t_1 + 2 \leq (t_1 + 2t_2 + 2)R$.

Case 3. The other possible shapes of rectilinear convex paths are those like $q_1' q_2' \cdots q_m'$ and $q_1 q_2 \cdots q_m$ as shown in Fig.5.16(b) (maybe in other directions). Otherwise, we can have a shorter tree by flipping some corner or Steiner point contradicting that T_{opt} is a shortest optimal Steiner tree for minimal Steiner points. For those convex paths like $q_1' q_2' \cdots q_m'$, it is easy to show that $|q_1' q_m'| \leq (t_1 + 2t_2 + 2)R$ holds. For those like $q_1 q_2 \cdots q_m$, when the turning point of path $q_i q_{i+1} \cdots q_j$ is a Steiner point q_k shown in Fig 5.16(b), it can be verified that $|q_1 q_m| \leq (t_1 + 2t_2 + 2)R$ also holds. When the turning point is a corner, $|q_1 q_m| = (t_1 + 3)R$ and $t_2 = 0$ will be obtained. This violates the inequality $|q_1 q_m| \leq (t_1 + 2t_2 + 2)R$. However, since there must exist another rectilinear convex path $q_1'' q_2'' \cdots q_n''$ which shares $q_i q_{i+1} \cdots q_m$ and it is not difficult to show that $|q_1'' q_m''| \leq (t_1'' + 2t_2'' + 2)R$ is true, where t_1'' is the number of angles of $180°$ and t_2'' is the number of angles of $270°$ among $n - 2$ angles along this convex path. Thus, we can charge cost 1 from $|q_1 q_m|$ to $|q_1'' q_m''|$. Therefore, overall we have $|q_1 q_m| \leq (t_1 + 2t_2 + 2)R$. The proof is then finished. □

The following lemma follows directly from the definition of optimal Steiner trees for the rectilinear Steiner tree problem for minimal Steiner points.

Lemma 5.18 *In the shortest optimal Steiner tree for minimal Steiner points, there are at most two big angles at a terminal with degree two, one big angle at a vertex with degree three, and no big angle with degree four.*

Note that every leaf in a shortest optimal Steiner tree is a terminal. A rectilinear Steiner tree is called *full* if every terminal is a leaf. When a rectilinear Steiner tree is not full, we can always find a terminal with degree more than one where we can break the tree into parts. In such a way, every rectilinear Steiner tree can be broken into several small full rectilinear Steiner trees, which are called *full components* of the rectilinear Steiner tree.

Lemma 5.19 *Let T be a full rectilinear Steiner tree interconnecting n terminals, and s_i denote the number of Steiner points with degree i in T. Then $2s_4 + s_3 = n - 2$.*

Proof. Since T has $s_4 + s_3 + s_2 + n - 1$ edges, and the total degrees of vertices in T is $4s_4 + 3s_3 + 2s_2 + n = 2(s_4 + s_3 + s_2 + n - 1)$. Hence, $2s_4 + s_3 = n - 2$. □

Given a shortest optimal Steiner tree for for the rectilinear Steiner tree problem for minimal Steiner points on n terminals, if it is a full rectilinear Steiner tree, then we can find a set of n rectilinear convex paths in the tree that satisfy the following three properties:

(1) each path connects two terminal points,
(2) each terminal point appears in exactly two paths, and
(3) each angle at a Steiner point appears in the paths exactly once.

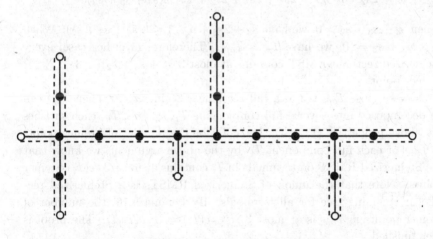

Fig. 5.17   Rectilinear convex paths having desired properties.

Fig.5.17 shows an example where terminals and Steiner points are marked by white and black nodes, respectively, and rectilinear convex paths by dashed lines. Now, we are ready to show the main theorem.

Theorem 5.6 *The Steinerized rectilinear minimum spanning tree is a* *polynomial-time approximation for the rectilinear Steiner tree problem for minimal Steiner points with performance ratio exactly 3.*

Proof.   Denote by $c(T_{opt})$ the number of Steiner points in T_{opt}. First, we assume that T_{opt} is a full rectilinear Steiner tree. Let s_i denote the number of Steiner points with degree i in T_{opt}. Let s_2' denote the number of Steiner points of degree two with angle $180°$, and s_2'' denote the number of Steiner points of degree two with angle $270°$. Clearly, $s_2 = s_2' + s_2''$. By Lemma 5.19, $n = 2s_4 + s_3 + 2$. Consider a rectilinear spanning tree T_s consisting of $n-1$ edges each connecting two terminals at endpoints of a rectilinear convex path in T_{opt}. By Lemma 5.17, each edge in T_s has length upper-bounded by $(t_1 + 2t_2 + 2)R$ where t_1 and t_2 are the numbers of big angles of $180°$ and $270°$ on the rectilinear convex path connecting two terminals, respectively. Hence we need at most $(t_1 + 2t_2 + 1)$ Steiner points to Steinerize the edge. By Lemma 5.18, the rectilinear spanning tree T_s can be Steinerized by at most $s_3 + 2s_2' + 2s_2'' + n - 1 = s_3 + 2s_2 + n - 1$ Steiner points. By Lemma 5.16, any Steinerized rectilinear MST contains at most $s_3 + 2s_2 + n - 1$ Steiner points. Clearly

$$s_3 + 2s_2 + n - 1 = 2s_4 + 2s_3 + 2s_2 + 1 = 2(s_4 + s_3 + s_2) + 1.$$

When $s_4 + s_3 + s_2 > 0$, we have $s_3 + 2s_2 + n - 1 \leq 3(s_4 + s_3 + s_2)$. When $s_4 + s_3 + s_2 = 0$, we have $T_s = T_{opt}$. Therefore, in either case, every Steinerized rectilinear MST contains at most $3(s_4 + s_3 + s_2)(= 3 \cdot c(T_{opt}))$ Steiner points.

Now suppose T_{opt} is not a full rectilinear Steiner tree. Then T_{opt} can be decomposed into several full components $T_1, T_2, \cdots, T_k$, each satisfies the above properties (1, 2, 3). Let $c(T_i)$ be the number of Steiner points in T_i. For each full component T_i, by the above argument, we know that the Steinerized RMST on terminals in T_i contains at most $3 \cdot c(T_i)$ Steiner points. Note that the union of Steinerized RMSTs is a Steinerized rectilinear spanning tree for all terminals. By Lemma 5.16, the number of Steiner points in T_{opt} is at most $3 \sum_{i=1}^{k} c(T_i) = 3 \cdot c(T_{opt})$. The proof is then finished. $\square$

5.2.2   Greedy Algorithm

Suppose that T_{opt} is a shortest optimal Steiner tree for minimal Steiner points in the rectilinear plane that has k full components $T_1, T_2, \cdots, T_k$. In the proof of Theorem 5.6, we have showed that the Steinerized RMST on terminals in T_i contains at most $(2 \cdot c(T_i) + 1)$ Steiner points. In the following we will study when this upper bound can be improved.

Lemma 5.20 *Let T_i' be the Steinerized RMST on terminals in T_i. Then*
(i) T_i' *contains at most $2 \cdot c(T_i) + 1$ Steiner points;*
(ii) T_i' *contains at most $2 \cdot c(T_i)$ Steiner points when T_i contains a Steiner point with degree at most three;*
(iii) T_i' *contains at most $2 \cdot c(T_i)$ Steiner points when T_i' contains an edge between two terminals.*

Proof. Note that conclusions (i-ii) follow immediately from the proof of Theorem 5.6. To show (iii), let n_i be the number of terminals in T_i. Note that there are exactly n_i paths in the forms shown in Lemma 5.17 in T_i. Choose any $(n_i - 1)$ of them and connect two endpoints of each path. We will obtain a rectilinear spanning tree. Its Steinerization is denoted by T_s. Now, assume that u is a Steiner point with degree at most three. When there is a big angle at u, we choose $(n_i - 1)$ rectilinear convex paths not containing the big angle. When there is no big angle at u, we can choose any $(n_i - 1)$ rectilinear convex path. In such a way, we can obtain $c(T_s) \leq s_3 + 2s_2 - 1 + (n_i - 1) \leq 2(s_4 + s_3 + s_2) = 2 \cdot c(T_i)$. The proof is then finished. $\qquad\square$

To deal with the case of the rectilinear plane, we could modify the greedy algorithm 5.2 for minimal Steiner points in the Euclidean plane as follows: At Step 2, if there exists a point within (rectilinear) distance R from a, b, and c, then $T_G := T_G \cup \{sa, sb, sc\}$.

Theorem 5.7 *The greedy algorithm for minimal Steiner points in the rectilinear plane returns a Steiner tree T_g that satisfies $c(T_g) \leq 2 \cdot c(T_{opt})$.*

Proof. Denote by T_i the T_g at the beginning of the i-th Step in the greedy algorithm for $i = 1, 2, 3$. Suppose that $T_3 - T_2$ contains m 3-stars. Then

$$c(T_g) \leq c(T_{smst}) - m,$$

where $c(T_{smst})$ is the number of Steiner points in T_{smst} which is the Steinerized RMST on all terminal points. We construct a Steinerized rectilinear spanning tree T_s as follows: Initially, put T_2 into T_s. For each full component T_s^j ($1 \leq j \leq k$), add to T_s the Steinerized rectilinear spanning tree H_j for terminals in T_s^j. If T_s has a cycle, then destroy the cycle by deleting some edges along with Steiner points of H_j. An important fact is that if H_j

does not contain an edge between two terminals, then at least one Steiner point must be deleted when destroying a cycle in $H_j \cup T_2$. From this fact and Lemma 5.19, we have

$$c(T_{smst}) \leq 2 \cdot c(T_{opt}) + h,$$

where h is the number of full components T_s^j's with properties that every Steiner point in T_s^j has degree four and $T_s^j \cup T_2$ has no cycle. Hence we have

$$c(T_g) \leq 2 \cdot c(T_{opt}) + h - m.$$

Now to prove the theorem it suffices to show $h \leq m$.

Suppose that T_2 has p connected components. Then T_3 has $(p - 2m)$ connected components $C_1, C_2, \cdots, C_{p-2m}$. We now construct a graph H with vertex set consisting of n terminals and the edge set defined as follows: First, we put all edges of T_2 into H. Then consider every full component T_s^j $(1 \leq j \leq k)$ with properties that every Steiner point in T_s^j has degree four and $T_s^j \cup T_2$ has no cycle. If T_s^j has only one Steiner point, then this Steiner point connects four terminals which must lie in at most two C_i's. Hence, among them there are two pairs of terminals; each pair lies in the same C_i. Connect the two pairs with two edges and put the two edges into H. If T_s^j has at least two Steiner points, then there must exist at least two Steiner points each connecting three terminals. We can also find two pairs of terminals among them such that each pair lies in the same C_i. Connect the two pairs with two edges and put the two edges into H.

Clearly, H has at most $(p - 2h)$ connected components. Since every connected components of H is included in a C_i, we have $p - 2m \leq p - 2h$. Therefore, $h \leq m$. The proof is then finished. $\square$

5.3 In Metric Spaces

In this section we present an elegant analysis [201] for the Steinerized minimum spanning tree algorithm for Steiner tree problem for minimal Steiner points in metric spaces.

Given a set P of terminal points in a metric space M, let $T_M(P)$ denote the set of all Minimum Spanning Trees (MSTs) for P in metric space M. Define

$$d_M \equiv \sup_P \min_{T \in T_M(P)} \max_{v \in P} deg_T(v),$$

where the supremum is taken over all finite subsets P in M. Observe that, if d_M is finite, then every set P of points in M admits an MST with maximum degree at most d_M.

Măndoiu and Zelikovsky [201] proved the following theorem and corollary giving an upper bound on the approximation performance ratio for Steinerized minimum spanning tree algorithm in metric space M in terms of d_M.

Theorem 5.8   *Steinerized minimum spanning tree algorithm is a $(d_M - 1)$-approximation algorithm for Steiner tree problem for minimal Steiner points in metric space M whose d_M is finite.*

Proof.   Let P be a set of terminal points in a metric space M, and let $T_{opt}^M(P)$ be an optimal solution to Steiner tree problem for minimal Steiner points in metric space M. Then denote the Steiner points in $T_{opt}^M(P)$ by $s_1, s_2, \cdots, s_k$, which are numbered in the order of the breadth-first traversal starting from an arbitrarily terminal point $p \in P$. Note that since all edges of $T_{opt}^M(P)$ have length at most R, s_i is within a distance of R of at least one point from $P \cup \{s_1, s_2, \cdots, s_{i-1}\}$ for each $i = 1, 2, \cdots, k$.

Suppose that each T_i for $1 \leq i \leq k$ is an MST for $P \cup \{s_1, s_2, \cdots, s_i\}$ with maximum degree at most d_M, and T_0 is an MST for P. For a Steiner tree T for P, let $c_2(T)$ be the set of all degree-2 Steiner points needed to place on the edges in T in order to ensure that T satisfies the length bound. Clearly, $|c_2(T_k)| = 0$.

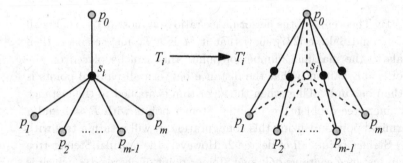

Fig. 5.18   Remove a Steiner point and make a new Steiner tree.

For each T_i, let $\{p_0, p_1, \cdots, p_m\}$ be the set of nodes adjacent to s_i in T_i, where $m + 1 \leq d_M$. Assume, without loss of generality, that p_0 is the closest neighbor of s_i in $P \cup \{s_1, s_2, \cdots, s_{i-1}\}$. Now we remove s_i and make all edges that are incident to s_i in T_i incident to p_0. As a result, we obtain a new Steiner tree T_i'. See Fig.5.18.

Note that $d(s_i, p_0) \leq R$ since the ordering of $\{s_1, s_2, \cdots, s_i\}$ ensures that s_i is within distance R of at least one point from $P \cup \{s_1, s_2, \cdots, s_{i-1}\}$ and p_0 is the point from this set closes to s_i. By the triangle inequality, any edge (p_j, p_0) needs at most one more degree-2 Steiner point than edge (p_j, s_i) to satisfy the length bound. Thus we have

$$|c_2(T_i')| \leq |c_2(T_i)| + m \leq |c_2(T_i)| + (d_M - 1).$$

Moreover, since T_i' does not contain s_i, it is an MST for the same set of points $P \cup \{s_1, s_2, \cdots, s_{i-1}\}$ as T_{i-1}. Hence we obtain $|c_2(T_{i-1})| \leq |c_2(T_i')|$, which, together with the above inequality, implies that

$$|c_2(T_{i-1})| \leq |c_2(T_i)| + (d_M - 1).$$

Taking the sum of above inequalities over $i = 0, 1, \cdots, k$ yields $|c_2(T_0)| \leq k \cdot (d_M - 1)$. Therefore, from Lemma 5.4 we deduce that the Steinerized MST for P contains at most $(d_M - 1)$ times number of Steiner points that $T_{opt}^M(P)$ contains. The proof is then finished. $\square$

Corollary 5.3 *Let M be a metric space L_p with finite d_M. Then the approximation performance ratio of Steinerized minimum spanning tree algorithm for minimal Steiner points in M is exactly $(d_M - 1)$.*

Proof.   By Theorem 5.8, the performance ratio is at most $(d_M - 1)$. Recall that Robins and Salowe [237] show that if M is a L_p metric space, then d_M equals to the maximal number of points that can be placed on the surface of a unit ball such that the distance between all pairs of points is greater than one unit. Thus when the algorithm is applied to the instance of the Steiner tree problem for minimal Steiner points with $R = 1$ and a set of terminal points realizing this configuration, it will return a tree with $(d_M - 1)$ Steiner points all of degree 2. However, the optimal Steiner tree to such an instance contains only one Steiner point of degree d_M, which is located at the center of the ball. The proof is then finished. $\square$

It is easy to know that for the Euclidean plane, $d_M = 5$ as shown in Fig.5.1, and for the rectilinear plane, $d_M = 4$ as shown in Fig.5.15. Thus Theorem 5.4 and Theorem 5.6 follow immediately from the above corollary.

5.4 Discussions

In this chapter we have studied the Steiner tree problem for minimal Steiner points. A closely related problem is the *bottleneck Steiner tree problem*. It asks for a Steiner tree interconnecting terminal points with at most k Steiner points such that the length of the longest edge in the tree is minimized. Clearly, this problem could be considered as the dual of Steiner tree problem for minimal Steiner points and will be discussed in the next chapter.

As one of the basic optimization models, the Steiner tree problem for minimal Steiner points finds an important application in *Wireless Sensor Networks* (WSNs). A *wireless sensor network* consists of many low-cost, low-power sensor nodes, which can perform sensing, simple computation, and transmission of sensed information. Since in many applications of WSNs, powers of sensors come from the equipped batteries that are usually not easy to maintain (e.g., recharge or replace), energy-efficiency is one of the most major concerns for the design of WSNs.

Lloyd and Xue [194] study the *relay node placement problem* in WSNs. Since energy consumption is a super linear function of the transmission distance, long distance transmission by sensor nodes is not energy efficient. One method to prolong network lifetime while preserving network connectivity is to deploy a small number of costly, but more powerful, relay nodes, which can communication with other sensor or relay nodes. They assume that sensor nodes have communication/transmission range $r > 0$, while relay nodes have range $R \geq r$. They study two versions of relay node placement problems under this assumption. In the first version, the objective is to deploy minimal relay nodes so that, between each pair of sensor nodes, there is a connecting path consisting of relay and/or sensor nodes. In the second version, the objective is to deploy the minimal relay nodes so that, between each pair of sensor nodes, there is a connecting path consisting solely of relay nodes. They propose a 7-approximation algorithm for the first version and a polynomial time $(5 + \epsilon)$-approximation algorithm for the second version, respectively, where $\epsilon > 0$ can be any given constant.

Min et al. [209] study the design of virtual backbone for supporting multicast communication in WSNs. A *virtual backbone* is a *connected dominating set* in the network, which is a subset of sensors such that they form a connected subnetwork and every sensor is either in the subset or adjacent to a sensor in the subset. Since multicasting can be performed first within virtual backbone and then to others, it has been recommended to manage and update the topology of virtual backbone instead of the topology of the whole network, which reduces both storage and message complexities. Clearly, the smaller virtual backbone gives the better performance. However, computing the connected dominating set of minimal size is NP-hard even in Unit Disk Graphs[1] (UDGs). Min et al. [209] propose a 6.8-approximation algorithm for the connected dominating set problem in UDGs, which can be implemented distributedly (Cheng [57] proposed a polynomial time approximation scheme for the problem, which, however, is not a distributed algorithm). Their algorithm consists of two steps. At the first step, it construct a maximal independent set[2], which is also a dominating set. At the second step, it produces a 3-approximate Steiner tree for minimal Steiner points in the maximal independent set using greedy algorithm 5.2.

---

[1]A *unit disk* is a disk with radius one. A *unit disk graph* is associated with a set of unit disks in the Euclidean plane. Each node is the center of a unit disk. An edge exists between two nodes u and v if and only if the Euclidean distance between u and v is at most one.

[2]An *independent set* of graph $G(V, E)$ is a subset S of V such that there is no edge in E between any two nodes in S.

Chapter 6

Bottleneck Steiner Tree Problem

Bottleneck Steiner tree problem arises from the design of Wireless Sensor Networks (WSNs). In WSNs, due to budget limits, suppose that we are only allowed to put totally $(n + k)$ base stations in the plane, where n of them must be located at given points. Clearly, one would like to have the distance between stations as small as possible. The problem is how to choose locations for other k stations to minimize the longest distance between stations. The problem can be more formally formulated as the following.

Problem 6.1 *Bottleneck Steiner Tree Problem* [88]

Instance A set P of n terminals in a plane and an integer $k > 0$.
Solution A Steiner tree T for P with at most k Steiner points.
Objective Minimizing the length of the longest edge in T.

Observe that the above problem can be considered as the dual of the Steiner tree problem for minimal Steiner points. Indeed, in the former problem the constraint is put on the number of Steiner points and the objective is on the length of the longest edge, while in the latter the constraint is put on the length of the longest edge in Steiner tree and the objective is about the number of Steiner points. So optimal solutions to these two problems have some common structural properties, and some basic ideas for designing approximation algorithms are similar. In this chapter, the *optimal Steiner tree* means the optimal Steiner tree for bottleneck Steiner tree problem unless specified otherwise.

In Section 6.1 we will present a negative result on approximation algorithms for bottleneck Steiner tree problem. In Sections 6.2 and 6.3 will study two approximation algorithms for bottleneck Steiner tree problem respectively. In the end we will discuss some related problems.

6.1   Complexity Study

The following theorem [262] gives a lower bound for the performance ratio of approximation algorithms for the bottleneck Steiner tree problem. The proof is based on the known result due to Gary and Johnson [111] who showed that the following problem is NP-complete.

Problem 6.2 *Planar Vertex Cover Problem*

Instance   A planar graph $G(V, E)$ with all vertices of degrees at most 4, and a positive integer $k < |V|$.

Question   Decide if there exists a *connected vertex cover* C of size k, i.e., a subset $C \subseteq V$ with $|C| = k$ such that for C contains at least one of end vertices of each edge in E and the subgraph induced by C is connected.

Theorem 6.1   *The bottleneck Steiner tree problem in the rectilinear plane does not admit any approximation algorithm with performance ratio less than 2 unless $P = NP$.*

Proof.   Suppose, by contradiction argument, that there exists a polynomial-time algorithm A_ϵ with performance ratio $(2-\epsilon)$ for some $\epsilon > 0$. Then using A_ϵ we will be able to design a polynomial-time algorithm for the planar vertex cover problem, which is impossible unless $P = NP$.

Given an instance of planar vertex cover problem, we first embed G into the rectilinear plane in such a way that all edges consists of some horizontal or vertical segments of lengths at least $(2k + 2)$ and every two edges meet at an angle of either 90° or 180°. And then we place some terminals on the interior of each edge in G in such a way that each edge $e \in E$ becomes a path P_e of many edges in the rectilinear plane each having length at most 1 while the first and the last edges have length of exactly 1. Denote by V' the set of all terminals. See Fig.6.1 where V has five vertices with a connected vertex cover containing two (black) vertices as shown in (a) and all terminals in V' are white nodes as shown in (b) with $k = 2$.

In the following we will prove that G has a connected vertex cover of size k if and only if when algorithm A_ϵ is applied to The bottleneck Steiner tree problem with input G' and k, it will produce a Steiner tree T (with at most k Steiner points) such that the rectilinear length of each edge in T is at most $(2 - \epsilon)$.

Clearly, if G has a connected vertex cover of C of size k, then putting k Steiner points at the k corresponding nodes contained in C at G', we can construct a Steiner tree on V' with k Steiner points such that the rectilinear length of each each in the tree is at most 1. This means that the rectilinear length of each edge in any optimal solution on the input G' is at most 1. Therefore, when A_ϵ is applied to G' it will return a Steiner tree T with k Steiner points such that the rectilinear length of each edge in T is at most $(2 - \epsilon)$. Refer to Fig.6.1 where T contains two (black) Steiner points as shown in Fig.6.1(c).

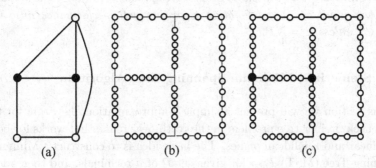

Fig. 6.1 (a) An instance of the planar vertex cover problem, (b) an instance of the bottleneck Steiner tree problem in the rectilinear plane with $k = 2$, and (c) a Steiner tree of $k = 2$ Steiner points.

Conversely, assume that when algorithm A_ϵ is applied to G' and k, it will return a Steiner tree T with k Steiner points such that the rectilinear length of each edge in the tree is at most $(2 - \epsilon)$. Note that by the way of constructing G', G' has two properties:

(1) Any two terminals in V' at two different edges of $E(G)$ have distance at least 2, and

(2) Any two terminals in V' at two non-adjacent edges of G have distance at least $2k + 2$, which means that they cannot be connected through k Steiner points.

From property (2) we deduce that in any full Steiner component of T, every two terminals lie on either the same edge or two adjacent edges. From property (1) we deduce that if a full Steiner component of T contains two terminals lying on two different edges, then it must contain at least one Steiner point. Thus we may move a Steiner point to the location of the vertex which covers all edges containing terminals in the full Steiner component and remove other Steiner points in the full Steiner component. The result is a Steiner tree T' with at most k Steiner points such that the rectilinear length of each edge in T' is at most 1. In addition, all Steiner points in T' lie at vertices of G. They form a connected vertex cover C of size at most k for G. (Add more vertices to V' to increase its size equal to k if necessary.) The proof is then finished.   □

Using almost the same argument, we are able to show the parallel result for the case of Euclidean plane, where the first and the last edges have lengths exactly $\sqrt{2}$ in Fig.6.1(b).

Theorem 6.2 *The bottleneck Steiner tree problem in the Euclidean plane does not admit any approximation algorithm with performance ratio less than $\sqrt{2}$ unless $P = NP$.*

6.2 Steinerized Minimum Spanning Tree Algorithm

In this section we will present a simple 2-approximation algorithm for the bottleneck Steiner tree problem in the rectilinear plane that works for both rectilinear and Euclidean planes. The basic idea is to construct a Minimum Spanning Tree (MST) T_{mst} for given set P of n terminals, and then add k Steiner points to the edges in T_{mst} to cut some long edges short, in the end we obtain a Steiner tree, which is called a *Steinerized minimum spanning tree*. Similarly, adding some degree-2 Steiner points to a spanning tree results in a *Steinerized spanning tree*.

In the following, for the simplicity of presentation, we allow a spanning tree to have crossing edges so that for any topology each edge of a spanning tree is a straight line segment between two vertices. Thus its length is the shortest distance between the two vertices. Note, however, that MSTs do not have crossing edges. The following lemma lays the basis for the Steinerized minimum spanning tree algorithm.

Lemma 6.1 *For any set P of terminals in the rectilinear plane, there exists a spanning tree T for P such that adding k Steiner points to T can make the longest edge in the resulting Steinerized spanning tree is at most twice that of the optimal Steiner tree T_{opt} to the bottleneck Steiner tree problem in the rectilinear plane.*

Proof. Since every Steiner tree can be decomposed into an edge-disjoint union of full Steiner subtrees, we may assume that T_{opt} is a full Steiner tree with k Steiner points. We arbitrarily select a Steiner point as the root of T_{opt}. In the following we will describe how to construct a Steiner tree T with at most k Steiner points of degree-2 such that the length of the longest edge in T is at most twice the length of the longest edge in T_{opt}.

See Fig.6.2. The construction of T is bottom-up by induction on the *height* of subtrees of T_{opt}, which is defined as the number of edges in the longest path from the root to a leaf in a rooted tree.

First, if the height of a subtree of T_{opt} is 1, i.e., there is one Steiner point in the subtree, then we can directly connect its leaves, which are all terminals in P, without any Steiner points. By the triangular inequality, the lengths of edges in the tree are at most twice that of the length of the longest edge in the subtree.

Next we assume that for any subtree T' of T_{opt} with height less than or equal to h, if T has s Steiner points, then we have: (1) there exists an *unused* path from the root of T' to a leaf containing l Steiner points, and (2) there exists a Steiner tree T'' with s' Steiner points of degree-2 such that the length of the longest edge is at most twice the length of the longest edge in T' and $l + s' \leq s$.

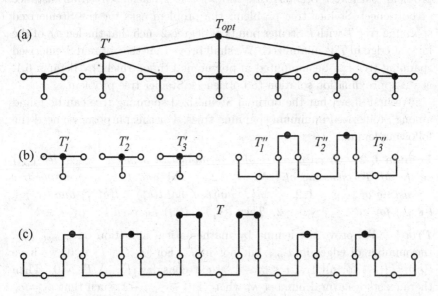

Fig. 6.2   Construct a Steiner tree T with at most k Steiner points of degree-2.

We now consider a subtree T' with height $(h + 1)$. Suppose that the degree of the root of T' is r. See Fig.6.2(a). Eliminating the root of T' results in r subtrees $T'_1, T'_2, \cdots, T'_r$, all of them have heights at most h. See Fig.6.2(b). Let k_i be the number of Steiner points in T'_i. By induction hypothesis (1), for each T'_i there exists an unused path from the root of

T_i' to a leaf of it with l_i Steiner points in T_i', and by induction hypothesis (2), there exists a Steiner tree T_i'' with at most $(k_i - l_i)$ Steiner points of degree-2 such that the length of the longest edges in T_i'' is at most twice the length of the longest edge in T_i'. Without loss of generality, we assume that $l_i \geq l_{i+1}$ for each $i = 1, 2, \cdots, r$. Now we connect these r subtrees $T_1'', T_2'', \cdots, T_r''$ with $(r - 1)$ edges and add $l_1, l_2, \cdots, l_{r-1}$ Steiner points to $(r - 1)$ new edges, respectively. See Fig.6.2(c). Note that after this process the unused paths for T_r remain unused, but those unused paths for $T_1', T_2', \cdots, T_{r-1}'$ have been already used. In the end, adding the root, which is a Steiner point, for T', we will obtain $(l_r + 1)$ Steiner points which are the Steiner points on the unused path for T'.

By the induction argument and construction, we obtain the desired Steiner tree T, and the proof is then finished.   □

In the following we shall study a variation of bottleneck Steiner tree problem, *bottleneck Steinerized spanning tree problem*. Given an instance of bottleneck Steiner tree problem, the problem asks for the Steinerized spanning tree T with k Steiner points of degree-2 such that the length of the longest edge in T is minimized. We shall prove that the optimal Steinerized spanning tree can be computed in polynomial time, which, by Lemma 6.1, is a 2-approximation solution to bottleneck Steiner tree problem.

We first show that the optimal Steinerized spanning tree can be found among Steinerized minimum spanning trees. For this purpose, we need the following lemma.

Lemma 6.2   *Given a set P of n terminals in a metric space, let T_{mst} be the MST consisting of $\{e_i^* | i = 1, 2, \cdots, n\}$ and T be a spanning tree consisting of $\{e_i | i = 1, 2, \cdots, n\}$. Suppose that $l(e_i^*) \leq l(e_{i+1}^*)$ and $l(e_i) \leq l(e_{i+1})$ for all $1 \leq i \leq n - 2$. Then $l(e_i^*) \leq l(e_i)$ for each $1 \leq i \leq n - 1$.*

Proof.   We prove the lemma by mathematical induction on $|T_{mst} \setminus T|$, the number of edges in T_{mst} but not in T. For $|T_{mst} \setminus T| = 0$, we have $l(e_i) = l(e_i^*)$ for all $1 \leq i \leq n - 1$. Next consider $|T_{mst} \setminus T| > 0$. Then there exists a natural number k, where $1 \leq k \leq n - 1$, such that $e_i = e_i^*$ for $1 \leq i \leq k - 1$ but $e_k \neq e_k^*$.

We now show that $l(e_k^*) \leq l(e_k)$. By contradiction argument, suppose that $l(e_k^*) > l(e_k)$. Then by the assumption of the lemma, we have $l(e_i^*) > l(e_k)$ for $i = k, \cdots, n-1$. Note that adding e_k to T_{mst} will results in a cycle containing e_k and an edge e_h^* for some h where $k \leq h \leq n - 1$. Deleting e_h^* yields a spanning tree whose total length is smaller than that of T_{mst}, contradicting that T_{mst} is an MST. Hence $l(e_k^*) \leq l(e_k)$.

We next add e_k^* to T. The resulting graph $T \cup \{e_k^*\}$ contains a cycle which contains e_k^* and an edge e_g for some g where $k \leq g \leq n-1$. Removing e_g from $T \cup \{e_k^*\}$ will produce a spanning tree T' with $|T_{mst} \backslash T'| = |T_{mst} \backslash T| - 1$. Note that all edges in T' satisfy the following order

$$l(e_1) \leq \cdots \leq l(e_{k-1}) \leq l(e_k^*) \leq (e_k) \leq \cdots \leq l(e_{g-1}) \leq l(e_{g+1}) \leq \cdots \leq l(e_{n-1}).$$

By the induction hypothesis, we obtain

$$l(e_{k+1}^*) \leq l(e_k), \cdots, l(e_g^*) \leq l(e_{g-1}) \text{ and}$$

$$l(e_{g+1}^*) \leq l(e_{g+1}), \cdots, l(e_{n-1}^*) \leq l(e_{n-1}).$$

Therefore, we have $l(e_i^*) \leq l(e_i)$ for all $1 \leq i \leq n-1$. $\qquad\square$

It immediately follows from Lemma 6.3 that when we use the same number of Steiner points to Steinerize a spanning tree and an MST, the resulting tree from the latter has the longest edge of length not exceeding that from the former. This implies that the optimal Steinerized spanning tree can be found among Steinerized minimum spanning trees.

We now describe how to add k Steiner points to an MST in order to obtain an optimal Steinerized spanning tree. A simple algorithm is as follows: For each edge $e_i = (u, v)$ in the MST T_{mst}, if we use $s(e_i)$ Steiner points to Steinerize it, then the length of the longest edge in the resulting path from u to v has the minimum value $l(e_i)/(s(e_i)+1)$, which is achieved when e_i is divided evenly by $s(e_i)$ Steiner points. Denote $l'(e_i) \equiv l(e_i)/(s(e_i)+1)$, which equals the length of each segment in path e_i after $s(e_i)$ is placed on it. Initially, set $s(e_i)$ to be zero. The basic idea is to add a degree-2 Steiner point to the edge e_i with the largest $s(\cdot)$ at each time. After e_i receives one more degree-2 Steiner point, update $l'(e_i)$ by increasing it by one. The process is repeated until k Steiner points of degree-2 are added.

Algorithm 6.1 *Steinerized Minimum Spanning Tree*

Step 0 Compute an MST $T_{mst} = \{e_1, e_2, \cdots, e_{n-1}\}$

        Set $s(e_i) := 0$ and $l'(e_i) := l(e_i)/(s(e_i)+1)$ for each $e_i \in T_{mst}$

        Set $s := 0$

Step 1 while $s < k$ **do begin**

        Sort all edges in T_{SMT} in a non-increasing order of $s(\cdot)$

        Choose e_j such that $s(e_j) = \max\{s(e_i) | i = 1, 2, \cdots, n-1\}$

        Add a degree-2 Steiner point to e_j

        Update $s(e_j) := s(e_j) + 1$ and $l'(e_i) := l(e_i)/(s(e_i)+1)$

Set $k := k + 1$
    end-while
Step 2 Place $s(e_i)$ Steiner points at each $e_i \in T_{smt}$
    Return the Steinerized minimum spanning tree T'_{smt}

Lemma 6.3   *Steinerized minimum spanning tree algorithm 6.1 returns an optimal Steinerized spanning tree with k Steiner points.*

Proof.   We will prove the lemma by mathematical induction on k. For $k = 0$, the lemma is trivially true. For any $k > 0$, let T_k be the Steinerized minimum spanning tree with k Steiner points returned by algorithm 6.1 and T_k^* be an optimal Steinerized spanning tree with k Steiner points. By the induction hypothesis, the length of the longest edge in T_k equals that of the longest edge in T_k^*. In the following we show that the length of the longest edge in T_{k+1} equals that the longest edge in T_k^*.

Denote by $l_{max}(T_k)$ the length of a longest edge in T_k for any $k > 0$. Then we have

$$l_{max}(T_k) = \max \left\{ \frac{l(e_i)}{s_k(e_i) + 1} \; \middle| \; 1 \le i \le n - 1 \right\},$$

where $s_k(e_i)$ is the numbers of Steiner points on the edge e_i of T_k. In addition, denote by l_{k+1}^* the length of a longest edge in an optimal Steinerized spanning tree T_{k+1}^* with $(k + 1)$ Steiner points. Clearly, $l_{k+1}^* \le l_{max}(T_{k+1}) \le l_{max}(T_k)$. If $l_{k+1}^* = l_{max}(T_k)$, then $l_{k+1}^* = l_{max}(T_{k+1})$, which means T_{k+1} is an optimal Steinerized spanning tree with $k+1$ Steiner points. Thus, we may assume that $l_{k+1}^* < l_{max}(T_k)$. Without loss of generality, suppose that e_j is an edge of T_{mst} satisfying $l_{max}(T_k) = l(e_j)/(s_k(e_j) + 1)$. Let $s_{k+1}^*(e_i)$ denote the number of Steiner points of T_{k+1}^* on the edge e_i of T_{mst}. Clearly, $s_{k+1}^*(e_j) \ge s_k(e_j) + 1$. In fact, if $s_{k+1}^*(e_j) \le s_k(e_j)$, then $l_{k+1}^* = l_{max}(T_{k+1}^*) \ge l(e_j)/(s_k(e_j) + 1) = l_{max}(T_k)$, contradicting our assumption.

We now show that $s_{k+1}^*(e_j) = s_k(e_j) + 1$. Suppose, by contradiction argument, that $s_{k+1}^*(e_j) > s_k(e_j) + 1$. Then, removing one Steiner point from e_j, we obtain a Steinerized spanning tree with k Steiner points which has the length of the longest edge not exceeding

$$\max \left\{ l_{k+1}^*, \frac{l(e_j)}{s_k(e_j) + 2} \right\} < l_{max}(T_k) = \frac{l(e_j)}{s_k(e_j) + 1},$$

which contradicts the induction hypothesis that the length of the longest edge in T_k equals that of the longest edge in T_k^*.

Next, we show that $s_k(e_i) \leq s_{k+1}^*(e_i)$ for each $i \neq j$. Suppose, by contradiction argument, that $s_k(e_i) > s_{k+1}^*(e_i)$ for some $i \neq j$. Then we have $s_k(e_i) > 0$, which implies that edge e_i receives some Steiner points at some step in the algorithm since $l(e_i)/s_k(e_i)$ achieves the largest $l'(\cdot)$-value. Since the $l'(\cdot)$-value does not increase for each edge in T_{mst} during the computation of the algorithm, we have

$$\frac{l(e_i)}{s_k(e_i)} \geq \frac{l(e_j)}{s_k(e_j) + 1} = l_{max}(T_k).$$

Hence we obtain

$$l_{k+1}^* = l_{max}(T_{k+1}^*) \geq \frac{l(e_i)}{s_{k+1}^*(e_i)} \geq \frac{l(e_i)}{s_k(e_i) + 1} \geq l_{max}(T_k),$$

which contradicts the assumption that $l_{k+1}^* < L_{max}(T_k)$.

We now have $s_k(e_i) \leq s_{k+1}^*(e_i)$ for each $i \neq j$. Moreover,

$$s_k(e_j) + \sum_{i \neq j} s_k(e_i) = k \text{ and } s_{k+1}^*(e_i) + \sum_{i \neq j} s_{k+1}^*(e_i) = k + 1,$$

which, together with $s_{k+1}^*(e_j) = s_k(e_j) + 1$, implies

$$\sum_{i \neq j} s_k(e_i) = \sum_{i \neq j} s_{k+1}^*(e_i).$$

Thus we obtain $s_k(e_i) = s_{k+1}^*(e_i)$ for each $i \neq j$, which implies $s_{k+1}(e_i) = s_{k+1}^*(e_i)$ for each $i = 1, 2, \cdots, n$. Therefore, we have $T_{k+1}^* = T_{k+1}$, the proof is then finished. $\square$

The following theorem follows directly from Lemmas 6.2-3.

Theorem 6.3 *Steinerized minimum spanning tree algorithm is a 2-approximation algorithm for bottleneck Steiner tree problem.*

6.3 3-Restricted Steiner Tree Algorithm

In this section we will study another algorithm for the bottleneck Steiner tree problem in the Euclidean plane. It is based on the following theorem due to Wang and Li [264].

Theorem 6.4 *There exists a 3-restricted Steiner tree T of k Steiner points whose longest edges have the length at most 1.866 times the length of* *the longest edges in the optimal Steiner tree T_{opt} to the bottleneck Steiner tree problem in the Euclidean plane.*

Proof.   Consider T_{opt} as a tree rooted at a Steiner point. We will modify T_{opt} bottom up into the desired 3-restricted Steiner tree T. Without loss of generality, we assume that T_{opt} is a full component.

We organize the nodes in T_{opt} level by level (ignoring degree-2 Steiner points). The $1-th$ level is the lowest level and $i-th$ level is above the $(i-1)-th$ level. Let s_3 be a node at the 3-rd level that has some grandchildren. Let s_2 be a child of s_3. If s_2 is a Steiner point, we can assume that s_2 has exactly two children that are terminals. Suppose, by contradiction argument, that s_2 has three (or more) children that are terminals, t_1, t_2, and t_3. Assume that they are clockwise around s_2. Then at least one of three angles $\angle t_1 s_2 t_2$, $\angle t_2 s_2 t_3$, and $\angle t_3 s_2 t_1$ is at most $120°$. Without loss of generality, assume that $\angle t_1 s_2 t_2 \leq 120°$ and $t_1 s_2$ is not shorter than $t_2 s_2$ as shown in Fig.6.3(a). Then $|t_1 t_2| \leq \sqrt{3}|t_1 s_2|$. Assume further that there are m Steiner points of degree-2 on the path from t_1 to s_2. Then we can directly connect t_1 with t_2 while equally placing m Steiner points of degree-2 on the segment $t_1 t_2$ such that each edge in the segment has length at most $\sqrt{3} < 1.866$. As a result, degree of s_2 is reduced by one.

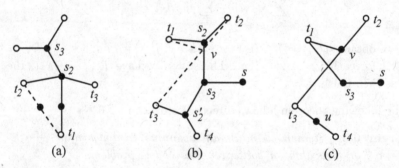

Fig. 6.3   (a) Degree of s_2 could be reduced by one, (b, c) degree of s_3 could be reduced by one.

Now we assume that s_2 has degree three, and consider two cases as follows.

Case 1. Every edge below s_3 in T_{opt} between its children and s_3 and its grandchildren has length at most 1. Suppose that s_3 has four grandchildren (the case of three grandchildren can be handled in a similar way). We deal with two cases separately in the following.

Case 1.1. s_3 has degree three. In this case, s_3 has two children s_2 and s_2', each of them has two children t_1, t_2 and t_3, t_4, respectively. Assume that $\angle t_1 s_2 t_2 > 120°$ and $\angle t_3 s_2' t_4 > 120°$ as shown in Fig.6.3(b). (Otherwise, we can directly connect t_1 and t_2, t_4 and t_5, respectively, and get $|t_1 t_2| \leq \sqrt{3}$ and $|t_4 t_5| \leq \sqrt{3}$. As a result, the degree s_2 is reduced by one. After that we can repeat the modification with fewer number of Steiner points whose degrees are greater than two.) Since one of angles $\angle s_3 s_2 t_1$ and $\angle t_2 s_2 s_3$ is at most 120°, and one of angles $\angle s_3 s_2' t_5$ and $\angle t_4 s_2' s_3$ is at most 120°. Without loss of generality, we may assume that $\angle s_3 s_2 t_1 \leq 120°$ and $\angle s_3 s_2' t_4 \leq 120°$. Thus we obtain $|t_1 s_3| \leq \sqrt{3}$ and $|t_4 s_3| \leq \sqrt{3}$. Let v be a node on edge $s_2 s_3$ with $|v s_3| = (2 - \sqrt{3})/2$. Then we have $|v t_2| \leq 1.866$, $v t_1 \leq 1.866$ and $|v t_4| \leq 1.866$. Note that we can replace s_2 and s_2' with two Steiner points v and u to connect those four terminals t_1, t_2, t_3, and t_4, and use s_3 to directly connect t_1 and its parent s in T^* as shown in Fig.6.3(c). Note that the length of every new edge $t_1 v, t_2 v, t_3 v, t_1 s_3$, and $t_3 u$, and $t_4 u$ is at most 1.866, and the length of $s_3 s$ is at most 1. As a result, the degree of s_3 has reduced by one. Hence we can continue the modification process with $(n-3)$ terminals in $P \cup \{s_3\} \setminus \{t_1, t_2, t_3, t_4\}$. Observe that v connects three terminals t_1, t_2, and t_3 forming a full Steiner component. Thus t_1, t_2, t_3, and t_4 are not included while s_3 is treated as a new terminal.

Case 1.2. s_3 has degree more than three. This case can be handled in a similar way except that more than one full component that spans three terminals will be generated.

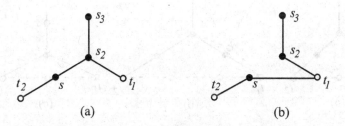

(a) (b)

Fig. 6.4   Case 2: (a) edge $s_2 t_2$ has length greater than 1 and it contains degree-2 Steiner points, and (b) degree of s_2 is reduced to two after modification.

Case 2. Some edges below s_3 in T_{opt} between its children and s_3 and its grandchildren has length greater than 1. Again let s_3 be a node at the third level that has some grandchildren and s_2 be one of its children. From the previous discussion, we can assume that s_2 has exactly two children t_1 and

t_2. Without loss of generality, we assume that $|s_2t_1| \le |s_2t_2|$ and $|s_2t_1| = m$ where m is an integer. See Fig.6.4(a). Note s_2t_2 can be divided into two segments, s_2s and st_2 where s is a degree-2 Steiner point in T_{opt} such that s_2s contains m edges where t_2s contains m' Steiner points of degree-2 including s. Note that the length of each edge on s_2s and st_2 is at most 1. Moreover, the total number of degree-2 Steiner points (not including s_2 and s) in s_2s and t_1s_2 is $2(m-1)$. Thus we can directly connect t_1 with s using $(\lceil 1.0713m \rceil - 1)$ Steiner points of degree-2, and connect t_1 with t_2 via s_2s using $(\lceil 1.0713m \rceil - 1 + m')$ Steiner points of degree-2 as shown in Fig.6.4(b). By the triangle inequality, we have $|t_1s| \le 2m$. Thus the length of each edge on t_1s is at most $2m/\lceil 1.0713m \rceil \le 1.866$. After than we can connect t_1 with s_2 using $2(m-1) - (\lceil 1.0713m \rceil - 1) = \lfloor 0.9287m \rfloor - 1$ Steiner points of degree-2. Thus t_1s_2 has $\lfloor 0.9287m \rfloor$ edges and each edge on it has length at most

$$\frac{m}{\lfloor 0.9287m \rfloor - 1} \le \frac{m}{0.9287m - 1},$$

which clearly is smaller than 1.68 if $m \ge 3$. Thus when $m \ge 3$, the degree of s_2 is reduced by one after modification as shown in Fig.6.4(b). Therefore, we can continue the modification process with $n-1$ terminals in $P \cup \{s_2\} \setminus \{t_1, t_2\}$, that is, t_1 and t_2 are not included while s_2 is treated as a new terminal.

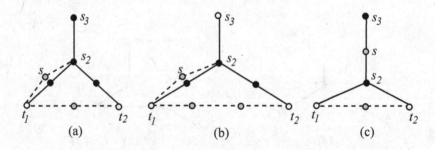

Fig. 6.5 Case of $m \le 2$: (a) $m = 2$, (b) $m = 2$, and (c) $m = 1$.

We now consider the cases of $m = 2$ and $m = 1$ as shown in Fig.6.5(a-b) and (c), respectively. Suppose that $|t_1s_2| < |t_2s_2|$. Then we can directly connect t_1 and t_2 using the same number of degree-2 Steiner points lying in t_2s_2. Note that the lengths of edges in t_1t_2 is at most 5/3 and 3/2 for (a,b) and (c), respectively. After modification, the degree of s_2 is reduced by one. Therefore, we can continue the modification process with $(n-1)$ terminals $P \cup \{s_2\} \setminus \{t_1, t_2\}$. In the following we assume that $|t_1 s_2| < |t_2 s_2|$ and consider two cases separately.

Case 2.1 $m = 2$. In this subcase, there are three Steiner points including s_2 in path $t_1 s_2 t_2$ in T^*. If $\angle t_1 s_2 t_2 \leq 120°$, then we have $|t_1 t_2| \leq 2\sqrt{3}$. Thus we can use one Steiner point of degree-2 to connect t_1 and t_2 directly, and then continue the process with $(n-1)$ terminals, that is, t_1 and t_2 are not included while s is treated as a new terminal as shown in Fig.6.5(a).

If one of angles $\angle t_1 s_2 s_3$ and $\angle t_2 s_2 s_3$ is at most 120°. Assume, without loss of generality, that $\angle t_1 s_2 s_3 \leq 120°$. Then we have $|t_1 s_3| \leq 1 + \sqrt{3}$. Thus we can use two degree-2 Steiner points to connect t_1 and t_2 and one Steiner point s to connect t_1 and s_3. Note that $|ss_3| \leq 1$ and s_2 is no longer in the resulting tree. Therefore, we can continue the process with $(n-1)$ terminals, that is, t_1 and t_2 are included while s is treated as a new terminal as shown in Fig.6.5(b).

Case 2.2 $m \leq 1$. We assume that $s_2 s_3 > 1$ since the case of $s_2 s_3 \leq 1$ has been discussed in Case 1. We can further assume that there exists a degree-2 Steiner point s in the path between s_2 and s_3. If $\angle t_1 s_2 t_2 \leq 120°$, then we can directly connect t_1 and t_2 with an edge of length at most $\sqrt{3}$. The modification process can still continue; Otherwise, one of angles $\angle t_1 s_2 s_3$ and $\angle t_2 s_2 s_3$ is at most 120°. Thus one of edges $t_1 s$ and $t_2 s$, say $t_1 s$, has length at most 1.732. Thus, we directly connect t_1 and s and use one Steiner point to connect t_1 and t_2. Note that in the resulting tree, Steiner point s_2 is not included any more as shown in Fig.6.5(c). In this subcase, we also can continue the modification process with $(n-1)$ terminals, that is, t_1 and t_2 are no longer included while s is treated as a new terminal.

We have studied all possible cases, and the proof is then finished. $\quad\square$

By Theorem 6.4, to design a good approximation algorithm for the bottleneck Steiner tree problem in the Euclidean plane, we may focus on how to solve the problem of producing a 3-restricted Steiner tree T_k^* with k Steiner points such that the length of the longest edge in T_k^* is minimal. Wang and Li [264] transformed this problem into the minimum spanning tree problem for 3-hypergraphs.

A *hypergraph* $H(V, E)$ is a generalization of a graph where the edge-set E is an arbitrary family of subsets of the vertex-set V (not just a family of subsets of two vertices in V as a graph). A *weighted hypergraph* $H(V, E; w)$ is a hypergraph such that each edge $e \in E$ has a weight $w(e)$. An *r-hypergraph* $H_r(V, E; w)$ is such a weighted hypergraph that each edge in E

has cardinality at most r. A *path* of length $l \geq 2$ in H is an alternative sequence of distinct vertices and edges $v_1, e_1, \cdots, v_{l-1}, e_{l-1}, v_l, e_l$ such that $v_i \in e_{i-1} \cap e_i$ for all $i = 2, 3, \cdots, l$. A hypergraph H is *connected* if there exists a path in H between any two vertices in V. A *cycle* of length $l \leq 2$ in H is an alternative sequence of distinct vertices and edges $v_1, e_1, \cdots, v_{l-1}, e_{l-1}, v_l, e_l$ such that $v_1 \in e_1 \cap e_l$ and $v_i \in e_{i-1} \cap e_i$ for all $i = 2, 3, \cdots, l$. A subhypergraph $T \subset E$ of H is a *tree* if and only if T is connected and contains no cycles. A tree T of a 3-hypergraph $H_3(V, E; w)$ is a *spanning tree* of H_3 if it contains every vertex in V. A *minimum spanning tree* T_{mst} for a 3-hypergraph $H_3(V, E; w)$ is a spanning tree of minimum weight (total weight of edges included in T_{mst}).

The transformation from the bottleneck Steiner tree problem in the Euclidean plane into the minimum spanning tree problem for 3-hypergraphs can be done as follows: construct a weighted hypergraph $H_3(V, E; w)$ with vertex-set $V = P$ and

$$ F = \{(a, b) \mid a, b \in P\} \cup \{(a, b, c) \mid a, b, c \in P\}. $$

The weight of an edge represents the smallest number of Steiner points that should be added to the edge (corresponding a set of two or three terminals) such that the length of Steiner tree interconnecting those two or three terminals is no greater than the length of the longest edge in the optimal solution to the bottleneck Steiner tree problem.

However, to assign a desired weight to each edge in E is not easy since the length of the longest edge in an optimal solution, denoted by $L_{max}(T_k^*)$, is not known. Thus we must refer to its upper bound $\beta \leq (1 + \epsilon)L_{max}(T_k^*)$ for any $\epsilon > 0$. Such a bound can be estimated in the following way: First, run the 2-approximation algorithm to get an upper bound β' of $L_{max}(T_k^*)$, and then use one of $\frac{\beta'}{2}, \frac{\beta'}{2}(1 + \frac{1}{p}), \frac{\beta'}{2}(1 + \frac{2}{p}), \cdots, \beta'$ as β, where p is an integer such $1/p \leq \epsilon$.

With an estimated upper bound β on $L_{max}(T_k^*)$, for each two-vertex edge $(a, b) \in E(H)$ we assign it a weight of the smallest number of Steiner points that should be added to the edge such that the length of each Steiner-ized edge in G is at most β, i.e.,

$$ w(a, b) = \left\lceil \frac{|ab|}{\beta} \right\rceil - 1. $$

Similarly, for each three-vertex edge $(a, b, c) \in E(H)$, we assign it a weight of the smallest number of Steiner points required to connect three terminals (possibly via a degree-3 Steiner points s) such that the length of each Steinerized edge in G is at most β. Such a number can be determined as follows: Suppose $|ab| \leq |bc| \leq |ca|$. Then

$$\beta' = \min\left\{\left\lceil\frac{|ab|}{\beta}\right\rceil + \left\lceil\frac{|bc|}{\beta}\right\rceil - 2, k\right\}$$

gives an upper bound on the number, where k is the bound on the number of Steiner points in an instance of the bottleneck Steiner tree problem. Note that we do not need to consider those edges that satisfy

$$\left\lceil\frac{|ab|}{\beta}\right\rceil + \left\lceil\frac{|bc|}{\beta}\right\rceil - 2 > k.$$

Now let i, j and l be the numbers of degree-2 Steiner points in the segments as, bs and cs such that $(i + j + l + 1)$ is minimized. Then to compute the weight $w(a, b, c)$ of three-vertex edge $(a, b, c) \in E(H)$, we just need to guess the values of i, j and l. This can be done by simply trying at most $O(k^3)$ possibilities. For each guessed triple $\{i, j, j\}$, we can determine if the three circles centered at terminals $a, b,$ and c with radius $i\beta'$, $j\beta'$ and $l\beta'$, denoted by $\odot a, \odot b$ and $\odot c$, have a point in common. Wang and Li [264] proved the following lemma, which shows this can be done in a constant time.

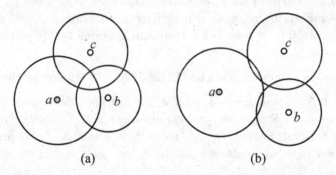

(a)   (b)

Fig. 6.6 There are six intersecting points: (a) one is located within all three circles and (b) none is located within all three circles.

Lemma 6.4 *Testing whether any three circles contain one common point can be done in a constant time.*

Proof. Note first that it is easy to determine whether two of $\odot a, \odot b$ and $\odot c$ are disjoint or one completely contains the other. In the former case, they do not contain any common point; while in the latter case, it is straightforward to test if the smaller of two circle is disjoint with the third circle.

Now we only consider other cases where two of three circles intersect with each other. Since there are only six intersecting points and these three circles contain one common point if and only if they contain at least one of the intersecting points, it is easy to determine whether one of the six points is included in all three circles as shown in Fig.6.6(a) or not as shown in Fig.6.6(b). The proof is then finished.   □

In the above we have described how to, given an instance of the bottleneck Steiner tree problem consisting of n terminals in the Euclidean plane and a natural number $k > 0$, find an instance of the minimum spanning tree problem in weighted 3-hypergraphs $H_3(V, E, w)$ such that there is a 3-restricted Steiner tree T for n terminals whose longest edge has length L_{max} if and only if there exists a spanning tree of $H_3(V, E, w)$ with weight equal to k and the length of longest corresponding edges at most L_{max}. To find the MST of $H_3(V, E, w)$, we can apply an existing algorithm for this problem. In particular, Prömel and Steger [227] proposed a fully polynomial randomized scheme for 3-uniform hypergraphs (where each edge contains exactly 3 vertices) and proved its performance as follows.

Lemma 6.5   *There exists a randomized algorithm that, given an edge-weighted 3-uniform hypergraphs of n vertices and maximum weight w_{max}, finds with probability at least 1/2 a minimum spanning tree in polynomial time of $(n + w_{max})$.*

From the above lemma, Wang and Li [264] proved the following theorem.

Theorem 6.5   *There exists a randomized algorithm that, given an instance of bottleneck Steiner tree problem and any $\epsilon > 0$, finds with probability at least 1/2 and running time of $\frac{1}{\epsilon} poly(|P|, k)$ a solution to the problem whose longest edge has length at most $(1.866 + \epsilon)$ times that of the optimal Steiner tree to the problem.*

Proof.   Note that there are $O(|P|^3)$ edges in the hypergraph, and computing the weight of an edge (a triple of three terminals) in the hypergraph needs time of $O(k^3)$ to guess the radii of three circles. Thus the weighted 3-hypergraph can be constructed in time of $O(n^3k^3)$. In addition, we can add an extra vertex into each two-vertex edge in the hypergraph to modify it into a 3-uniform hypergraph. Hence by Lemma 6.5 and Theorem 6.4 we deduce that there exists a desired algorithm for the bottleneck Steiner tree problem. The proof is then finished.   □

6.4 Discussions

Berman and Zelikovsky [30] study the bottleneck Steiner tree problem in graphs as follows.

Problem 6.3 *Bottleneck Steiner Tree Problem in Graphs*

Instance A graph $G(V, E)$ with a length $l(e)$ on each edge $e \in E$ and a terminal-set $P \subset V$.

Solution A Steiner tree T that spans P in the distance graph $G_D(P)$ and each Steiner point has degree at least 3.

Objective Minimizing the bottleneck length of T, $\max\{l(e) \mid e \in T\}$.

For the hardness of the above problem, they prove the following theorem using the reduction from the problem of *exact cover by 3-sets* (X3C)[1].

Theorem 6.6 *The bottleneck Steiner tree problem in graphs does not admit any approximation algorithm with performance ratio less than 2 unless $P = NP$.*

For the approximability of the problem, Ganley and Salowe [106] first prove that the minimum spanning tree algorithm has performance ratio of $2 \log_2 |P|$, and then Berman and Zelikovsky [30] get a better algorithm reducing the ratio by half.

Theorem 6.7 *The bottleneck Steiner tree problem in graphs admits an approximation algorithm with performance ratio no greater than $\log_2 |P|$.*

---

[1]Given a universal set X with $|X| = 3k$ and a collection C of subsets of X each containing exactly three elements, the problem is to determine if there is a subcollection C' of C that is a partition of X.

Chapter 7

Steiner k-Tree and k-Path Routing Problems

Steiner k-tree and k-path problems arises from the multicast communication in all-optical networks. *Multicast* is a point-to-multipoint communication that enables a node to send or forward data to multiple recipients [242]. In order to perform multicast communication in a network, the source and all destination nodes must be interconnected by a tree. Thus, the problem of *multicast routing* in networks is usually treated as finding a tree in a network that spans source and all destination nodes [242]. One of the objectives of multicast routing is to minimize the *network cost* of produced trees, which is defined as the sum of costs of all the links in the tree. This can be considered as the Steiner tree problem in graphs (Problem 2.1).

To implement multicast routing in a wavelength-routed optical network [22], the concept of a *light-tree* and the cross-connect architecture of *splitter-and-delivery* were proposed in [5; 243]. A light-tree interconnecting the source and all destination nodes uses a dedicated wavelength on all of its branches. Each intermediate node in a light-tree must have a splitter so that copies of data in the optical domain can be made and delivered to each of its children. An n-way splitter is an optical device which splits an input signal among n outputs, thus reducing the power of each output to $\frac{1}{n}$-th of that of the original signal [5]. As a result, while the power budget may allow data on a given wavelength to be "dropped off" (or "delivered to") more than one destination node, it many not be possible to drop off data at an arbitrary number of destination nodes using a single light-tree.

So Hadas [125] proposed the *multi-tree routing model*. Under this model only some specified maximum number of light splitting are allowed per transmission, and then multicast routing is to find a set of light-trees such that each of them includes at most k destination nodes which can receive data and every destination node is designated to receive data in one of the light-trees.

However, the application of multi-tree routing model is not easy because most networks are designed to mainly support *unicast* (point-to-point) communication and in a hetero-geneous high-speed network, some switches or routers may not have multicast capability [242]. Even if a switch has multicast capability, duplicating the data and forwarding it to its neighbor nodes in the multicast tree causes heavy load at the switch and makes routing complicated. In addition, it is impossible to store multicast routing information in a network of large size.

To overcome those difficulties, Hadas [125] proposed *multi-path routing model*, which could be considered as a generalization of point-to-point connection. Under this model the data is sent from the source node to a destination node in a light-path. During the data transmission along the path, if an intermediate node itself is a destination node, then the data is stored (dropped) and a copy of the data is forwarded to its adjacent neighbor in the path. In each path some destination nodes are designated where the data is stored (dropped). Accordingly, *multicast routing* is to find a set of such paths so that every destination node is designated to receive the data in a path. Compared with the tree model of multicast routing, this simple model makes multicast easier and more efficient to implement, but at the expense of increasing the network cost (since the cost of a multicast tree is generally less than that of a set of paths rooted a fixed node).

In this chapter, we will study how to establish a multicast connection under multi-tree and multi-path routing models in all-optical networks [122; 123][1]. In Section 7.1 we shall formulate the problem as Steiner k-path and k-tree routing problems and study their computational complexity. In Section 7.2 and Section 7.3 we will study some approximation algorithms for these two problems, respectively.

7.1   Problem Formulation and Complexity Study

7.1.1   *Problem Formulation*

We model the network under consideration as an arc-weighted digraph $\vec{G}(V, A, c)$, where vertex-set V is the set of nodes in the network representing switches/routers and arc-set A is the set of links between nodes representing wires. For arc $(u, v) \in A$, cost function $c : A \to \Re^+$ mea-

---

[1]Although they were proposed for wavelength routed optical networks, the basic idea may also applicable for packet switching networks [252].

sures the desirability of using a particular arc (a lower cost means more desirable). We assume that $\vec{G}(V, A, c)$ is totally symmetric, that is, there is an arc $(u, v) \in A$ from u to v if and only if there is an arc $(v, u) \in A$ from v to u, and the costs on two arcs between u and v are the same, i.e., $c(u, v) = c(v, u)$. For the simplicity of presentation, we denote by $G(V, A, c)$ the underlying graph of $\vec{G}(V, A, c)$, that is a undirected graph that has the same vertex-set V and two arcs between u and v is replaced an edge between them whose cost is the same as the cost of one of the arcs.

As usual we also assume that the cost function c is additive over the links in a path $p(u, v)$ between u and v, i.e.,

$$c(p(u, v)) \equiv \sum_{a \in p(u, v)} c(a).$$

For the simplicity of presentation, we denote by $P_{G'}(u, v)$ the shortest path from u to v in subgraph G' of G. These notations will be used for both directed and indirected graphs.

We define a *k-path* as directed trail in $\vec{G}$ such that in the trail at most k nodes in D are designated to receive the data (all other nodes in the trail, including destination nodes, can only forward the data to their neighbors in the trail), where an arc can appear in the trail at most once (but a node may appear more than once). In addition, we define a *k-path routing* of $< s, D >$, denoted by $R_p(s, D; k) = \{P_i \,|\, i\}$, as a set of k-paths such that every destination node in D must be designated to receive the data in one of the k-paths in $R(s, D; k)$. Two k-paths in $R_p(s, D; k)$ may share an arc, which will not cause any trouble during data transmission under time division multiplexing [252]. Clearly, $m \equiv \lceil |D|/k \rceil \leq |R_p(s, D; k)| \leq |D|$. Since, data is transmitted through each arc in a k-path exactly once, the cost of multicasting data is then defined as the total costs of k-paths in $R_p(s, D; k)$, i.e.,

$$c\big(R_p(s, D; k)\big) \equiv \sum_{P_i \in R_p(s, D; k)} c(P_i).$$

Multicast routing problem under k-path routing model is defined as follows.

Problem 7.1 *Steiner k-Path Routing Problem* [123]

Instance	A graph $\vec{G}(V, A, c)$, a source node $s \in V$ and a subset $D \subset V$ of destination nodes, and an integer $k \geq 1$.
Solution	A k-path routing $R_p(s, D; k)$.
Objective	Minimizing the cost of $R_p(s, D; k)$.

Similarly, a *k-tree* is defined as a tree in G such that in the tree at most k nodes in D are designated to receive data (all other nodes in the tree, including destination nodes, can only forward data to their neighbors in the tree). A *k-tree routing* of $< s, D >$, denoted by $R_t(s, D; k) = \{T_i \,|\, i\}$, is defined as a set of k-trees such that every destination node in D must be designated to receive data in one of the k-trees in $R(s, D; k)$. Two k-trees in $R_t(s, D; k)$ may also share an edge under the wavelength division multiplexing. Clearly, $m \equiv \lceil |D|/k \rceil \leq |R(s, D; k)| \leq |D|$. Since, data is transmitted through each edge in a k-tree exactly once, the cost of multicasting data is then defined as the total costs of k-trees in $R_t(s, D; k)$, i.e.,

$$c\big(R_t(s, D; k)\big) \equiv \sum_{T_i \in R_t(s, D; k)} c(T_i).$$

Multicast routing problem under k-tree routing model is defined as follows.

Problem 7.2 *Steiner k-Tree Routing Problem* [122]

Instance	A graph $G(V, E)$, a source node $s \in V$ and a subset $D \subset V$ of destination nodes, and an integer $k \geq 1$.
Solution	A k-tree routing $R_t(s, D; k)$.
Objective	Minimizing the cost of $R_t(s, D; k)$.

It is worthy of emphasizing that under k-tree routing model, a k-tree may contain more than k nodes in $\{s\} \cup D$ (but at k of them need to be assigned to receive data), but in the discussion of k-Steiner ratios in Chapter 2, a k-tree contains exactly k terminals in P. In addition, Steiner tree problem in graphs (Problem 2.1) could be considered as a special case of Steiner k-tree routing problem (Problem 7.2) with $k \geq |V|$ and $P = \{s\} \cup D$.

7.1.2 Complexity Study

Let us first consider Steiner k-tree routing problem. Clearly, when $k = 1$, the optimal k-tree routing consists of $|D|$ shortest paths from source s to each of $|D|$ destination nodes. Thus it can be found in polynomial time. The following theorem shows that this is also true when $k = 2$. The proof is based on a polynomial-time reduction from the *minimum weighted matching* problem: given a complete graph H of even $|V(H)|$ with a weight function: $w(,) : V(H) \times V(H) \to Q^+$, find a complete *matching* $M \subseteq E(H)$ such that every node in $V(H)$ is incident to exactly one edge in M and the total weight of edges in M is minimal. This problem can be solved in polynomial time (refer to Chapter 11 of [224]).

Theorem 7.1 *The Steiner k-tree routing problem with $k = 2$ is polynomial-time solvable.*

Proof. Given a multicast connection request $< s, D >$ on network G with $D = \{d_1, \cdots, d_{|D|}\}$, the reduction can be done as follows (see Fig.7.1, where D contains six destination nodes in black).

Step 1. Compute the shortest path $p_G(u, v)$ for each node pair $u, v \in V$.

Step 2. Compute SMT $T(d_i, d_j, s)$ of $\{d_i, d_j\}$ for each pair $d_i, d_j \in D$, i.e., find $u \in V$ such that $c(p_G(d_i, u)) + c(p_G(d_j, u)) + c(p_G(s, u)) = \min\{c(p_G(d_i, v)) + c(p_G(d_j, v)) + c(p_G(s, v)) \mid v \in V\}$.

Step 3. Construct an auxiliary graph $G'(D \cup \{s_1, \cdots, s_{|D|}\}, E')$ such that there is an edge between d_i and d_j for $i \neq j$ with weight $w(d_i, d_j) = c(T(d_i, d_j, s))$, there is an edge between s_i and s_j for $i \neq j$ with weight zero, there is an edge between s_i and d_i for each i with weight $w(s_i, d_i) = c(p_G(s, d_i))$, and there is no edge between d_i and s_j for $i \neq j$ (or equivalently, edges between them have very large costs).

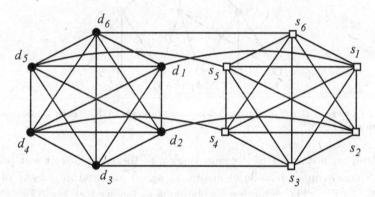

Fig. 7.1   The reduction from the Steiner k-tree routing problem with $k = 2$ to minimum weight matching problem.

Clearly, Steps 1-3 can be done in polynomial time. Moreover, given a minimum weight matching $M \subset E'$ of G', we can produce an optimal k-tree routing $R_t(s, D; k)$ on G as follows: (1) For each edge $(d_i, d_j) \in M$, produce a k-tree that is an SMT of $\{s, d_i, d_j\}$; (2) For each edge $(s_i, d_i) \in M$, produce a k-tree that is the shortest path $p_G(s, d_i)$ from s to d_i. The proof is then finished. □

However, Lin [187] proved the following theorem showing that the problem becomes NP-hard when $k = 3$. His proof is based on a polynomial reduction from the decision version of *exact 3-set cover problem*: Given a universal set S and a collection of 3-sets $\mathcal{C} = \{C_1, C_2, \cdots, C_m\}$, where every C_i is a set containing 3 elements from S, decide if there exists a subcollection of disjoint 3-sets whose union is S. This was proved to be NP-hard [110].

Theorem 7.2　*The Steiner k-tree routing problem with $k = 3$ is NP-hard.*

Proof.　It suffices to show that there exists a polynomial-time reduction from the decision version of the exact 3-set cover problem to the decision version of the Steiner k-tree routing problem with $k = 3$. The latter is as follows: Given a graph $G(V, E)$, a node $s \in V$ and a subset $D \subset V$, and a constant $B > 0$, decide if there exists a 3-tree routing with cost at most B.

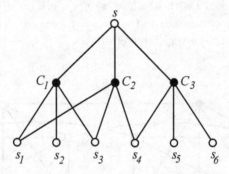

Fig. 7.2　The reduction from the exact 3-set cover problem to the Steiner k-tree routing problem with $k = 3$.

Given an instance I of decision version of the exact 3-set cover problem: S consisting of $n = 3q$ elements, $s_1, s_2, \cdots, s_n$, and m subsets of S, $C_1, C_2, \cdots, C_m$. The reduction can be done as follows (see Fig.7.2):

Step 1.　Create one destination node for every element s_i, $1 \le i \le n$, and create one Steiner point for every 3-set C_i, and there is an edge connecting this point to every destination node inside the set.

Step 2.　Create a source node s, which is adjacent to every Steiner point.

Step 3.　Assign a unit weight to all edges and set $B = 4q$.

Clearly we can produce in polynomial time such an instance I' of the decision version of the Steiner k-tree routing problem with $k = 3$. Moreover, it is easy to verify that instance I' has a 3-tree routing of cost at most $4q$ if and only if instance I has a subcollection of q 3-sets $\mathcal{C} = \{C_{i1}, C_{i2}, \cdots, C_{iq}\}$ such that $\cup_{j=1}^{q} C_{ij} = S$ and $C_{is} \cap C_{it} = \emptyset$ for $s \neq t$ and $1 \leq s, t \leq q$. The proof is then finished. $\qquad\square$

We now consider Steiner k-path routing problem. When $k = 1$, the problem is equivalent to the shortest path problem, thus it is polynomial-time solvable. When $k = 2$, using almost the same argument for proving Theorem 1.1, we can prove the following theorem.

Theorem 7.3 *The Steiner k-path routing problem is polynomial-time solvable when $k = 2$.*

Proof. Again we will reduce the problem to minimum weight matching problem. Given a multicast connection request $< s, D >$ on network $\vec{G}$ with $D = \{d_1, \cdots, d_{|D|}\}$, the reduction can be done as follows.

Step 1. Compute the shortest path $p_G(d_i, d_j)$ between each node pair $d_i, d_j \in D$ for $i \neq j$, and the shortest path $p_G(s, d_i)$ from source s to each d_i in D.

Step 2. Construct an auxiliary graph $G'(D \cup \{s_1, \cdots, s_{|D|}\}, E')$: there is an edge between d_i and d_j for $i \neq j$ with weight $w(d_i, d_j) = \min\{c(p_G(s, d_i)) + c(p_G(d_i, d_j)), c(p_G(s, d_j)) + c(p_G(d_j, d_i))\}$, there is an edge between s_i and s_j for $i \neq j$ with weight zero, and there is an edge between s_i and d_i for each i with weight $w(s_i, d_i) = c(p_G(s, d_i))$.

Clearly, Steps 1-2 can be done in polynomial time. Moreover, given a minimum weight matching of G' that is a perfect matching M of G' whose weight is minimal, we can produce an optimal k-path routing $R_p(s, D; k)$ on G as follows: (1) For each edge $(d_i, d_j) \in M$, produce a k-path from s to d_j via d_i that consists of $P_G(s, d_i)$ and $P_G(d_i, d_j)$ if $w(d_i, d_j) = c(p_G(s, d_i)) + c(p_G(d_i, d_j))$ and a k-path from s to d_i via d_j that consists of $P_G(s, d_j)$ and $P_G(d_j, d_i)$ if $w(d_i, d_j) = c(p_G(s, d_j)) + c(p_G(d_i, d_j))$; (2) For each edge $(s_i, d_i) \in M$, produce a k-path from s to d_i.

The correctness of the reduction follows from two facts, which can be easily verified: (a) Each of possible shortest k-paths is associated with exactly one edge in G' whose cost is the weight of the edge, and (b) each of destination nodes is incident to exactly one edge in M. That means, every destination node is designated in a produced k-path. The proof is then finished. $\qquad\square$

The following theorem, however, shows that the problem is NP-hard for general k.

Theorem 7.4 *The Steiner k-path routing problem is NP-hard.*

Proof. We will consider the decision version of Steiner k-path routing problem. Given a multicast connection $< s, D >$ on network G, an integer $k > 2$ and a bound $B > 0$, the problem asks if there is a k-path routing for $< s, D >$ whose cost is at most B.

It was proved (refer to Theorem 16.7 of [224]) that *Hamilton circuit problem* for graphs with all nodes of degree three (that is a 3-regular graph) is NP-complete. It was also proved (refer to Corollary 1 of Section 15.6 of [224]) that *Hamilton path problem* is NP-complete through a simple reduction as follows. Given a 3-regular graph $G(V, E)$, construct a new graph $G(V', E')$ where $V' = V \cup \{x, y, z\}$ and $E' = E \cup \{(y, z), (x, v_0)\} \cup \{(y, v) \mid (v, v_0) \in E\}$, for some fixed $v_0 \in V$. See Fig.7.2. It can be verified that $G(V, E)$ has a Hamilton circuit[2] if and only if $G(V', E')$ has a Hamilton path[3].

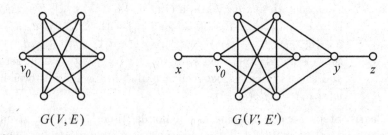

$$G(V, E) \qquad\qquad G(V', E')$$

Fig. 7.3 Constructing a new graph $G(V', E')$ from given 3-regular graph $G(V, E)$.

We now show that Hamilton path problem for above defined graph $G(V', E')$ can be reduced in polynomial time to Steiner k-path routing problem. First, we construct $\vec{G}(V', A, c)$ by substituting each edge $(u, v) \in E'$ with a pair of arcs (u, v) and (v, u) whose costs c are equal to one. Secondly, we set $s := x$, $D := V' \setminus \{x\}$, $B := |V'| - 1$, and $k := |V'|$. It is easy to verify that $G(V', E')$ has a Hamilton path if and only if $\vec{G}(V', A, c)$ has a k-routing $R_p(s, D; k)$ whose cost is at most B. The proof is then finished. $\qquad\square$

---

[2] A *Hamilton circuit* (also called cycle) of graph $G(V, E)$ is a closed path that consists of $|V|$ edges and contains every vertex in V.

[3] A *Hamilton path* of graph $G(V, E)$ is a path that consists of $(|V| - 1)$ edges and contains every vertex in V.

7.2 Algorithms for k-Path Routing Problem

As in the previous section we have proved that Steiner k-path routing problem is NP-hard in general, in this section we will study two approximation algorithms for the problem that both have guaranteed performances.

7.2.1 *Steiner Tree Based Algorithm*

The basic idea of this algorithm is to simply produce a Steiner tree of $D \cup \{s\}$ first and then transverse the obtained tree. See Fig.7.4-5.

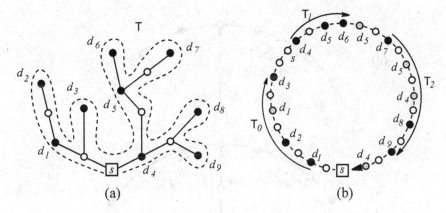

Fig. 7.4 (a) Traverse along the Steiner tree T_S, and (b) Partition the obtained trail $\mathcal{T}$.

Algorithm 7.1 *Steiner Tree Based Approximation Algorithm*

Step 1 Construct an auxiliary complete edge-weighted graph G_a of $D \cup \{s\}$. For each node pair $u, v \in D \cup \{s\}$, the weight of edge (u, v) in G_a is the cost of the shortest path in G between u and v.

Step 2 Construct an MST T_{mst} of G_a and substitute each edge in T_{mst} by the shortest path between two endpoints in G. And then as a result, obtain a Steiner tree T_S of $D \cup \{s\}$ in G.

Step 3 Produce a directed trail of $D \cup \{s\}$ by traversing each vertex of $D \cup \{s\}$ along T_S, and denote it by $\mathcal{T} = (s \to d_1 \to \cdots \to d_{|D|} \to s)$.

Step 4 Partition $\mathcal{T}$ into m subtrails $\mathcal{T}_i$ for $i = 0, 1, \cdots m - 1$ such that $d_{ik+1}, d_{ik+2}, \cdots, d_{ik+k}$ are designated in $\mathcal{T}_i$ to receive the data.

Step 5 Find v_i in $\mathcal{T}_i$ which is closest to source s for each i.

Step 6 Construct two k-paths via v_i to include k designated destination

nodes in subtrail T_i: $P_i' = \{s \to d_{i'} \to d_{i'-1} \to \cdots \to d_{ik+1}\}$,
$P_i'' = \{s \to d_{i'+1} \to d_{i'+2} \to \cdots \to d_{ik+k}\}$, where v_i lies between
$d_{i'}$ and $d_{i'+1}$ in T_i.

Return k-path routing $R_{ST} = \{P_i' \mid i=0,1,\cdots,m{-}1\} \cup \{P_i'' \mid i=0,1,\cdots,m{-}1\}$.

Fig.7.4(a-b) illustrate Steps 3-4, and Fig.7.5(a-b) illustrate Steps 5-6, respectively. Note that in Step 4, T_i may contain more than k destination nodes. We now study the guaranteed performance of Algorithm 7.1 in worst case analysis. To do this, we need the following lemma. Given a multicast connection $< s, D >$, let R_{opt} be an optimal k-path routing and $c(R_{opt})$ be its cost.

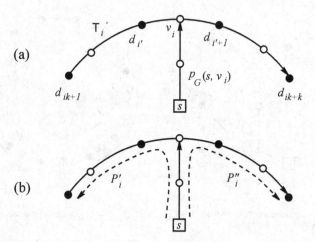

Fig. 7.5   (a) Find node $v_i \in T_i$ closest to source s, (b) Produce two k-paths P_i' and P_i''.

Lemma 7.1 *Let $d_{i'}$ be the destination node in trail T_i that is designated to receive the data and is closest to s. Then*

$$\sum_{i=0}^{m-1} c(p_G(s, d_{i'})) \leq c(R_{opt}). \qquad (7.1)$$

Proof. Consider an optimal routing R_{opt} that has N k-paths $P_1^*, P_2^*, \cdots, P_N^*$, where $N \geq m$. We construct an auxiliary weighted bipartite graph $B(X, Y)$, where $X = \{T_i \mid i = 0, 1, \cdots, m-1\}$ and $Y = \{P_i^* \mid i = 1, \cdots, N\}$. There exists an edge (T_i, P_j^*) in $B(X, Y)$ if and only if T_i and P_j^* designate $d \geq 1$ destination nodes in common and the weight of the edge is $w(T_i, P_j^*) = d$. See Fig.7.6.

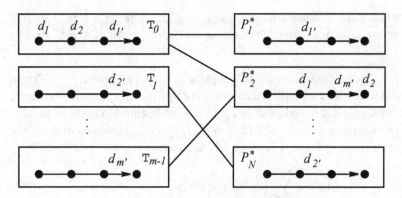

Fig. 7.6 Construct bipartite graph $B(X, Y)$.

Now we prove, by using Hall's Theorem [224], that $B(X, Y)$ has a (perfect) matching such that each T_i is incident to an edge in the matching. Suppose, by contradiction, that there exists a subset $X_0 \subseteq X$ such that X_0's neighbor set $Y_0 \subseteq Y$, which consists of vertices adjacent with some vertices in X_0, satisfies $|Y_0| \le |X_0| - 1$. Since each T_i designates at most k destination nodes and each of them is designated in exactly one optimal k-path, then the total weight of edges incident to T_i is at most k. For each P_j^* we have the same result. Now for $X' \subseteq X$ and $Y' \subseteq Y$, let $w(X')$ and $w(Y')$ denote the total weights of edges incident to some $T_i \in X'$ and $P_j^* \in Y'$, respectively. Then $w(Y') \le k|Y'|$, this implies $w(X_0) \le w(Y_0) \le k|Y_0|$. In addition, we have

$$w(X \setminus X_0) \le k|X \setminus X_0| = k(m - |X_0|).$$

Hence we obtain the following contradiction:

$$|D| = w(X_0) + w(X \setminus X_0) \le k|Y_0| + k(m - |X_0|) \le k(m - 1) < |D|.$$

Therefore, there exists a desired matching. Without loss of generality, we denote this matching by $M = \{(T_i, P_i^*) \mid i\}$. This means that for each i there exists a destination node designated in both T_i and P_i^*. Thus the cost of P_i^* is not less than the cost of the shortest path from s to that common designated destination node, which, by the definition of $d_{i'}$, is not less than the cost of the shortest path from s to $d_{i'}$, i.e., $c(P_i^*) \ge c(p_G(s, d_{i'}))$. To sum up this inequality over i, we obtain the desired inequality (7.1). The proof is then finished. $\square$

Theorem 7.5 *Steiner tree based algorithm 7.1 returns a k-path routing R_{ST} whose cost is at most four times that of an optimal k-path routing R_{opt}.*

Proof.   Notice that an optimal k-routing $R_{opt} = \{P_i^* \,|\, i = 1, \cdots, N\}$ can be transformed into a (undirected) spanning tree of G_a, that can be done by substituting each directed path in P_i^* from one destination node to another with the shortest path in G. Thus $c(R_{opt}) \geq c(T_{mst})$. According to the rules of Algorithm 7.1 and Lemma 7.1, we have

$$c(R_{ST}) = \sum_{i=0}^{m-1} \Big(c(P_i') + c(P_i'') \Big) \tag{7.2}$$

$$\leq c(T) + 2 \sum_{i=0}^{m-1} c\big(p_G(s, v_i)\big)$$

$$\leq 2 \cdot c(T_{mst}) + 2 \cdot c(R_{opt}) \leq 4 \cdot c(R_{opt}),$$

which proves the theorem. $\square$

Cai et al. [42] gave an example showing the bound 4 in Theorem 7.5 is tight. In their example, the optimal k-path routing consists of m k-paths, $\{P_1^*, P_2^*, \cdots, P_m^*\}$, where

$$P_1^* = \{s \rightarrow d_{mk-1} \rightarrow d_{mk} \rightarrow d_1 \rightarrow \cdots \rightarrow d_{k-2}\},$$
$$P_2^* = \{s \rightarrow d_{k-1} \rightarrow d_k \rightarrow d_{k-1} \rightarrow \cdots \rightarrow d_{2k-2}\},$$
$$\cdots = \cdots$$
$$P_m^* = \{s \rightarrow d_{(m-1)k-1} \rightarrow d_{(m-1)k} \rightarrow d_{(m-1)k+1} \rightarrow \cdots \rightarrow d_{mk-2}\}.$$

The costs of the edges on the optimal k-path routing are defined as follows:

$$c(s, d_{ik-1}) = M, i = 1, 2, \cdots, m;$$
$$c(d_j, d_{j+1}) = 1, \; j \neq ik - 2 \text{ for some } i.$$

The underlying network G is the completion of the routing tree. It is easy to verify that the optimal k-path routing and MST both have cost of $m(M + k - 1)$ while the Hamilton circuit has cost of $2m(M + k - 1)$. According to the partitioning scheme of Algorithm 7.1, $d_1, d_2, \cdots, d_k$ are on a segment and among them d_{k-1} is the closest to source s. Hence,

Algorithm 7.1 returns a k-path routing has a cost $m(4M + 2k - 3)$, which is asymptotically $4m(M + k - 1)$.

In addition, Steiner tree based algorithm 7.1 can be easily modified so that the bound on its approximation ratio could be reduced to 2 for $m = 1, 2$ and 3 for $m = 3$, respectively.

Applying the same idea of Algorithm 7.1, Cai et al. [42] proposed a better approximation algorithm by introducing a different technique for partitioning the Hamilton circuit.

Suppose, without loss of generality, that the destination nodes on the Hamilton circuit are indexed consecutively from d_1 to $d_{|D|}$, with source node s lying in between d_1 and $d_{|D|}$. Assume $lk < |D| \leq (l+1)k$ for some l. They make a new Hamilton circuit as follows:

(1) Make $((l+1)k - |D|)$ copies of s, $d_{|D|+1}, \cdots, d_{(l+1)k}$;
(2) Connect those new nodes in a path and replace s with them;
(3) Set $w(d_{|D|}, d_{|D|+1}) = w(d_{|D|}, s)$, $w(d_i, d_{i+1}) = 0$,
    for $i = |D| + 1, ..., (l+1)k - 1$, and $w(d_{(l+1)k}, d_1) = w(s, d_1)$.

Clearly, the new Hamilton circuit contains exactly $(l+1)k$ nodes while its weight remains unchanged.

Now in stead of dividing the Hamilton circuit into $(l+1)$ segments starting from d_1 each containing exactly k nodes, they choose an index j such that $\sum_{i=0}^{l} w(s, d_{ik+j})$ is minimal, denote it by j^*, and then divide the Hamilton circuit into $(l+1)$ segments starting from d_{j^*}, i.e.,

$$(d_{j^*}, d_{j^*+1}, d_{j^*+k-1}), \cdots, (d_{j^*+lk}, d_{j^*+lk+1}, d_{j^*+(l+1)k-1}),$$

where the subscripts take module of $(l+1)k$. In the end they append each i-th segment (path) with edge $(s, d_{(i-1)k+j^*})$ for $i = 0, 1, \cdots, l$, and obtain a k-path routing. Cai et al. [42] proved the following theorem.

Theorem 7.6 *Given a multicast connection $< s, D >$ and $k \geq 2$, the improved Algorithm 7.1 produces a k-path routing R_{IST} whose cost is at most three times that of an optimal k-path routing R_{opt}.*

Proof. Let $R_{opt} = \{P_i^* \mid i = 1, \cdots, m\}$ be an optimal k-path routing for $< s, D >$. Let d_j be a destination node in path P_i^*, then the distance from d_j to s in graph G is no greater than the distance from d_j to s in P_i^*, i.e., $w(d_j, s) \leq c(p_{P_i^*}(d_j, s))$. Now suppose that all destination nodes in P_i^* are $d_{j_1}, d_{j_2}, \cdots, d_{j_l}$, where $l \leq k$. Then we have

$$\sum_{j=1}^{l} w(d_{j_j}, s) \leq \sum_{j=1}^{l} c\big(p_{P_i^*}(d_{j_j}, s)\big) \leq l \cdot w(P_i^*) \leq k \cdot w(P_i^*).$$

Hence we have

$$\sum_{i=1}^{|D|} w(d_i, s) \leq k \sum_{i=1}^{|D|} w(P_i^*) = k \cdot c(R_{opt}). \qquad (7.3)$$

According to the method that we choose j^*, we deduce from the inequality (7.3) that

$$\sum_{i=1}^{l} w(d_{ik+j^*}, s) \leq c(R_{opt}),$$

otherwise we will have

$$\sum_{i=1}^{l} w(d_{ik+j}, s) > c(R_{opt}), \quad j = 1, 2, \cdots, k.$$

Recall that the Hamilton circuit has weight no greater than two times of $c(R_{opt})$, which, together with $(l+1)$ paths from s to d_{ik+j^*} for $i = 0, 1, \cdots, l$, makes a k-path routing whose weight is at most three times that of the optimal k-path routing R_{opt}. The proof is then finished. $\square$

7.2.2 *Set Cover Based Algorithm*

The basic idea of the second algorithm is to transform Steiner k-path routing problem to the *set cover problem*, which can be solved approximately by a greedy algorithm [65]. An instance of set cover problem consists of a finite set S and a collection $\mathcal{C}$ of subsets $S_i \subset S$ with a weight $w(S_i) > 0$ for $i = 1, 2, \cdots, m$ that satisfies $\cup_{i=1}^{m} S_i = S$. A subcollection $\mathcal{C}' \subseteq \mathcal{C}$ is called a *set cover* if each element in S is in at least one subset of $\mathcal{C}'$. The objective is to find a set cover whose weight, the total weights of subsets in the set cover, is minimal.

Algorithm 7.2 *Set Cover Based Approximation Algorithm*

Step 1 For each subset of j destination nodes $\{d_{i_1}, \cdots, d_{i_j}\} \subseteq D$, where $j = 1, \cdots, k$, produce a j-path (as in the proof of Corollary 7.1):

Step 1.1 Produce an auxiliary graph $G(d_{i_1}, \cdots, d_{i_j})$. It is a complete edge-weighted graph of $\{d_{i_1}, \cdots, d_{i_j}\} \cup \{s\}$. For $u, v \in \{d_{i_1}, \cdots, d_{i_j}\} \cup \{s\}$ the weight of edge (u, v) is the cost of the shortest path in G between u and v.

Step 1.2 Construct an MST $T_{mst}(d_{i_1}, \cdots, d_{i_j})$ of $G(d_{i_1}, \cdots, d_{i_j})$ and substitute each edge in $T_{mst}(d_{i_1}, \cdots, d_{i_j})$ by the shortest path between two endpoints in G. Obtain a Steiner tree $T_S(d_{i_1}, \cdots, d_{i_j})$ of $\{d_{i_1}, \cdots, d_{i_j}\} \cup \{s\}$.

Step 1.3 Obtain a j-path $P(d_{i_1}, \cdots, d_{i_j})$ by traversing each vertex of $\{d_{i_1}, \cdots, d_{i_j}\} \cup \{s\}$ along $T_S(d_{i_1}, \cdots, d_{i_j})$ whose edges are replaced by two arcs between two endpoints of the edges.

Step 2 Assign weight $w(d_{i_1}, \cdots, d_{i_j}) = c(P(d_{i_1}, \cdots, d_{i_j}))$ for each subset $\{d_{i_1}, \cdots, d_{i_j}\}$. Find a set cover $\mathcal{C}'$ of D, by using a subcollection of $\mathcal{C} = \{\{d_{i_1}, \cdots, d_{i_j}\} \mid$ all $\binom{|D|}{j}$ j-tuples $\{d_{i_1}, \cdots, d_{i_j}\}, j = 1, \cdots, k\}$.

Step 2.1 Choose a subset $S_g = \{d_{i_1}, \cdots, d_{i_g}\}$ that satisfies the equality $w(d_{i_1}, \cdots, d_{i_g})/g = \min\{w(d_{i_1}, \cdots, d_{i_j})/j \mid \{d_{i_1}, \cdots, d_{i_j}\} \in \mathcal{C}\}$.

Step 2.2 Put $\{d_{i_1}, \cdots, d_{i_g}\}$ into set cover $\mathcal{C}'$, and then remove every element in $\{d_{i_1}, \cdots, d_{i_g}\}$ from D, $\mathcal{C}$ and each $\{d_{i_1}, \cdots, d_{i_j}\}$.

Step 2.3 Repeat (Step 2.1-2) until D is empty.

Step 3 Return k-path routing R_{SC} that contains k-path $P(d_{i_1}, \cdots, d_{i_j})$ if $\{d_{i_1}, \cdots, d_{i_j}\}$ is in $\mathcal{C}'$.

Theorem 7.7 *Given a multicast connection $< s, D >$ and $k \geq 2$, Algorithm 7.2 returns a k-path routing R_{SC} whose cost is $2\mathcal{H}(k)$ times that of an optimal k-path routing R_{opt}, where $\mathcal{H}(k) = \sum_{i=1}^{k} 1/i$.*

Proof. Let $\mathcal{C}_{opt}$ be an optimal set cover of the set cover problem generated at Step 2. Then it can be proved (refer to [65]) that

$$c(R_{SC}) = \sum_{\{d_{i_1}, \cdots, d_{i_j}\} \in \mathcal{C}'} c(P(d_{i_1}, \cdots, d_{i_j}))$$

$$\leq \mathcal{H}(k) \sum_{\{d_{i_1}, \cdots, d_{i_j}\} \in \mathcal{C}_{opt}} c(P(d_{i_1}, \cdots, d_{i_j})). \qquad (7.4)$$

Now consider an optimal k-path routing $R_{opt} = \{P^*(d_{i_1}, \cdots, d_{i_j})\}$, where $P^*(d_{i_1}, \cdots, d_{i_j})$ is an optimal k-path designating all destinations in $\{d_{i_1}, \cdots, d_{i_j}\}$. Let $C^* = \{\{d_{i_1}, \cdots, d_{i_j}\} \mid P^*(d_{i_1}, \cdots, d_{i_j}) \in R_{opt}\}$. Since C^* is a set cover, then we have

$$\sum_{\{d_{i_1}, \cdots, d_{i_j}\} \in C_{opt}} c\big(P(d_{i_1}, \cdots, d_{i_j})\big) \leq \sum_{\{d_{i_1}, \cdots, d_{i_j}\} \in C^*} c\big(P(d_{i_1}, \cdots, d_{i_j})\big). \quad (7.5)$$

Notice that by using the same argument as in the proof of Theorem 7.5, we can prove that each k-path $P(d_{i_1}, \cdots, d_{i_j})$ produced by Algorithm 7.2 satisfies the following inequality

$$c\big(P(d_{i_1}, \cdots, d_{i_j})\big) \leq 2 \cdot c\big(P^*(d_{i_1}, \cdots, d_{i_j})\big). \quad (7.6)$$

Therefore, to combine above three inequalities we have

$$c(R_{SC}) \leq 2\mathcal{H}(k) \sum_{\{d_{i_1}, \cdots, d_{i_j}\} \in C^*} c\big(P^*(d_{i_1}, \cdots, d_{i_j})\big) = 2\mathcal{H}(k)c(R_{opt}).$$

This is the desired bound, and the proof is then finished. □

For the case of small $k = 3$, Algorithm 7.2 can be modified so that its approximation ratio can be reduced to $11/6$. In addition, it is easy to verify that the time-complexity of Algorithm 7.2 is $O(k|D|^k|V|^2)$.

7.3   Algorithms for k-Tree Routing Problem

As in Section 7.1 we have proved that Steiner k-tree routing problem in general is NP-hard, in this section we will propose two approximation algorithms that have guaranteed performance ratios.

7.3.1   *Hamilton Circuit Based Algorithm*

The first algorithm is similar to Steiner tree based algorithm 7.1. It first produces a directed trail of low cost including all nodes in $D \cup \{s\}$, and then break it into m small trails on which at most k nodes in D are specified, in the end for each small trail make a k-tree constituting of s and those specified nodes in D. The directed trail can be obtained by constructing a Hamilton circuit of low cost in an auxiliary graph whose vertex-set is $D \cup \{s\}$.

Algorithm 7.3 *Hamilton Circuit Based Algorithm*

Step 1 Construct an auxiliary graph G_a as in Step 1 of Algorithm 7.1.
Step 2 Produce a Hamilton circuit H_c of G_a using Christonfides' method[61].

Step 3 Obtain a directed trail T of $D \cup \{s\}$ in G by substituting each edge in H_c by the shortest path between its two endpoints of the edge in G, $T = (s \to d_1 \to \cdots \to d_{|D|} \to s)$.

Step 4 Partition T into m subtrails T_i for $i = 0, 1, \cdots m - 1$ such that $d_{ik+1}, d_{ik+2}, \cdots, d_{ik+k}$ are designated in T_i to receive the data. For each i, find v_i in T_i which is closest to source s.

Step 5 Construct a k-tree T_i consisting of two paths $(d_{ik+1} \to d_{ik+2} \to \cdots \to d_{ik+k})$ and $(s \to v_i)$, where $d_{ik+1}, d_{ik+2}, \cdots, d_{ik+k}$ in subtrail T_i are designated to receive the data in T_i.

Step 6 Return k-tree routing $R_{HC} = \{T_i \mid i = 0, 1, \cdots m - 1\}$.

Notice that as the auxiliary graph G_a produced at Step 1 is a complete graph and the cost function defined on its edges satisfies triangular inequality, in Step 2 Christonfides' method can be employed to construct a Hamilton circuit H_c of G_a. The following lemma comes directly from the well-known result due to Christonfides [61].

Lemma 7.2 *For any given multicast connection $< s, D >$ on G, the Hamilton circuit H_c of G_a produced at Steps 1-2 of Algorithm 7.3 has cost at most 3/2 times that of the minimum Hamilton circuit of G_a.*

To study the performance of Algorithm 7.3, we also need Lemma 7.1, which, it is easy to see, is also applicable here.

Theorem 7.8 *Given a multicast connection $< s, D >$ and $k > 2$, Hamilton circuit based algorithm 7.3 returns a k-tree routing R_{HC} whose cost is at most four times that of the optimal k-tree routing R_{opt}.*

Proof. Let H_{opt} be the minimum Hamilton circuit of G_a. Then we have $2 \cdot c(R_{opt}) \geq c(H_{opt})$ since two R_{opt}s correspond a Hamilton circuit of G_a. In addition, by Lemma 7.2 we have $\frac{3}{2}c(H_{opt}) \geq c(H_c)$. Thus according to the rules of Algorithm 7.3 and Lemma 7.1, we have

$$c(R_{HC}) = \sum_{i=0}^{m-1} c(T_i)$$

$$= \sum_{i=0}^{m-1} c(T_i) + \sum_{i=0}^{m-1} c(p_G(s, v_i))$$

$$< c(T) + c(R_{opt}) \leq c(H_c) + c(R_{opt})$$

$$\leq \frac{3}{2} \cdot 2 \cdot c(R_{opt}) + c(R_{opt}) = 4 \cdot c(R_{opt}),$$

which proves the theorem. □

When $m = 2$, Hamilton circuit based algorithm 7.3 can be modified as follows: After partitioning $\mathcal{T}$ into two subtrails $\mathcal{T}_o$ and $\mathcal{T}_1$ at Step 4, construct a k-tree routing consisting of two k-trees,

$$T_1 = \{s \to d_1 \to d_2 \to \cdots \to d_{k-1} \to d_k\},$$
$$T_2 = \{s \to d_{|D|} \to d_{|D|-1} \to \cdots \to d_{k+2} \to d_{k+1}\}.$$

Applying the same argument as used in the proof of Theorem 7.8, we can deduce $c(R_{HC}) \leq 3 \cdot c(R_{opt})$.

7.3.2 Steiner Tree Based Algorithm

The Steiner tree based algorithm is proposed by Cai et al. [42]. In stead of constructing a Hamilton circuit of auxiliary graph G_a, it first produces a low cost of Steiner tree T in G interconnecting all nodes in $\{s\} \cup D$. Here we can use an α-approximation algorithm for Steiner tree problem in graphs.

Suppose that there are l branches of T rooted at s. If each of l branches contains at most k destination nodes in D, then T constitutes a k-tree routing with cost $c(T) \leq \alpha \cdot c(R_{opt})$ since the cost of SMT $c(T_{smt}) \leq c(R_{opt})$ and $c(T) \leq \alpha \cdot c(T_{smt})$.

In the following we consider the case that there exists (at least) one branch rooted at s, denote it by T_b, that contains more than k destination nodes. We will demonstrate how to divide it into some small sub-branches so that each of them contains at most k destination nodes in D while keeping the total cost of distances from the roots of these sub-branches to s within two times of the cost of the optimal k-tree routing. Note that those branches which contain no more than k destination nodes in D do not need dividing operations. In such a way, we will obtain a k-tree routing whose cost is at most $(2 + \alpha)$ times that of the optimal k-tree routing.

We now assume that Steiner tree T_b contains $d \ (> k)$ destination nodes in D. Cai et al. [42] proved the following two lemmas that deal with two cases separately, $d \geq 2k/3$ and $2k/3 > d > k$.

Lemma 7.3 T_b *could be partitioned into two subtrees such that they have one node in common and the number of destination nodes they contain is at least $d/3$ but at most $2d/3$.*

Lemma 7.4 *Suppose $k < d \leq 3k/2$ and $k \geq 3$. Let D_0 be a set of $(d - k/2 + 1)$ destination nodes in T_b. Then T_b could be partitioned into two* subtrees T_1 and T_2 such that they have one node in common and contain a set of destination nodes $D_1 \subset D_0$ and $D_2 \subset D_0$ with $0 < |D_1|, |D_2| \leq k$, respectively.

Proof. Consider T_b as a Steiner tree rooted at a node b in T_b. Denote by $d(v)$ be the number of destination nodes in the subtree of T_b rooted at v. Let r denote the farthest node from b that has $d(r) \geq d - k/2$. Note that r is well-defined and unique since $k < d \leq 3k/2$. Now rooting tree T_b at node r instead of b will guarantee that r is the only node with $d(r) \geq d - k/2$. Clearly, the degree of node r is at least 2.

In order to obtain a desired pair of partition of T_b, we first arbitrarily partition T_b into two subtrees T_1 and T_2 both rooted at r. Suppose that T_1 and T_2 contain a set D_1 and D_2 of destination nodes, respectively. If $0 < |D_1|, |D_2| \leq k$, then we get the desired partition.

Now we assume that the current partition does not satisfy $0 < |D_1|, |D_2| \leq k$, and assume further, without loss of generality, that $|D_2| \geq |D_1|$. Then we have $|D_1| \leq 1/k$ and $|D_2| > d - k/2$. If not, we will obtain either $|D_2| \leq d - k/2$ or $|D_1| > 1/k$ (which implies $|D_2| \leq d - k/2$). Since $d \leq 3k/2$, we obtain $|D_1| \leq |D_2| \leq k$. Moreover, notice that $|D_0| = d - k/2 + 1$, thus there must be at least one node in D_0 included in T_1, and at least one distinct node in D_0 included in T_2, i.e., $|D_1| > 0$ and $|D_2| > 0$, contradicting that the current partition is not a desired partition.

As we have proved in the above, $|D_1| \leq 1/k$ and $|D_2| > d - k/2$, if the current partition is not a desired partition, we now show how to modify the current partition into a desired partition. For this purpose, we consider subtree T_2. Since r is the only node with $d(r) \geq d - k/2$, T_2 must have at least two branches rooted at r and each of these branches contains at most $(d - k/2)$ destination nodes (including r if r is a destination node). Denote these branches by T_{2i} for $i = 1, 2, \cdots l$ and the set of destination nodes in T_{2i} by D_{2i}. Thus $|D_{2i}| \leq d - k/2$ for each i. In the following we consider two cases.

Case 1. There is a branch T_{2i} for some i with $|D_{2i}| > k/2$. In this case, we have $|D_{2i}| \leq d - k/2 \leq k$. We can repartition T in such a way that T_2 has only one branch T_{2i} while T_1 consists of the rest. Clearly, this produces a desired partition since T_1 is not empty but contains at most $(d - k/2 + 1 - (k/2 + 1)) \leq d - k \leq k/2$ destination nodes.

Case 2. Every branch contains at most $k/2$ destination nodes, i.e., $|D_{2i}| \leq k/2$ for $i = 1, 2, \cdots, l$. For ease presentation, we relabel T_1 as T_{20} and D_1 as D_{20}. Recall that D_0 has $d - 2/k + 1 > k/2 + 1$ destination nodes, thus there are at least two branches in $\{T_{20}, T_{21}, \cdots, T_{2l}\}$, say T_{2i} and T_{2j}, that contain distinct destination nodes in D_0. We repartition the current Steiner tree T by grouping T_{2i} with some other subtrees as T_1 and T_{2j} with the rest of subtrees as T_2 in such a way that T_1 and T_2 both contain at most k destination nodes. This can certainly be done since every subtree in $\{T_{20}, T_{21}, \cdots, T_{2l}\}$ contains at most $k/2$ destination nodes. It is easy to verify that the new partition is a desired partition. The proof is then finished.   □

Algorithm 7.4 *Steiner Tree Based Algorithm*

Step 0 Produce an α-approximation of SMT T of $\{s\} \cup D$.

   Consider T as a tree rooted at s, and set k-tree routing $R := \emptyset$.

Step 1 if a branch of T contains at most k destination nodes in D,

   then put it into R;

   else delete the edge incident to s from the branch

      and obtain a subtree.

   go to either **Step 2** or **Step 3**.

Step 2 For each subtree T' that contains more than $3k/2$ destination nodes, apply the partitioning process given in the proof of Lemma 7.6 to get two subtrees of T'.

   if it contains more than $3k/2$ destination nodes,

   then repeatedly apply the process to the subsubtree

   (When no further partition is needed, we obtain a set of subtrees each containing no more than $3k/2$ destination nodes.)

Step 3 For each produced subtree of branch T'' that is not a k-tree, apply the partitioning process given in the proof of Lemma 7.6 and get two subtrees of T'' each containing at most k destination nodes.

Step 4 For each obtained k-tree, find the destination node closest to source s in the tree and connect it to s via shortest path between them, that produces a k-tree rooted at s.

Step 5 Return k-tree routing R_{ST} that consists of all k-trees produced.

For the simplicity of presentation, we will call a resultant subtree in Step 2 a *type-1 k-tree* if it contains at most k destination nodes, and the resultant subtrees at the end of Step 3 *type-2 k-trees*.

Theorem 7.9   *Given a multicast connection request $< s, D >$ and $k > 2$, the Steiner tree based algorithm 7.4 returns a k-tree routing R_{ST} whose cost is at most $(2 + \alpha)$ times that of the optimal k-tree routing R_{opt}.*

Proof.   Let $R_{opt} = \{T_1^*, T_2^*, \cdots, T_m^*\}$ be an optimal k-tree routing, where each k-tree T_i^* has cost $c(T_i^*)$. Then $c(R_{opt}) = \sum_{i=1}^m c(T_i^*)$. Let d_j be a destination node in tree T_i^*, then the distance from d_j to s in graph G is no greater than the distance from d_j to s in T_i^*, i.e., $w(d_j, s) \leq c(p_{T_i^*}(d_j, s))$. Applying a similar argument used for k-path routing in the proof of Theorem 7.6, we have

$$\sum_{i=1}^{|D|} w(d_i, s) \leq k \sum_{i=1}^{|D|} w(P_i^*) = k \cdot c(R_{opt}). \tag{7.7}$$

Note that to prove the theorem it suffices to show that the total lengths of the shortest paths from source s to the destination node closest to s in each obtained k-tree in Step 3 is at most two times cost of R_{opt}.

First suppose that there are g type-1 k-trees, $T_1', T_2', \cdots, T_g'$, then for each k-tree T_i', let $k/2$ closest destination nodes to s be the representatives for T_i', and denote them by $d_{i,j}'$ for $j = 1, 2, \cdots, k/2$, which are labelled in the order of non-decreasing distance from source s.

Secondly, consider every pair of type-2 k-trees T_1'' and T_2''. If T_1'' (T_2'', respectively) contains no less than $k/2$ destination nodes, then let $k/2$ closest destination nodes to s be the representatives for T_i' (T_2'', respectively); otherwise it contains $l < k/2$ destination nodes, let these l nodes along with $(k/2 - l)$ destination nodes farthest to s in its sibling tree be the representatives. At the same time for its sibling k-tree, there are still $k/2$ destination nodes closest to s that can be selected to be its own representatives. Now each type-2 k-tree also has exactly $k/2$ representatives and no two k-trees share one common representative. Note that such an assignment of representatives to type-2 k-trees could be implemented since the total number of destination nodes in one pair of type-2 k-trees is greater than k. Suppose that there are h pairs of type-2 k-trees, $\{T_{11}, T_{12}\}, \{T_{21}, T_{22}\}, \cdots, \{T_{h1}, T_{h2}\}$, then denote their representatives by

$$\{d_{j,1}'', d_{j,2}'', \cdots, d_{j,k/2}''\}, \{d_{j,1}'', d_{j,2}'', \cdots, d_{j,k/2}''\}, \quad \text{for } j = 1, 2, \cdots, g,$$

which are labelled in the order of non-decreasing distance from source s. For easy presentation, for every pair of type-2 k-trees T_{i1} and T_{i2}, denote by $d_{i^*,1}'', d_{i^*,2}'', \cdots, d_{i^*,k/2}''$ the $k/2$ destination nodes closest to s among all destination nodes in both of them, and denote by $d_{i^*,k/2+1}'', d_{i^*,k/2+2}'', \cdots, d_{i^*,k}''$ the $k/2$ destination nodes farthest to s among all destination nodes in both of them.

It follows from inequality (7.7) that

$$\sum_{i=1}^{g}\sum_{j=1}^{k/2} c(d'_{i,j}, s) + \sum_{i=1}^{h}\sum_{j=1}^{k} c(d''_{i,j}, s) \le \sum_{i=1}^{|D|} c(d_i, s) \le k \cdot c(R_{opt}).$$

Using the non-decreasing order of these destination nodes with respect to their distance from s, from the above inequalities we deduce

$$\sum_{i=1}^{g} c(d'_{i,j}, s) + \sum_{i=1}^{h} \left(c(d''_{i,1}, s) + c(d''_{i^*,k/2+1}, s) \right) \le 2 \cdot c(R_{opt}).$$

Clearly, destination node $d'_{i,1}$ connects type-1 k-tree to source s. Note also that $d''_{i,1}$ must be a representative for either type-2 k-tree T_{i1} or T_{i2}, thus it is connected to source s via the shortest path. Suppose, without loss of generality, that $d''_{i,1}$ is a representative for T_{i1}, then the closest destination node $d_{i2,1}$ in T_{i2}, which is a representative of T_{i2}, has the distance to s no larger than $d''_{i,k/2+1}$'s distance to s, i.e., $c(d_{i2,1}, s) \le c(d''_{i,k/2+1}, s)$. Hence we have

$$\sum_{i=1}^{g} c(d'_{i,j}, s) + \sum_{i=1}^{h} \left(c(d''_{i,1}, s) + c(d''_{i,k/2+1}, s) \right) \le 2 \cdot c(R_{opt}),$$

which means the total length of paths added to connect the source to the obtained k-trees is at most $2 \cdot c(R^*_{opt})$. Therefore the total cost of the obtained k-tree routing is at most $(2 + \alpha)c(R^*_{opt})$. The proof is finished. □

7.4 Discussions

A more general version of Steiner k-tree routing problem is as follows.

Problem 7.3 *Capacitated Steiner Minimum Tree Problem*

Instance	An undirected graph $G(V, E)$ with a non-negative cost $l(e)$ on each edge $e \in E$, a root vertex $r \in V$ and a subset $D \subseteq V$, each vertex $v \in D$ needs to transmit $d(v)$ units of flow to r, and a positive integer k.
Solution	A sets of Steiner trees rooted at r that span all vertices in D and each tree carries at most k units of flow.
Objective	Minimizing the total length of Steiner trees.

When $D = V$, the above problem is the well-known *Capacitated Minimum Spanning Tree* (CMST) Problem, which finds an application in telecommunication network design [158]: When designing a minimum cost tree network, we need to consider how to instal expensive (e.g., fiber-optic) cables along its links. Each cable has a certain cost and a prespecified capacity on the amount of traffic flow it can handle. Every source node in the network has some demand that needs to be transmitted to the sink node. The objective is to construct a minimum cost tree network for simultaneous routing of all demands from the source nodes to the sink node.

Capacitated minimum Steiner tree problem is NP-hard since Steiner tree problem is a special case of this problem with $k = \infty$. In fact, capacitated minimum spanning tree problem is NP-hard even when vertices have unit weights and $k = 3$ (but it is polynomial-time solvable for $k = 2$ [110]). Moreover, the geometric version of the problem, where costs of edges are the Euclidean distance between endpoints of them, remains NP-hard.

Jothi and Raghavachari [158] proved, among other results, that capacitated minimum Steiner and spanning tree problems admit a $(\rho^{-1}\alpha + 2)$-approximation and a $(\rho^{-1} + 2)$-approximation algorithms, respectively, where ρ is the Steiner ratio and α is the best achievable approximation ratio for Steiner tree problem. Their algorithm adopts the same basic idea as that of Algorithm 7.4 due to Cai et al. [42]: First construct an α-approximation of SMT T of $D \cup \{r\}$ and root T at the root vertex r. Next, prune subtrees of cost at most k in a bottom-up way and add edges to connect r to the closest node in each of the pruned subtrees.

In addition, we also notice that Steiner k-path and k-tree routing problems (Problem 7.1-2) are similar to the following two well-known problems, respectively.

Problem 7.4 *k-Travelling Salesman Problem* (k-TSP)

Instance   A graph $G(V, E)$ with a cost $l(e)$ on each edge $e \in E$, and a positive integer $k \leq |V|$.

Solution   A tour (closed path) T that contains k vertices and k edges, called k-tour.

Objective Minimizing the total length of the edges in T, $l(T) \equiv \sum_{e \in T} l(e)$.

Problem 7.5 *k-MST Problem* (or Minimum Weight k-Tree Problem)

Instance   A connected graph $G(V, E)$ with a cost $l(e)$ on each edge $e \in E$, and a positive integer $k \leq |V|$.

Solution   A tree T that interconnects k vertices, called k-tree.

Objective Minimizing the total length of the edges in T, $l(T) \equiv \sum_{e \in T} l(e)$.

Problem 7.5 is also known as *quorum-cast problem* [59]. It finds applications in data updating and replicated data management of distributed database systems.

Notice that in the above two problems, when the set of k vertices is determined, the shortest k-tour and the k-MST on that set are shortest tour and SMT of the set, respectively. In particular, when $k = |V|$, they are reduced to the travelling salesman problem and Steiner tree problem in graphs, respectively. Notice that unlike the classical travelling salesman problem and Steiner tree problem, the main difficulty arises from two tasks, one needs to determine not only the order in which to visit the cities (or edges to connect the vertices), but also which cities should be visited (or which points should be interconnected).

Arora and Karakostas [14] proposed a $(2 + \epsilon)$-approximation algorithm for k-MST problem, the ratio holds for both the rooted and unrooted versions of the problem. (In the rooted version there is a specified root vertex that must be in the tree produced.) Recently, Garg [112] improved the result by removing ϵ. His result also leads to a 2-approximation algorithm for k-travelling salesman problem. Their algorithms apply an *linear programming relaxation* and the *primal-dual approach*. For a geometrical version of Problem 7.4: V contains a set of points in the Euclidean plane where the cost of an edge is the Euclidean distance between its two endpoints, Arora [10] and Mitchell [214] independently proposed a polynomial time approximation scheme for this case. Their algorithms use techniques of *portals* and *guillotine cuts* that we discussed in Chapter 3.

Another problem similar to Problems 7.1-5 is *minimal k-broadcasting network problem* [131; 184]. *k-broadcasting* is a special scheme to disseminate a single message, originated at any node in a network, to all other nodes of the network by letting each informed node transmit the message to at most k neighbors simultaneously. A *minimal k-broadcast network* is a communication network in which k-broadcasting can be completed in minimum time from any node. An *optimal k-broadcast network* is a minimal k-broadcast network with the minimal number of edges.

In this chapter we have studied the multicast routing problem under multi-path and multi-tree routing models for minimum network cost. In fact, to establish a multicast connection, we need also to consider the *wavelength assignment problem*: each of generated k-paths (or k-trees) should be assigned a wavelength in such a way that two wavelengths are needed if any two of them share a common link in the network. In an all-optical

Wavelength Division Multiplexing (WDM) network without wavelength conversions (every link in the path or tree should be assigned the same wavelength), wavelength assignment is the key to guarantee the quality of service and to reduce communication costs. In Chapter 8, we will study this problem. Recently, Wang et al. [265] considered wavelength assignment problem for WDM multicast with two criteria: find a subset of available wavelengths for each link such that, (1) the maximum number of destinations can be reached, and (2) the wavelength cost is minimized under the condition of (1). They studied the computational complexity of the problem and proposed some heuristic algorithms.

Chapter 8

Steiner Tree Coloring Problem

*Steiner tree coloring problem*s arise from the application of multicast communications in all optical *Wavelength Division Multiplexing* (WDM) networks. In a WDM network, nodes interested in some particular data make a *multicast group*, which requires a *multicast connection* for sending data from its source(s) to its destinations. Given a set of multicast connection requests, two steps are needed to set up the connections, routing and wavelength assignment. *Multicast routing* is to connect all members in each multicast group with a tree, which is called *light-tree*. *Wavelength assignment* is to assign a wavelength to each of generated light-trees in such a way that no two trees sharing a common link are assigned the same wavelength. Since the number of wavelengths can be used in a WDM network is very limited, how to make a good use of wavelengths becomes very important. This motivates extensive studies on the problem of multicast routing and wavelength assignment in WDM networks.

Clearly, how to set up a multicast routing with minimal network cost in WDM networks is equivalent to the Steiner tree problem in graphs (Problem 2.1) or Steiner k-tree problem (Problem 4.1). There are two basic versions of Steiner tree coloring problem [52; 51], which consider not only multicast routing but also wavelength assignment.

(1) *Maximum tree coloring problem.* It aims at finding an optimal way of multicast routing and wavelength assignment to maximize the throughput, where *throughput* is the number of requests that can be accepted/satisfied given a prespecified number of wavelengths.

(2) *Minimum tree coloring problem.* It aims at finding an optimal way of multicast routing and wavelength assignment to accept/satisfy all requests with the minimal number of wavelengths.

221

The remainder of this chapter is organized as follows. In Sections 8.1 and 8.2 we will study maximum and minimum tree routing and coloring problems, respectively. For each problem, we shall first prove some inapproximability results for some graphs, and then some positive results as well by proposing some approximation algorithms for general and special graphs. In Section 8.3, we will conclude the chapter with some remarks.

8.1 Maximum Tree Coloring

In this section we study maximum tree coloring problem. We will first formulate the maximum tree coloring problem by introducing a few notations, and then we will present some inapproximability results for the problem in trees, meshes and tori. After that we will propose a greedy algorithm for the problem in general graphs and two approximation algorithms for the problem in two special graphs.

8.1.1 *Problem Formulation*

Let G be a graph with vertex set $V(G)$ with $|V(G)| = n$ and edge set $E(G)$ with $|E(G)| = m$, and $\mathcal{S} = \{S_1, S_2, \cdots, S_g\}$ be a set of g groups, where each S_i is a subset of $V(G)$. A tree interconnecting S_i is a tree of G with $S_i \subseteq V(T_i)$. A family $\mathcal{T} = \{T_1, T_2, \cdots, T_g\}$ of trees is said to be a *tree family* of $\mathcal{S} = \{S_1, S_2, \cdots, S_g\}$ if there is a permutation ρ on $\{1, 2, \cdots, g\}$ such that $T_{\rho(i)}$ is a tree interconnecting S_i for each $1 \leq i \leq g$. A *coloring* $\{(T_i, c_i) \mid i = 1, 2, \cdots, t\}$ of a tree family $\{T_1, T_2, \cdots, T_t\}$ is called *proper* if tree T_i $(1 \leq i \leq t)$ receives color $c_i \in \{1, 2, \cdots, c\}$ such that $c_i \neq c_j$ whenever $E(T_i) \cap E(T_j) \neq \emptyset$ for each i, where c is a positive integer and $\{1, 2, \cdots, c\}$ is the set of available colors. Clearly, tree family $\mathcal{S}$ has a proper tree coloring with $c \geq g$, and it may have a proper tree coloring with $c < g$. The *maximum tree coloring problem* is to find the tree family of a maximum subset of $\mathcal{S}$ that has a proper coloring, which can be formulated as follows.

Problem 8.1 *Maximum Tree Coloring Problem*

Instance	A graph $G(V, E)$, a family $\mathcal{S} = \{S_1, S_2, \cdots, S_g\}$ of subsets of V, and a set of c colors.
Solution	A proper tree coloring $\{(T_i, c_i) \mid i = 1, 2, \cdots, t\}$, where $\mathcal{T} = \{T_1, T_2, \cdots, T_t\}$ is a tree family of $\mathcal{S}$.
Objective	Maximize the size of tree coloring, denoted by $s(G, \mathcal{S}, c)$.

When every group has only two members, a tree interconnecting a group is simply a path connecting the two members. The maximum tree coloring problem in this case is commonly known as *maximum path coloring problem*, and has been extensively studied for several topologies, such as trees, rings and meshes. Maximum path coloring problem is NP-hard for all these three topologies [180; 260; 220], and is approximable within 1.58 in trees [260], within 1.5 in rings [221], and within $O(1)$ in 2-dimensional meshes [173; 260]. A simplified version of maximum path coloring problem assumes that the set of paths is prespecified [222]. For this version it was also proved NP-hard in general (but polynomial-time solvable when the graph is a chain) and inapproximable within m^δ for some $\delta > 0$ unless $NP = P$.

When the number of available colors is exactly one, the maximum tree coloring problem reduces to *maximum edge-disjoint Steiner tree problem*, and its special case mentioned above is referred to as the *maximum edge-disjoint path problem*. The standard greedy approaches [260] guarantee that if maximum edge-disjoint Steiner tree problem is approximable within r, then maximum tree coloring problem is approximable within $1/(1 - e^{-1/r})$. Nevertheless, even maximum edge-disjoint path problem seems hard to approximate: the current-best approximation guarantee is $(\sqrt{m}+1)$ achieved through greedy selection of shortest paths [174] .

In addition, both the decision and optimization versions of maximum tree coloring problem are closely related to *maximum k-colorable induced subgraph problem* and its special case, *maximum independent set problem*. The former is to find a maximum subset of $V(G)$ that is the union of k independent sets in G while the latter is to find an independent set of largest cardinality $\alpha(G)$, where an *independent set* of G is a set of pairwise nonadjacent vertices in $V(G)$ (in particular, a single vertex forms an independent set). Bellare et al. [27] proved that maximum independent set problem is inapproximable within $n^{1/4-\varepsilon}$ for any $\varepsilon > 0$ assuming $NP \neq P$. Since the faith in the hypothesis $NP \neq ZPP^1$ is almost as strong as $NP \neq P$, the following negative result from [132], as well as positive result from [129], explains the lack of progress on good approximation for maximum independent set problem and maximum k-colorable induced subgraph problem.

Theorem 8.1 (i) *Maximum independent set problem is inapproximable within $n^{1-\varepsilon}$ for any $\varepsilon > 0$, unless $NP = ZPP$.* (ii) *Maximum k-colorable induced subgraph problem is approximable within $O(n(\log\log n/\log n)^2)$.*

---

[1]ZPP is the class of problems that can be solved in expected polynomial time by a probabilistic algorithm that never makes an error, i.e. only the running time is stochastic.

8.1.2   *Inapproximability Analysis*

In this subsection, we shall show that maximum tree coloring problem is as hard as maximum independent set problem. Then our inapproximability results follow from Theorem 8.1 one way or another. Roughly speaking, we assume existence of an r-approximation algorithm A for maximum tree coloring problem, and use A to design an r-approximation algorithm A' for maximum independent set problem.

Given a graph G, the set of edges in $E(G)$ incident with a vertex $v \in V(G)$ is denoted by $\delta(v)$. A tree family $\mathcal{T} = \{T_1, T_2, \cdots, T_g\}$ in graph G is usually associated with its *intersection graph* $G_\mathcal{T}$ with vertex set $V(G_\mathcal{T}) = \{v_1, v_2, \cdots, v_g\}$ and edge set $E(G_\mathcal{T}) = \{v_i v_j \mid T_i \text{ and } T_j \text{ share at least one edge in } G\}$. Clearly, an independent set S in $G_\mathcal{T}$ (resp. the union S of c independent sets in $G_\mathcal{T}$) corresponds to a set $\{T_i \mid v_i \in S\}$ of edge-disjoint trees in G (resp. a set $\{T_i \mid v_i \in S\}$ of trees in G that admits a coloring using colors in $\{1, 2, \cdots, c\}$).

When the underlying graph is a tree, the tree interconnection group S_i is uniquely determined by the group members in S_i. The first inapproximability result [52] is for the maximum tree coloring problem in star graphs. A *star graph* is a tree with at most one vertex (called *center*) of degree greater than one.

Theorem 8.2   *The maximum tree coloring problem in trees is inapproximable within* $\max\{g^{1-\varepsilon}, m^{1/2-\varepsilon}\}$ *for any* $\varepsilon > 0$, *unless* $NP = Z\mathbb{P}P$.

Proof.   Suppose for a contradiction that for some $\varepsilon > 0$, there is an approximation algorithm A with ratio $\max\{g^{1-\varepsilon}, m^{1/2-\varepsilon}\}$ for the maximum tree coloring problem in trees. Consider an arbitrary graph H with $V(H) = \{v_1, v_2, \cdots, v_n\}$ and $E(H) = \{e_1, e_2, \cdots, e_m\}$. We construct a star graph G with $m + 1$ vertices and m edges by setting $V(G) := \{a, b_1, b_2, \cdots, b_m\}$ and $E(G) := \{ab_i \mid i = 1, 2, \cdots, m\}$, where a is the center. Define $\mathcal{S} := \{S_1, S_2, \cdots, S_n\}$ by $S_i := \{a\} \cup \left(\cup_{e_j \in \delta(v_i)} b_j\right)$, $1 \leq i \leq n$. Let T_i be the unique tree in G interconnecting S_i for each $1 \leq i \leq n$. Then $\mathcal{T} = \{T_1, T_2, \cdots, T_n\}$ is the unique tree family of $\mathcal{S}$, and $G_\mathcal{T} = H$. Now algorithm A' runs A on the instance $(G, \mathcal{S}, 1)$ and outputs $\{v_i \mid$ algorithm A outputs $T_i\}$. It is easy to see that A' is an approximation algorithm for maximum independent set problem and has performance ratio, $\max\{g^{1-\varepsilon}, m^{1/2-\varepsilon}\} = \max\{n^{1-\varepsilon}, m^{1/2-\varepsilon}\} = n^{1-\varepsilon}$, the same as that of A, a contradiction to Theorem 8.1. The proof is then finished. $\square$

Notice that the star graph used in the above reduction has a center of a very large degree. A natural question is: whether low degrees make maximum tree coloring problem easier? The following inapproximability result [52] on 2-dimensional meshes and tori gives a negative answer.

Theorem 8.3 *The maximum tree coloring problem in meshes (tori) is inapproximable within* $\max\{g^{1-\varepsilon}, \frac{1}{3}m^{1/4-\varepsilon}\}$ *for any* $\varepsilon > 0$, *unless* $NP = ZPP$.

We will only prove the case of meshes since the proof for tori is similar. For graph H with $V(H) = \{v_1, v_2, \cdots, v_n\}$ and $E(H) = \{e_1, e_2, \cdots, e_m\}$, we define groups $\mathcal{S} = \{S_1, S_2, \cdots, S_n\}$ on a $(5m \times 5m)$-mesh G as follows. Assume the vertices in G are labelled as in the Cartesian plane with its corners located at $(0,0)$, $(0, 5m - 1)$, $(5m - 1, 0)$, and $(5m - 1, 5m - 1)$, respectively. Associate each edge e_j $(1 \leq j \leq m)$ in H with two vertex sets in G: $R_j := \{(\ell, 5j - k) \mid \ell = 0, 1, \cdots, 5m - 1; k = 1, 2, 3, 4, 5\}$ and $R'_j := \{(5j - k, \ell) \mid \ell = 0, 1, \cdots, 5m - 1; k = 1, 2, 3, 4, 5\}$. Notice that R_j (resp. R'_j) consists of vertices located on five consecutive rows (resp. columns) of G, and satisfies the following two properties:

Each of $\{R_1, \cdots, R_m\}$ and $\{R'_1, \cdots, R'_m\}$ is a partition of $V(G)$, (8.1)

$R_j \cap R'_k$ induces a $(5 \times 5)-$ submesh G_{jk} of G, for $1 \leq j, k \leq m$. (8.2)

Now corresponding to vertex v_i in G, the i-th group

$$S_i := \bigcup_{e_j \in \delta(v_i)} (R_j \cup R'_j) \tag{8.3}$$

in G is defined as the union of $R_j \cup R'_j$ for all e_j incident with v_i. See Fig.8.1-2 for an example.

Lemma 8.1 *Let* $\mathcal{S} = \{S_1, S_2, \cdots, S_n\}$ *consist of n groups in $(5m \times 5m)$-mesh G as defined in (8.3), and let* $\mathcal{T}' = \{T'_1, T'_2, \cdots, T'_n\}$ *be a tree family in G such that T'_i is a tree interconnecting S_i, $1 \leq i \leq n$. Then*
(i) *there is a tree family* $\mathcal{T} = \{T_1, T_2, \cdots, T_n\}$ *such that T_i is a tree spanning S_i, $1 \leq i \leq n$, and the intersection graph of $\mathcal{T}$ is H; and*
(ii) *there does not exist distinct $i, j, k \in \{1, 2, \cdots, n\}$ such that $v_i v_j \in E(H)$ and T'_i, T'_j, T'_k are pairwise edge-disjoint in G.*

Proof. To justify claim (i), let us first construct a tree family $\mathcal{T} = \{T_1, T_2, \cdots, T_n\}$ so that each T_i $(1 \leq i \leq n)$ is a tree obtained from its vertex set $V(T_i) := S_i$ by two steps. In the first step, for every e_j incident with v_i, we add five rows each connecting all vertices in $\{(\ell, 5j - k) \mid \ell = 0, 1, \cdots, 5m - 1\}$, $1 \leq k \leq 5$. Then the horizontal edges on the five rows span R_j. Summing over all $e_j \in \delta(v_i)$, in total $5|\delta(v_i)|$ rows are added. In the second step, we use vertical edges with both ends in R_j' for some $e_j \in \delta(v_i)$ to connect the $5|\delta(v_i)|$ rows and the rest vertices in S_i under the condition that the resulting graph is a tree. (Though there are many possible T_i's, it is not a hard task to pick any one of them. Fig.8.3 gives an illustration for the example depicted in Fig.8.1-2.)

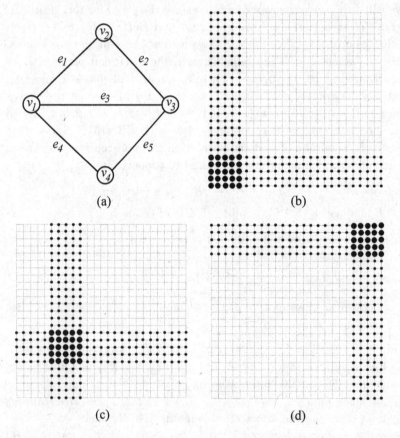

Fig. 8.1   Reduction: (a) Graph G, (b) $R_1 \cup R_1'$, (c) $R_2 \cup R_2'$, and (d) $R_5 \cup R_5'$.

By the construction, it suffices to show that the intersection graph G_T of T is identical with H. Indeed, for every edge $v_h v_i = e_j$ in H, trees T_h and T_i in G share common edges on the rows that span R_j. On the other hand, for every pair of nonadjacent vertices v_h and v_i in H, since $\delta(v_h) \cap \delta(v_i) = \emptyset$, we deduce from (8.1) that $R_j \cap R_k = \emptyset = R_j' \cap R_k'$ for all

$e_j \in \delta(v_h), e_k \in \delta(v_i)$. Therefore, combining the definitions of S_h and S_i (recall property (8.3) and the constructions of T_h and T_i we see that T_h and T_i shares neither a common horizontal edge nor a common vertical edge. In other words, T_h and T_i are edge-disjoint. Thus $G_{\mathcal{T}} = H$ as desired.

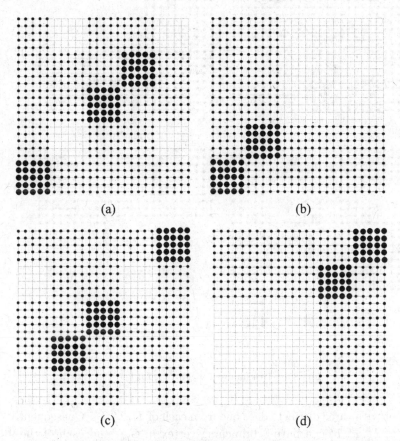

(a)   (b)

(c)   (d)

Fig. 8.2 Reduction: (a) $S_1 = \bigcup_{i=1,3,4}(R_i \cup R'_i)$, (b) $S_2 = \bigcup_{i=1,2}(R_i \cup R'_i)$, (c) $S_3 = \bigcup_{i=2,3,5}(R_i \cup R'_i)$, and (d) $S_4 = \bigcup_{i=4,5}(R_i \cup R'_i)$.

We now prove claim (ii). Suppose on the contrary that $v_i v_j = e_p \in E(H)$ and T'_i, T'_j, T'_k are pairwise edge-disjoint. Since $e_p \in \delta(v_i) \cap \delta(v_j)$, by property (8.3), both T'_i and T'_j contain $(R_p \cup R'_p) \subseteq S_i \cap S_j$. Take $e_q \in \delta(v_k)$. Obviously $e_p \neq e_q$. Recalling property (8.2), we have a (5×5)- submesh G_{pq} in G induced by $R_p \cap R'_q$. Note that the 25 vertices of G_{pq} are all contained in $S_i \cap S_j \cap S_k \subseteq V(T'_i) \cap V(T'_j) \cap V(T'_k)$, and hence every vertex in G_{pq} is

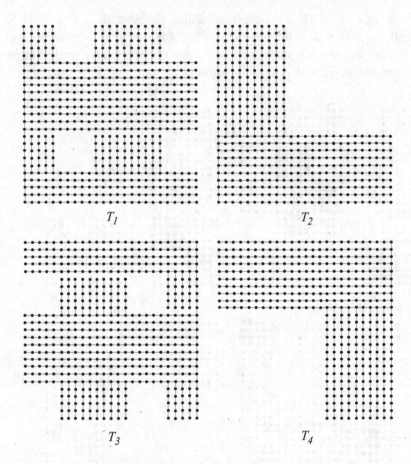

Fig. 8.3   Tree family. $\mathcal{T} = \{T_1, T_2, T_3, T_4\}$.

incident with three distinct edges one from each of T'_i, T'_j, T'_k. Consequently none of T'_h, T'_i, T'_j can have a branching vertex in G_{pq}, and each of the 9 internal vertices of G_{pq} is a leaf of at least two of T'_i, T'_j, T'_k. Therefore, there are in total at least 18 different paths in $T'_i \cup T'_j \cup T'_k$ connecting these leaves to the boundary of G_{pq} because every of T'_i, T'_j, T'_k has vertices outside G_{pq}. Two of those paths must have a common edge in G_{pq} as G_{pq} has only 16 boundary vertices. The two different paths are contained in exactly one tree in $\{T'_h, T'_i, T'_j\}$. It follows that this tree has a branching vertex in G_{pq}. The contradiction establishes claims (ii). The proof is then finished.   □

Proof of Theorem 8.3. Suppose that for some $\varepsilon > 0$, there exists an approximation algorithm A for the maximum tree coloring problem in 2-dimensional meshes with performance ratio $\max\{g^{1-\varepsilon}, \frac{1}{3}m^{1/4-\varepsilon}\}$. By Theorem 8.1, it suffices to present a polynomial time algorithm A' which always finds an independent set of size at least $\alpha(H)/n^{1-\varepsilon}$ in any given graph H on n vertices. If H is a complete graph, then A' returns an arbitrary vertex of H. So we assume that $\alpha(H) \geq 2$, and by examining all pairs of vertices in H, A' can find two nonadjacent vertices v_1 and v_2 in H in square time,

Suppose $V(H) = \{v_1, v_2, \cdots, v_n\}$ and $E(H) = \{e_1, e_2, \cdots, e_{m'}\}$, respectively. Let groups $\mathcal{S} = \{S_1, S_2, \cdots, S_n\}$ on a $(5m' \times 5m')$-mesh G be defined for H as in (8.3). Then the number of the edges in G is $m = 10m'(5m' - 1) < 50n^4$, and by Lemma 8.1(i) there is a tree family $\mathcal{T} = \{T_1, \cdots, T_n\}$ in G such that T_i interconnects S_i $(1 \leq i \leq n)$ and $G_{\mathcal{T}} = H$. This implies that there is a subset $\mathcal{S}$ of $\mathcal{T}$ consisting of $\alpha(H)$ pairwise edge-disjoint trees. Observe that $\mathcal{S}$ is a solution to the instance $(G, \mathcal{S}, 1)$ of the maximum tree coloring problem with only one color available. Hence the optimal value $s_{opt}(G, \mathcal{S}, 1) \geq |\mathcal{S}| = \alpha(H)$. Now running algorithm A on $(G, \mathcal{S}, 1)$, algorithm A' yields a solution $\{T'_{i_1}, T'_{i_2}, \cdots, T'_{i_\beta}\}$ consisting of β pairwise edge-disjoint trees interconnecting multicast groups $S_{i_1}, S_{i_2}, \cdots, S_{i_\beta}$. As a result, algorithm A' outputs $S = \{v_1, v_2\}$ if $\beta = 2$ and $S = \{v_{i_1}, v_{i_2}, \cdots, v_{i_\beta}\}$ if $\beta \geq 3$. Notice that S is an independent set in G of size β (recall Lemma 8.1(ii)). Moreover, we have

$$\frac{\alpha(H)}{|S|} \leq \frac{s_{opt}(G, \mathcal{S}, 1)}{\beta} = \frac{s_{opt}(G, \mathcal{S}, 1)}{s_A(G, \mathcal{S}, 1)}$$

$$\leq \max\left\{g^{1-\varepsilon}, \frac{1}{3}m^{1/4-\varepsilon}\right\} = \max\left\{n^{1-\varepsilon}, \frac{1}{3}m^{1/4-\varepsilon}\right\} = n^{1-\varepsilon},$$

which shows that $|S|$ approximates $\alpha(H)$ within $n^{1-\varepsilon}$. It follows that A' is an $n^{1-\varepsilon}$-approximation algorithm for maximum independent set problem. The proof is then finished. □

8.1.3 *Approximation Algorithms*

In this subsection, we will first propose a simple greedy algorithm for the maximum tree routing problem in general graphs, and then two approximation algorithms for the problem in trees and rings.

8.1.3.1 *For General Graphs*

The main philosophy of our greedy strategy is to produce trees of fewer edges whenever possible. This is based on a natural intuition: a tree of fewer edges potentially has more chances to use the same color with others, and therefore coloring more trees.

In order to carry out the greedy strategy, it is worth noting that finding a tree for each given group S_i with minimal number of edges is the Steiner tree problem in graphs (Problem 2.1). The current best approximation algorithm for this problem has performance ratio of 1.55 [238].

For a given instance $(G, \mathcal{S}, c)$ of maximum tree coloring problem, the implementation of greedy strategy consists of a number of iterations. In the $(i + 1)$-th iteration, set $\mathcal{S}_i$ contains all currently unrooted multicast groups. For every $j = 1, 2, \cdots, c$, let G_j^{i+1} be the subgraph of G obtained by removing all edges in the trees already colored with color j. Clearly, G_j^{i+1} contains a Steiner tree of S, for every $S \in \mathcal{S}_i$ whose connection can be established using color j. Subsequently, for every such S, compute an α-approximate Steiner Minimum Tree (SMT) of S in G_j^{i+1}; all these α-approximations are put into a set $\mathcal{T}_i$ (Steps 5-7). When all js have been considered, every group in $\mathcal{S}_i$ whose connection can be established has at least a tree of $\mathcal{T}_i$, and every tree in $\mathcal{T}_i$ can be colored with an appropriate color. If $\mathcal{T}_i = \emptyset$, then no more connection can be established and the algorithm terminates; else among all produced trees in $\mathcal{T}_i$, the algorithm selects the one with the minimum number of edges and colors it with an appropriate color (Steps 9-10), and then proceeds to the next iteration.

Algorithm 8.1 *Greedy Tree Coloring*

---

(1) $i \leftarrow 0$, $\mathcal{C}_0 \leftarrow \emptyset$, $\mathcal{S}_0 \leftarrow \mathcal{S}$.

(2) **while** $\mathcal{S}_i \neq \emptyset$ **do begin**

(3) $\quad$ $\mathcal{T}_i \leftarrow \emptyset$

(4) $\quad$ **For** $1 \leq j \leq c$ **do**

(5) $\quad\quad$ **while** G_j^{i+1} contains a tree spanning $S \in \mathcal{S}_i$ **do begin**

(6) $\quad\quad\quad$ $\mathcal{T}_i \leftarrow \mathcal{T}_i \cup \{\alpha\text{-approximate SMT of } S \text{ in } G_j^{i+1}\}$

(7) $\quad\quad$ **end-while**

(8) $\quad$ **If** $\mathcal{T}_i \neq \emptyset$ **then**

(9) $\quad\quad$ Pick $T_{i+1} \in \mathcal{T}_i$ and $j \in \{1, 2, ..., c\}$ such that

$$|E(T_{i+1})| = \min_{T \in \mathcal{T}_i} |E(T)| \text{ and } E(T_{i+1}) \cap \left(\bigcup_{T:(T,j) \in \mathcal{C}_i} E(T) \right) = \emptyset$$

(10) $\quad\quad$ $\mathcal{C}_{i+1} \leftarrow \mathcal{C}_i \cup \{(T_{i+1}, j)\}$, $\mathcal{S}_{i+1} \leftarrow \mathcal{S}_i - \{S \mid T_{i+1} \text{ spans } S\}$

(11) $\quad$ **else** $\mathcal{S}_{i+1} \leftarrow \emptyset$

(12) $\quad$ $t \leftarrow i$, $i \leftarrow i + 1$

(13) **end-while**

(14) **return** $\mathcal{C} \leftarrow \mathcal{C}_t$

Theorem 8.4 *Greedy tree coloring algorithm 8.1 returns a $(\sqrt{\alpha \cdot m} + 1)$-approximate solution to maximum tree coloring problem.*

Proof. Clearly, greedy tree coloring algorithm 8.1 terminates after at most g iterations (Steps 2-14) since $|\mathcal{S}_{i+1}| \leq |\mathcal{S}_i| - 1$ for every i (see Steps 10-12). Additionally, Step 11 implies that at most one tree is output for one group. The correctness follows from Steps 4-11 which guarantee inductively that every $\mathcal{C}_i$ $(1 \leq i \leq t)$ is a solution to $(G, \mathcal{S}, c)$.

We now turn to estimate the performance ratio of greedy tree coloring algorithm. It is obvious that $t \geq \min\{c, g\}$. If $c \geq g$, then the algorithm solves $(G, \mathcal{S}, c)$ optimally. So we assume $c < g$ and therefore $t \geq c$. Consider an optimal solution to $(G, \mathcal{S}, c)$ and let $\mathcal{T}_{opt}$ consist of the trees of the optimal solution for the groups unrouted by the algorithm. Therefore, we have

$$s_{opt}(G, \mathcal{S}, c) \leq |\mathcal{T}_{opt}| + t, \text{ and} \tag{8.4}$$

$$\text{no } (c+1) \text{ trees in } \mathcal{T}_{opt} \text{ can share the same edge in } G. \tag{8.5}$$

If $\mathcal{T}_{opt} = \emptyset$, then $t = s_{opt}(G, \mathcal{S}, c)$ and we are done. So we assume $\mathcal{T}_{opt} \neq \emptyset$ and consider an arbitrary $T \in \mathcal{T}_{opt}$. Suppose that T is a tree interconnecting group S. Observe that in Step 11 S is contained in every of $\mathcal{S}_1, \mathcal{S}_2, \cdots, \mathcal{S}_t$. Since $\mathcal{S}_{t+1}$ must be empty (otherwise the algorithm should return at least $(t+1)$ trees), by Step 5, we have

$$T \nsubseteq G \setminus \left(\bigcup_{(T', j) \in \mathcal{C}_t} E(T') \right) \quad \text{for every } 1 \leq j \leq c.$$

On the other hand, it is clear that $\mathcal{C}_{c-1}$ does not use color k for some $k \in \{1, 2, \cdots, c\}$, so the first c rounds of **while**-loop (Steps 2-14) always find

$$T \subseteq G \setminus \left(\bigcup_{(T', k) \in \mathcal{C}_{h-1}} E(T') \right) = G \quad \text{for } 1 \leq h \leq c.$$

Hence we may take an integer $i(T) \geq c$ such that

$$T \subseteq \bigcup_{j=1}^{c} \left(G \setminus \left(\bigcup_{(T', j) \in \mathcal{C}_{h-1}} E(T') \right) \right) \text{ for each } 1 \leq h \leq i(T), \text{ and} \tag{8.6}$$

$$E(T) \bigcap \left(\bigcup_{(T', j) \in \mathcal{C}_{i(T)}} E(T') \right) \neq \emptyset \text{ for each } 1 \leq j \leq c. \tag{8.7}$$

Subsequently, for every tree $T \in \mathcal{T}_{opt}$, we can find an $h(T) \leq i(T)$ and *charge* T to a common edge $e_{h(T)}^T \in E(T) \cap E(T_{h(s)})$ of T and the tree $T_{h(T)}$

in a way that no two trees in $\mathcal{T}_{opt}$ are both charged to the same edge of the same tree in $\{T_1, T_2, \cdots, T_t\}$, i.e.,

$$\text{either } e^T_{h(T)} \neq e^{T'}_{h(T')} \text{ or } h(T) \neq h(T'), \text{ for any distinct } T, T' \in \mathcal{T}_{opt}. \ (8.8)$$

To establish such a correspondence between trees in $\mathcal{T}_{opt}$ and edges in $T_1, T_2, \cdots, T_t$, we shall make use of a matching in a bipartite graph H as follows: the vertex set of H is the disjoint union of independent set X and independent set Y for which $X := \mathcal{T}_{opt}$ consists of $|\mathcal{T}_{opt}|$ vertices one for a tree in $\mathcal{T}_{opt}$, while $Y := \{e_i \, | \, e \in E(T_i), i = 1, 2, \cdots, t\}$ is considered a multi-set of size $\sum_{i=1}^{t} |E(T_i)|$ so that every edge $e \in E(G)$ has a number of copies e_i in Y, each carrying an index i if and only if $e \in E(T_i)$. The edge-set of H contains an edge joining $T \in X$ and $e_h \in Y$ if and only if $h \leq i(T)$ and $e \in E(T) \cap E(T_h)$ in G. Since, by Step 10, $\mathcal{C}_{i(S)}$ does not route any of $T_{i(S)+1}, \cdots, T_t$, it follows from (8.7) that in H every $T \in X$ has at least c neighbors in Y. For any $X' \subseteq X$, we use $N(X')$ to denote the set of vertices in $H \setminus X'$ each having a neighbor in X'. Clearly, $N(X') \subseteq Y$. If $|N(X')| < |X'|$ for some $X' \subseteq X$, then there exists $e_i \in N(X')$ which has at least $(c+1)$ neighbors in X', so these $(c+1)$ neighbors are $(c+1)$ trees in $\mathcal{T}_{opt}$ sharing the same edge $e \in E(G)$, contradicting (8.5). Thus $|N(X')| \geq |X'|$ for every $X' \subseteq X$, and Hall's theorem [127] guarantees the existence of a matching in H that saturates every $T \in X$. Suppose e_h is the neighbor of T in this matching. We then define $h(S) := h$ and $e^S_{h(T)} := e$. From the structure of H, it is easy to see that (8.8) is satisfied.

Furthermore, since $h(T) \leq i(T)$, by (8.6) and by Steps 5-7, $T_{h(T)-1}$ contains a tree T' interconnecting S with $|E(T')| \leq \alpha |E(T)|$. In turn, from the choice of $T_{h(T)} \in \mathcal{T}_{h(T)-1}$ made in Step 9, we deduce that $|E(T_{h(T)})| \leq |E(T')| \leq \alpha |E(T)|$. Thus $|E(T)| \geq \frac{1}{\alpha} |E(T_{h(T)})|$ holds for every $T \in \mathcal{T}_{opt}$. Let $s_i := |\{T \, | \, T \in \mathcal{T}_{opt}, h(T) = i\}|$ denote the number of trees in $\mathcal{T}_{opt}$ that are charged to edges of T_i, $1 \leq i \leq t$, then $\sum_{i=1}^{t} s_i = |\mathcal{T}_{opt}|$, and by (8.8), $s_i \leq |E(T_i)|$. Recall from (8.5) that the total number of edges of all trees in $\mathcal{T}_{opt}$ does not exceed $c \cdot m$. This yields

$$c \cdot m \geq \sum_{T \in \mathcal{T}_{opt}} |E(T)| = \sum_{i=1}^{t} \sum_{T \in \mathcal{T}_{opt}, h(T)=i} |E(T)|$$

$$\geq \sum_{i=1}^{t} \sum_{T \in \mathcal{T}_{opt}, h(T)=i} \frac{1}{\alpha} |E(T_{h(T)})| = \frac{1}{\alpha} \sum_{i=1}^{t} |E(T_i)| s_i \geq \frac{1}{\alpha} \sum_{i=1}^{t} s_i^2.$$

Combining this with $t \geq c$, we have

$$|\mathcal{T}_{opt}|/t = \Big(\sum_{i=1}^{t} s_i\Big)/t \leq \sqrt{\Big(\sum_{i=1}^{t} s_i^2\Big)/t} \leq \sqrt{\alpha \cdot c \cdot m/t} \leq \sqrt{\alpha \cdot m}.$$

This, together with (8.4), yields the desired performance ratio $s_{opt}(G, \mathcal{S}, c)/t \leq \sqrt{\alpha \cdot m} + 1$. The proof is then finished. $\qquad\square$

One of the techniques used in the above proof [52] borrows an idea used in the work of [174] for maximum edge-disjoint path problem. However, combining those two approaches in [174] and [260] can only lead to a performance ratio of $1/(1 - e^{-1/(\sqrt{\alpha \cdot m}+1)})$ for the greedy tree coloring algorithm, which is greater than $(\sqrt{\alpha \cdot m} + 1)$ obtained in Theorem 8.4.

8.1.3.2 *For Special Graphs*

When the given graph G is a tree, the tree family $\mathcal{T}$ for any set $\mathcal{S}$ of groups is unique, and the maximum tree coloring problem is reduced to coloring as many trees in $\mathcal{T}$ as possible using c colors. Clearly, the algorithm proposed by Halldórsson [129] for the maximum k-colorable induced subgraph problem on $G_{\mathcal{T}}$ carries over to the maximum tree coloring problem on $\mathcal{S}$, and has an approximation ratio $O(g(\log \log g / \log g)^2)$.

When the size of each group in $\mathcal{S}$ is upper bounded by a constant k (as for *Steiner k-tree problem*, i.e., Problem 4.1), the maximum degree of any tree in $\mathcal{T}$ is no more than k. We call such a tree family a *k-tree family*. Notice that the maximum tree coloring problem on k-tree family in trees is NP-hard even when $k = 2$ [260].

Theorem 8.5 *The maximum tree coloring problem in tree graphs is approximable within $1/(1 - e^{-1/k})$ for any given k-tree family.*

To prove the above theorem [52], we apply the idea of iterative application of an algorithm for computing a maximal set of edge-disjoint trees. The standard iterative method by [260] goes as follows: First run the algorithm on the tree family $\mathcal{T}$ to get a maximal set of edge-disjoint trees. All trees in this set are colored with color 1, and then removed from the current tree family. And then run the algorithm on the remaining tree family to find the maximal set of edge-disjoint trees and color them with a new color. Repeat this process until either no color can be used or no more tree is left uncolored. Wan and Liu [260] proved that if this algorithm is a k-approximation algorithm for computing a maximum set of edge-disjoint trees, then this iterative method provides a $1/(1 - e^{-1/k})$-approximation for maximum tree coloring problem.

To present the algorithm, we root the tree G at an arbitrary vertex r. The *level* of a vertex $v \in V(G)$ is defined as the length of the path from r to v. We use ℓ to denote the highest level of vertices in G. Let T be a tree in G, the root of T is the vertex in T that has the lowest level, and the *level* of T is equal to the level of its root.

Algorithm 8.2 *Disjoint Tree Iterating*

---

(1) $i \leftarrow \ell - 1, \mathcal{T}_d \leftarrow \emptyset$
(2) **While** $i \neq -1$ **do begin**
(3)   Find a maximal set $\mathcal{T}_i$ of edge-disjoint trees
             in $G \backslash \underset{T \in \mathcal{T}_d}{\cup} E(T)$ for each level i in G
(4) $\mathcal{T}_d \leftarrow \mathcal{T}_d \cup \mathcal{T}_i, i \leftarrow i - 1$
(5) **end-while**
(6) **return** $\mathcal{T}_d$

Proof of Theorem 8.5 It suffices to show that there are at most $k|\mathcal{T}_d|$ edge-disjoint trees in any k-tree family $\mathcal{T}$. Note that $\mathcal{T}_d$ is the disjoint union of $\mathcal{T}_0, \mathcal{T}_1, \cdots, \mathcal{T}_{\ell-1}$. Denote by $\mathcal{T}_{max}$ the subset of $\mathcal{T}$ consisting of a maximum number of edge-disjoint trees, and set $\mathcal{T}'_j := \{T \in \mathcal{T}_{max} \mid T$ is edge-disjoint from every tree in $\cup_{i=j+1}^{\ell-1} \mathcal{T}_i$, and shares a common edge with some tree in $\mathcal{T}_j\}$, $\ell - 1 \geq j \geq 0$. Then the maximality in Step 3 implies $|\mathcal{T}_{max}| = \sum_{j=0}^{\ell-1} |\mathcal{T}'_j|$. Since every tree in $\mathcal{T}'_j$ is edge-disjoint from every tree in $\mathcal{T}_d$ of level higher than j, and shares a common edge with a tree in $\mathcal{T}_d$ of level j, every tree in $\mathcal{T}'_j$ has an edge that is incident with a vertex of level j and contained in a tree in $\mathcal{T}_j$. It is easy to see that $|\mathcal{T}'_j| \leq k|\mathcal{T}_j|$ for all $\ell - 1 \geq j \geq 0$. Thus we have $|\mathcal{T}_{max}| = \sum_{j=0}^{\ell-1} |\mathcal{T}'_j| \leq k \sum_{j=0}^{\ell-1} |\mathcal{T}_j| = k|\mathcal{T}_d|$. The proof is then finished.   □

When the given graph is a ring, a tree for a group is simply a path containing all vertices in the group. In this simple case, by matching pairs of groups that can be routed as two edge-disjoint paths, an easy extension of the algorithm due to Nomikos et al. [221] for *unicast* in rings can compute an 1.5-approximation for the maximum tree coloring problem in rings. Thus we have the following theorem.

Theorem 8.6 *The maximum tree coloring problem in rings is approximable within 1.5.*

8.2 Minimum Tree Coloring

In this section we study minimum tree coloring problem. We will first formulate the minimum tree coloring problem by introducing a few notations, and then we will present some inapproximability results for the problem in trees, meshes and tori. After that we will propose a greedy algorithm for the problem in general graphs

8.2.1 *Problem Formulation*

Let S be a set of groups in graph G and let T be a tree family over S, the *maximum load* of G with respect to T (or simply the *load* of T), denoted by L_T, refers to the maximum number of trees in T that share a common edge in G. The *minimum maximum load* of G is defined as $L(G, S) \equiv \min\{L_T \mid T$ is a tree family over $S\}$. Clearly, the minimum number of colors necessary for coloring any tree family of S is at least $L(G, S)$.

The *intersection graph* G_T of a tree family $T = \{T_1, T_2, \cdots, T_g\}$ in graph G is an undirected graph with vertex-set $V(G_T) = \{v_1, v_2, \cdots, v_g\}$ and edge-set $E(G_T) = \{v_i v_j \mid T_i$ and T_j share at least one edge in $G, 1 \leq i < j \leq g\}$. Evidently, a k-coloring of G_T gives rise to a coloring of T with no more than k colors, and vice versa. For an instance (G, S) of minimum tree coloring problem, the solution value is defined as the number of colors used to color the trees in the output tree family over S. The problem is to find the tree family for given set of groups that could be colored with minimal number of colors. The problem is more formally formulated as follows.

Problem 8.2 *Minimum Tree Coloring Problem*

Instance A graph $G(V, E)$, a family $S = \{S_1, S_2, \cdots, S_g\}$ of subsets of V.

Solution A proper tree coloring $\{(T_i, c_i) \mid i = 1, 2, \cdots, g\}$, where $T = \{T_1, T_2, \cdots, T_g\}$ is a tree family of S.

Objective Minimize the number of colors used for tree coloring, $c(G, S)$.

When every group has only two members, a tree over a group is simply a path connecting the two members, the problem in this case is known as *minimum path coloring problem*. This version has been extensively studied for several topologies, e.g., trees, rings, meshes, etc. In particular, the minimum path coloring problem is known to be NP-hard for all the three topologies [254; 116; 96], and is approximable within 4/3 for trees [218], within 2 for rings [231], and within poly$(\log \log m)$ for 2-dimensional meshes

[230]; but it is polynomial-time solvable when the underlying graph is a chain [124] or a bounded-degree tree [219].

In addition, minimum tree coloring problem is closely related to the *vertex coloring problem*. Given a graph H, a *k-coloring* of H is a function $\phi : V(H) \rightarrow \{1, 2, \cdots, k\}$ such that each color class $\{v \,|\, \phi(v) = i$ and $v \in V(H)\}$ contains no two adjacent vertices of H for each $1 \leq i \leq k$. We say that H is *k-colorable* if it admits a *k*-coloring. The *chromatic number* $\chi(H)$ is the minimum value of k for which H is *k*-colorable. The *vertex coloring problem* on H is to find a $\chi(H)$-coloring of H.

It was shown in [27] that the vertex coloring problem on graph H of n vertices cannot be approximated within $n^{\frac{1}{7}-\varepsilon}$ for any $\varepsilon > 0$ assuming $NP \neq P$. In addition, based on a little stronger assumption $NP \neq ZPP$, the following negative result [99] and positive results [128] prove the intractability of approximation for the vertex coloring problem, which will be used in our discussion.

Theorem 8.7   (i) *The vertex coloring problem on graph of n vertices is not approximable within $n^{1-\varepsilon}$ for any $\varepsilon > 0$, unless $NP = ZPP$, and* (ii) *it is approximable within $O(n(\log \log n)^2/(\log n)^3)$.*

8.2.2   *Inapproximability Analysis*

In this subsection, we will show that minimum tree coloring problem is as hard as the vertex coloring problem, and then deduce the inapproximability results directly from Theorem 8.7. We consider the minimum tree coloring problem for three special graph topologies including trees, meshes and tori, which occur in a variety of applications in computer communication networks.

To distinguish the graph in an instance of minimum tree coloring problem and that in an instance of the vertex coloring problem, we reserve symbol G for the former and symbol H for the latter. For ease of description, hereafter we assume $|V(H)| = g$ and $|E(H)| = m$. Let $v \in V(H)$, we use $\delta(v)$ to denote the set of edges in $E(H)$ incident with v.

The first inapproximability result [51] concerns with the minimum tree coloring problem in star graphs. A *star graph* is a tree with at most one node of degree greater than one, which is called the *center* of the star graph.

Theorem 8.8   *The minimum tree coloring problem in trees is not approximable within $\max\{g^{1-\varepsilon}, m^{\frac{1}{2}-\varepsilon}\}$ for any $\varepsilon > 0$, unless $NP = ZPP$.*

Proof. Given a graph H with vertex-set $V(H) = \{v_1, v_2, \cdots, v_g\}$ and edge-set $E(H) = \{e_1, e_2, \cdots, e_m\}$, the star graph G with $(m+1)$ vertices and m edges is defined by $V(G) := \{c, d_1, d_2, \cdots, d_m\}$ and $E(G) := \{(c, d_i) \mid i = 1, 2, \cdots, m\}$, and $\mathcal{S} := \{S_1, S_2, \cdots, S_g\}$ is defined by $S_i := \{c\} \cup \left(\bigcup_{e_j \in \delta(v_i)} d_j\right)$ for each $1 \leq i \leq g$. Let $\mathcal{T} = \{T_1, T_2, \cdots, T_g\}$ be the tree family over $\mathcal{S}$ such that T_i is the unique minimal tree in G over S_i for each $1 \leq i \leq g$. It is easy to verify that $G_{\mathcal{T}} = H$ and $c_{opt}(G, \mathcal{S}) = \chi(G_{\mathcal{T}}) = \chi(H)$.

Suppose that there is an r-approximation algorithm A for the vertex coloring problem, then A colors the vertices in $G_{\mathcal{T}}$, and therefore the corresponding trees in $\mathcal{T}$, with $c_A(G_{\mathcal{T}}) \leq r\chi(G_{\mathcal{T}}) = r c_{opt}(G, \mathcal{S})$ colors. Conversely, any r'-approximation algorithm A' for minimum tree coloring problem would color the trees in $\mathcal{T}$, and therefore the corresponding vertices in $G_{\mathcal{T}} = H$, with $c_B(G, \mathcal{S}) \leq r' c_{opt}(G, \mathcal{S}) = r'\chi(G_{\mathcal{T}}) = r'\chi(H)$. Thus under approximation-ratio-preserving reduction, the minimum tree coloring problem in star graphs is equivalent to the vertex coloring problem, and then the conclusion follows immediately from Theorem 8.7. The proof is then finished. $\square$

We now consider the minimum tree coloring problem in meshes and tori. As in the discussion of Section 8.1.2, let $k \geq 2$ be a positive integer. The 2-dimensional $(k \times k)$-mesh (resp. *torus*) is an undirected graph with vertex-set $\{(a_1, a_2) \mid a_1, a_2 \in \{0, \cdots, k-1\}\}$ and edge-set $\{(a_1, a_2)(b_1, b_2) \mid a_i = b_i$ and $a_j = b_j \pm 1$, for $\{i, j\} = \{1, 2\}\}$ (resp. $\{(a_1, a_2)(b_1, b_2) \mid a_i = b_i$ and $a_j \equiv b_j \pm 1 \pmod{k}$, for $\{i, j\} = \{1, 2\}\}$). We will only present the analysis for meshes since the same argument is applicable to tori unless otherwise noted.

Let H be an arbitrary graph without isolated vertices. Suppose $V(H) = \{v_1, v_2, \cdots, v_g\}$ and $E(H) = \{e_1, e_2, \cdots, e_m\}$, as in Section 8.1.2, we define the group set $\mathcal{S} = \{S_1, S_2, \cdots, S_g\}$ on a $(5m \times 5m)$-mesh G as

$$S_i := \bigcup_{e_j \in \delta(v_i)} \{(\ell, 5j - k), (5j - k, \ell) \mid \ell = 0, 1, \cdots, 5m - 1; k = 1, \cdots, 5\} \quad (8.9)$$

where $i = 1, 2, \cdots, g$, and we prove the the following lemma.

Lemma 8.2 *Let $\mathcal{S} = \{S_1, S_2, \cdots, S_g\}$ consist of g groups in $5m \times 5m$ mesh G as defined in (8.9). Then*
(i) *A tree family $\mathcal{T}$ over $\mathcal{S}$ can be computed in polynomial time such that*
   $G_{\mathcal{T}} = H$ *and* $L_{\mathcal{T}} = 2$; *and*

(ii) *there do not exist three distinct integers* $i, j, k \in \{1, 2, \cdots, g\}$ *such that* $v_i v_j \in E(H)$ *and* T_i', T_j', T_k' *are pairwise edge-disjoint, for any tree family* $\mathcal{T}' = \{T_1', T_2', \cdots, T_g'\}$ *in* G *where* T_i' *is a tree of* S_i *for each* g.

By a gap-preserving reduction (see [15] for an introduction to this concept) from the vertex coloring problem, we shall establish the following inapproximability threshold for the minimum tree coloring problem in meshes and tori [51].

Theorem 8.9   *The minimum tree coloring problem in 2-dimensional meshes (tori) is not approximable within* $g^{1-\varepsilon}$ *for any* $\varepsilon > 0$, *unless* $NP = ZPP$.

Proof.   The theorem will follow directly from Theorem 8.7 if we can prove that, given an instance of the vertex coloring problem on graph H, an instance $(G, \mathcal{S})$ of the minimum tree coloring problem in meshes can be constructed in polynomial time such that $\frac{1}{2}\chi(H) \leq c_{opt}(G, \mathcal{S}) \leq \chi(H)$.

We now show that the $(G, \mathcal{S})$ described in (8.9) is as desired. Since, by Lemma 8.2(i), $c_{opt}(G, \mathcal{S}) \leq \chi(G_{\mathcal{T}}) = \chi(H)$, it remains to exhibit a $2c_{opt}(G, \mathcal{S})$-coloring of H. Let $\{(T_1', c_1), (T_2', c_2), \cdots, (T_g', c_g)\}$ be an optimal solution in which T_i' is a tree in G over S_i and with color c_i for each $1 \leq i \leq g$. Considering the intersection graph $G_{\mathcal{T}'}$ of $\mathcal{T}' = \{T_1', T_2', \cdots, T_g'\}$, we deduce from Lemma 8.2(ii) that

$$v_i v_k \in E(G_{\mathcal{T}'}) \text{ or } v_j v_k \in E(G_{\mathcal{T}'}) \text{ for any distinct} \atop v_i, v_j, v_k \text{ such that } v_i v_j \in E(H) \text{ and } v_i v_j \notin E(G_{\mathcal{T}'}). \tag{8.10}$$

Note that $G_{\mathcal{T}'}$ has a $c_{opt}(G, \mathcal{S})$-coloring $\phi' : V \to \{1, 2, \cdots, c_{opt}(G, \mathcal{S})\}$ with $\phi'(v_i) = c_i$ for each $1 \leq i \leq g$. We now prove that graph H' with vertex-set $V(H') := V(H) = V(G_{\mathcal{T}'})$ and edge-set $E(H') = E(H) \cup E(G_{\mathcal{T}'})$ has a $2c_{opt}(G, \mathcal{S})$-coloring ϕ. If ϕ' is a proper coloring of H', then we are done since we can simply set $\phi := \phi'$; otherwise we can assume, without loss of generality, that $e_i = a_i b_i, i = 1, \cdots, \ell$, are all edges in $E(H) \setminus E(G_{\mathcal{T}'})$ with both ends assigned the same color in ϕ'. So $1 \leq \ell \leq m$, and $\phi'(a_i) = \phi'(b_i)$ for each $1 \leq i \leq \ell$. By (8.10), every vertex in $V(G_{\mathcal{T}'}) \setminus \{a_i, b_i\}$ is adjacent to a_i or b_i in $G_{\mathcal{T}'}$ and hence assigned by ϕ' a color different from $\phi'(a_i) = \phi'(b_i)$. It follows that all 2ℓ vertices $a_1, b_1, \cdots, a_\ell, b_\ell$ are distinct, $\ell \leq c_{opt}(G, \mathcal{S})$, and

$$\phi(v) \equiv \begin{cases} c_{opt}(G, \mathcal{S}) + i, & v = a_i, \text{ for some } 1 \leq i \leq \ell; \\ \phi(v), & v \in V(H) \setminus \{a_1, a_2, \cdots, a_\ell\}. \end{cases}$$

defines a $2c_{opt}(G, S)$-coloring of H' as claimed. Since H is a subgraph of H', ϕ is also a $2c_{opt}(G, S)$-coloring of H. The proof is then finished. □

Despite the same inapproximability threshold for trees (Theorem 8.8) and meshes (Theorem 8.9), the comparison of their proofs shows that a few more choices for routing in meshes/tori might bring a little bit of benefit in computing an approximate solution to the minimum tree coloring problem (though the benefit is negligible when g sufficiently large). Unfortunately, two popular routing strategies for the minimum tree coloring problem in meshes, the *shortest path tree* strategy and the *single path* strategy fail to exploit the benefit. Under the shortest path tree strategy, every group is connected by a tree, called *shortest path tree*, such that a distinguished group member, called *source*, is connected to every member in the group through a shortest path. Under the *single path strategy*, every tree over a group is a (single) path that spans all group members.

Theorem 8.10 *The minimum tree coloring problems in meshes under the shortest path tree and single path strategies are both equivalent to vertex coloring problem in terms of approximation-ratio-preserving reductions.*

Proof. To see the equivalence, it suffices to consider an arbitrary graph H with $V(H) = \{v_1, v_2, \cdots, v_g\}$ and $E(H) = \{e_1, e_2, \cdots, e_m\}$, and define a set $S = \{S_1, S_2, \cdots, S_g\}$ of g groups in a mesh G satisfying the properties
(a) H is the intersection graph of some tree family over S, and
(b) the intersection graph of any tree family over S contains H as a subgraph.

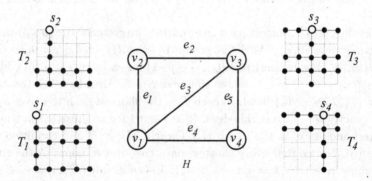

Fig. 8.4 Graph H and tree family $T = \{T_1, T_2, T_3, T_4\}$ under the shortest path tree routing.

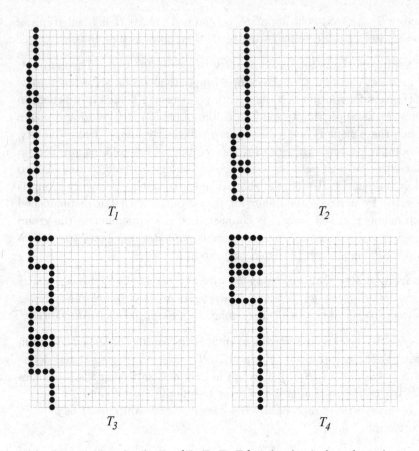

Fig. 8.5 Tree family $\mathcal{T} = \{T_1, T_2, T_3, T_4\}$ under the single path routing.

In the case of shortest path tree routing, we consider the $(p \times p)$-mesh G with $p = \max\{m, g\}$. We define groups by $S_i = \{(i-1, k) \mid k = 0, 1, \cdots, p-1\} \cup \{(h, j-1) \mid e_j \in \delta(v_i); h = 0, 1, \cdots, p-1\}$ for each $1 \leq i \leq g$. In addition, we set $s_i := (i-1, p-1)$ as the source of S_i. We then have the tree family $\mathcal{T} = \{T_1, T_2, \cdots, T_g\}$ in which each T_i is the shortest path tree over S_i that has exactly $(p-1)$ vertical edges. (See Fig.8.4 for an illustration, where T_i corresponds to v_i in the graph H.) Clearly, $G_{\mathcal{F}} = H$ and condition (a) is satisfied. Notice that every shortest path tree over S_i contains the column spanning $\{(i-1, k) \mid k = 0, 1, \cdots, p-1\}$. Hence for any $(v_h, v_i) = e_j \in E(H)$ with $h < i$, since $\{(h-1, j-1), (h, j-1), \cdots, (i-1, j-1)\} \subseteq V(S_h) \cap V(S_i)$, it is easy to see that every shortest path tree over S_h share at least one common edge with any shortest path tree over S_i. Thus condition (b) holds.

In the case of single path routing, we consider $(5m \times 5m)$-mesh G and define $S_i := \{(0, 5j - k) \mid e_j \in \delta(v_i); k = 1, 2, \cdots, 5\}$ for each $1 \leq i \leq g$. Then we have the tree family $\mathcal{T} = \{T_1, T_2, \cdots, T_g\}$ in which T_i is the path in G with vertex set $V(P_i) = S_i \cup \{(h, 5j - 5), (h, 5j - 1) \mid e_j \in \delta(v_i); h = 1, 2, \cdots, i\} \cup \{(i, 5j - k) \mid e_j \notin \delta(v_i); k = 1, 2, \cdots, 5\}$. See Fig.8.5. Observe that $G_{\mathcal{T}} = H$. Moreover, for any $v_h v_i = e_j \in E(H)$, any path over S_h and any path over S_i must share an edge incident with one of the five vertices $(0, 5j - 5), (0, 5j - 4), (0, 5j - 3), (0, 5j - 2), (0, 5j - 1)$ on the boundary of G since $V(S_h) \cap V(S_i)$ contains the five vertices and at least three of the five vertices are each incident with two edges from T_h and from T_i. Hence both conditions (a) and (b) are satisfied. The proof is then finished. □

We conclude this subsection with a brief discussion on the relationship between the minimum number of colors and the minimum maximum load. Based on the inequality $c_{opt}(G, \mathcal{S}) \geq L(G, \mathcal{S})$, a common approach (e.g. [24; 153]) to solve an instance $(G, \mathcal{S})$ of minimum tree coloring problem is to first lower the maximum load of the graph before coloring the trees. Nevertheless, the following theorem shows that it might not contribute much because the gap between $c_{opt}(G, \mathcal{S})$ and $L(G, \mathcal{S})$ can be arbitrarily large in general.

Theorem 8.11   *For any $\beta > 0$, there is an instance $(G, \mathcal{S})$ of the minimum tree coloring problem such that $c_{opt}(G, \mathcal{S}) \geq \beta \cdot L(G, \mathcal{S})$, where G can be a tree or a mesh or a torus.*

Proof.   Let H be a graph with $\chi(H) \geq 4\beta$. First let the star graph G and the group set $\mathcal{S}$ be constructed as in the proof of Theorem 8.8, then $L(G, \mathcal{S}) = 2$ while $c_{opt}(G, \mathcal{S}) = \chi(H) \geq 4\beta$. Second, let the mesh (torus) G, the group set $\mathcal{S}$, and the tree family $\mathcal{T}$ over $\mathcal{S}$ be as described in Lemma 8.2, then $L(G, \mathcal{S}) \leq L_{\mathcal{T}} = 2$, and by the proof of Theorem 8.10, we have $c_{opt}(G, \mathcal{S}) \geq \chi(H)/2 \geq 2\beta$. The proof is then finished. □

8.2.3   *Approximation Algorithms*

In this subsection, we first present a greedy algorithm for the minimum tree coloring problem in general graphs and analyze its performance, and then we study approximation algorithms for the problem in some special graph topologies including trees, tori, and rings.

8.2.3.1 *For General Graphs*

The main philosophy of the greedy strategy for minimal tree coloring problem is the same as that for maximum tree coloring problem (refer to Section 8.1.3): route trees using as few edges as possible. Under such a strategy, in order to save colors, one always tries to assign one color to as many groups as possible by constructing trees of fewer edges. It does not introduce a new color unless it has to. The algorithm iteratively finds a large maximal set of edge-disjoint trees over some currently unrouted groups, and assigns them a unique color. In the following algorithmic descriptions, $\mathrm{DIS\_TREES}(G, \mathcal{S}_i)$ is used as a subroutine that, given a set $\mathcal{S}$ of groups in graph G, returns a tree family $\mathcal{T}$ of edge-disjoint trees over a subset of $\mathcal{S}$. If $\mathcal{T}$ is a tree family over a subset of $\mathcal{S}$, then $\mathrm{GROUPS}(\mathcal{T}, \mathcal{S})$ denotes an arbitrary subset of $\mathcal{S}$ over which $\mathcal{T}$ is a tree family, and $E(\mathcal{T})$ stands for $\bigcup_{T \in \mathcal{T}} E(T)$. For convenience, we set $E(\emptyset) = \emptyset$.

Algorithm 8.3 *Greedy Tree Coloring*

---

(1) $i \leftarrow 0$, $\mathcal{B}_0 \leftarrow \emptyset$, $\mathcal{S}_0 \leftarrow \mathcal{S}$
(2) **while** $\mathcal{S}_i \neq \emptyset$ **do begin**
(3) $\mathcal{T}_i \leftarrow \mathrm{DIS\_TREES}(G, \mathcal{S}_i)$
(4) $\mathcal{S}_{i+1} \leftarrow \mathcal{S}_i \setminus \mathrm{GROUPS}(\mathcal{T}_i, \mathcal{S}_i)$
(5) $\mathcal{B}_{i+1} \leftarrow \mathcal{B}_i \cup \{(T, i+1) \mid T \in \mathcal{T}_i\}$ //assign $i+1$ to all trees in $\mathcal{T}_i$
(6) $i \leftarrow i + 1$
(7) **end-while**
(8) **return** $c \leftarrow i$ and $\mathcal{C} \leftarrow \mathcal{B}_c$

PROCEDURE DIS_TREES

(1) $\mathcal{T} \leftarrow \emptyset$
(2) **repeat**
(3) $\mathcal{S} \leftarrow \{\text{an } \alpha\text{-approx. of SMT of } S \text{ in } G \setminus E(\mathcal{T}) \mid S \in \mathcal{S} \setminus \mathrm{GROUPS}(\mathcal{T}, \mathcal{S})\}$
(4)   Take $T \in \mathcal{S}$ such that $|E(T)| = \min_{\dot{S} \in \mathcal{S}} |E(\dot{S})|$
(5) $\mathcal{T} \leftarrow \mathcal{T} \cup \{T\}$
(6) **until** $\mathcal{S} = \emptyset$
(7) **return** $\mathcal{T}$

Before proceeding to the performance analysis of greedy tree coloring algorithm 8.3, let us make some necessary preparations. First recall that greedy tree coloring algorithm 8.1, which is for maximum tree coloring problem, has approximation performance ratio bounded by $(\sqrt{\alpha \cdot m} + 1)$. It consists of a number of iterations. In the $(i+1)$-st iteration, set $\mathcal{S}_i$ contains all currently unrouted groups; for every $j = 1, 2, \cdots, c$, it obtains subgraph G_j^{i+1} of G by removing from G edges in the trees already colored with color j, and computes, for every $S \in \mathcal{S}_i$ whose connection can be established using color j, an α-approximate SMT over S in G_j^{i+1}; subsequently, all these α-approximations are put into a set $\mathcal{T}_i$; if $\mathcal{T}_i = \emptyset$, then the algorithm terminates, else

(∗) selects a tree T in $\mathcal{T}_i$ and an integer j with $T \subseteq G_j^{i+1}$ such that $|E(T)|$ *is minimum*, and colors T with j, and then proceeds to the next iteration.

It is obvious that for any $(G, \mathcal{S})$, the trees returned by procedure DIS_TREES on instance $(G, \mathcal{S})$ are exactly those returned by greedy tree coloring algorithm 8.1 on instance $(G, \mathcal{S}, 1)$. By this fact and the $(\sqrt{\alpha \cdot m} + 1)$-performance guarantee of greedy tree coloring algorithm 8.1, applying the techniques used in [21; 172] we can an approximation ratio of $\lceil \sqrt{\alpha \cdot m} + 1 \rceil (\log g + 1)$ for greedy tree coloring algorithm 8.3. Next, we improve this slightly by presenting a somewhat different analysis.

Let us consider an adaptation of greedy tree coloring algorithm 8.1, called algorithm 8.1', which does the same as the algorithm 8.1 except applies rule (∗') below in place of rule (∗).

(∗') Select a tree T in $\mathcal{T}_i$ and an integer j with $T \subseteq G_j^{i+1}$ such that j is *minimum, and subject to the minimality of j, $|E(T)|$ is minimum*, and color T with j, and then proceed to the next iteration.

Essentially the adaption of greedy tree coloring algorithm 8.1 has a pseudo-code description quite similar to that of greedy tree coloring algorithm 8.3.

Algorithm 8.4 *Adaption of Greedy Tree Coloring*

---

(1) $i \leftarrow 0$, $\mathcal{B}_0 \leftarrow \emptyset$, $\mathcal{S}_0 \leftarrow \mathcal{S}$.
(2) **while** $\mathcal{S}_i \neq \emptyset$ and $i \neq c + 1$ **do begin**
(3)   execute Steps 3-6 of greedy tree coloring algorithm 8.3
(4) **end-while**
(5) **return** $\mathcal{C} \leftarrow \mathcal{B}_i$

By similarity of the above algorithm 8.4 and greedy tree coloring algorithm 8.1, minor modification on the analysis of Algorithm 8.1 in Section 8.1.3 shows that Algorithm 8.4 remains a $(\sqrt{\alpha \cdot m} + 1)$-approximation algorithm for maximum tree coloring problem. In the following we include a short and a little different proof [51].

Lemma 8.3   *The adaption of greedy tree coloring algorithm 8.4 returns a $(\sqrt{\alpha \cdot m} + 1)$-solution to maximum tree coloring problem.*

Proof.   We may assume that $c < g$, $t \geq c$, and Algorithm 8.4 used all c colors. Consider an optimal solution to $(G, \mathcal{S}, c)$ and let $\mathcal{T}_{opt}$ consist of the trees in the optimal solution over groups unrouted by Algorithm 8.4. It suffices to show $|\mathcal{T}_{opt}|/t \leq \sqrt{\alpha \cdot m}$.

Notice that from rule $(*')$, for every tree T' in $\mathcal{T}_{opt}$ and every $j = 1, 2, \cdots, c$, T' shares a common edge with some tree $T \in \{T_1, \cdots, T_t\}$ such that $(T, j) \in \mathcal{C}$ and $|E(T)| \leq \alpha |E(T')|$. We employ similar construction to that used in the proof of Theorem 8.4. Let H be a bipartite graph with bipartition (X, Y) in which $X = \mathcal{T}_{opt}$ and $Y = \{e_i | e \in E(T_i), i = 1, 2, \cdots, t\}$, and an edge join $T' \in X$ and $e_i \in Y$ if and only if $e \in E(T')$ and $|E(T_i)| \leq \alpha |E(S)|$. Since every $T' \in X$ has at least c neighbors in Y, and no $e_i \in Y$ can have more than c neighbors in X, there is a matching in T' that matches every $T' \in X$ with an $e_{h(T')}^{T'} \in Y$ with $e^{T'} \in E(T') \cap E(T_{h(T')})$ and $|E(T')| \geq |E(T_{h(T')})|/\alpha$. Since $e_{h(T'')}^{T''} \neq e_{h(T')}^{T'}$ for distinct $T', T'' \in \mathcal{T}_{opt}$, we see that $|E(T_i)| \geq s_i := |\{T' | T' \in \mathcal{T}_{opt}, h(T') = i\}|$ holds for $i = 1, 2, \cdots, t$. Therefore, as in the proof of Theorem 8.4, we obtain the following two inequalities

$$c \cdot m \geq \frac{1}{\alpha} \sum_{i=1}^{t} s_i^2, \quad \text{and}$$

$$|\mathcal{T}_{opt}|/t \leq \sqrt{\alpha \cdot m}.$$

The proof is then finished.   □

We are now ready to estimate the approximation performance ratio of greedy tree coloring algorithm 8.3 [51].

Theorem 8.12   *Greedy tree coloring algorithm 8.3 for minimum tree coloring problem has approximation ratio at most $\lceil M \rceil (\log \lceil g/M \rceil + 2) + 2$, where $M = \sqrt{\alpha \cdot m} + 1$.*

Proof.   If $g \leq M$ or $m \leq 2$, then the algorithm uses at most M colors and has approximation ratio at most M. Thus we assume

$$g > M \geq \sqrt{\alpha \times 3} + 1. \tag{8.11}$$

For an instance (G, S) of minimum tree coloring problem, let $c^* = c_{opt}(G, S)$ denote the minimum number of colors needed, and let c denote the number of colors used by the algorithm, and $\mathcal{B}_\infty, \mathcal{B}_\in, \cdots, \mathcal{B}_\lrcorner$ denote the sets constructed in Step 5 of the algorithm.

Note that when the adaption of greed tree coloring algorithm 8.4 is applied on the instance (G, S_0, c^*), it returns $\bigcup_{i=0}^{c^*-1}\{(T, i+1) \mid T \in \mathcal{T}_i\} = \mathcal{B}_{c^*}$. Hence by Lemma 8.3, we have $\beta_1 := |\mathcal{B}_{c^*}| \geq g/M$. Subsequently, we see that there exists an execution of Algorithm 8.4 on the instance (G, S_{c^*}, c^*) whose output is the same as $\mathcal{B}_{2c^*} - \mathcal{B}_{c^*}$, and again by Lemma 8.3 we obtain

$$\beta_2 := |\mathcal{B}_{2c^*} - \mathcal{B}_{c^*}| \geq |S_{c^*}|/M = (g - |\mathcal{B}_{c^*}|)/M.$$

Continuing in this way, we get

$$\beta_{i+1} := |\mathcal{B}_{(i+1)c^*} - \mathcal{B}_{ic^*}| \geq \lceil (g - |\mathcal{B}_{ic^*}|)/M \rceil, i = 0, 1, \cdots, \lfloor c/c^* \rfloor, \quad (8.12)$$

where $\mathcal{B}_0 := \emptyset$. Notice that

$$|\mathcal{B}_{jc^*}| = \sum_{i=1}^{j} \beta_j, j = 1, 2, \cdots, \lfloor c/c^* \rfloor \text{ and } \sum_{i=1}^{\lfloor c/c^* \rfloor} \beta_i \leq g.$$

To estimate the value c, we consider the sequence $\beta'_0 = 0$, $\beta'_i = \lceil (g - \sum_{j=0}^{i-1} \beta'_j)/M \rceil$, $i = 1, 2, \cdots$, and take the maximum positive integer c' such that $\beta'_{\lfloor c'/c^* \rfloor} \geq 1$ and $\sum_{i=1}^{\lfloor c'/c^* \rfloor} \beta'_i \leq g$. It can be seen from (8.12) that $c \leq c'$. Moreover $\beta'_1 \geq \beta'_2 \geq \cdots \geq \beta'_{\lfloor c'/c^* \rfloor} \geq 1$, $\beta'_1 \geq 2$ (by (8.11), and

$$\frac{c'}{c^*} < \underbrace{\frac{1}{\beta'_1} + \cdots + \frac{1}{\beta'_1}}_{\beta'_1} + \underbrace{\frac{1}{\beta'_2} + \cdots + \frac{1}{\beta'_2}}_{\beta'_2} + \cdots + \underbrace{\frac{1}{\beta'_{\lfloor c'/c^* \rfloor}} + \cdots + \frac{1}{\beta'_{\lfloor c'/c^* \rfloor}}}_{\beta'_{\lfloor c'/c^* \rfloor}} + 1. \quad (8.13)$$

Note that the right hand side of (8.13) can have at most $(\lfloor M \rfloor + i)$ terms $1/i$ for $1 \leq i \leq \lceil M \rceil$. If $\beta'_1 \leq \lceil M \rceil$, then

$$\frac{c}{c^*} \leq \frac{c'}{c^*} < \lfloor M \rfloor \left(\frac{1}{\beta'_1} + \frac{1}{\beta'_1 - 1} + \cdots + \frac{1}{2} + 1 \right) + \beta'_1$$
$$\leq \lceil M \rceil (\log \beta'_1 + 1) + \lceil M \rceil$$
$$\leq \lceil M \rceil (\log \lceil g/M \rceil + 2)$$

giving the result. So we assume that for some integer k,

$$\beta'_1 > \cdots > \beta'_{k-1} > \beta'_k \geq \lceil M \rceil + 1 > \beta'_{k+1} \geq \cdots \geq \beta'_{\lfloor c'/c^* \rfloor}.$$

Therefore, by the definitions of β_i', we have

$$g - \sum_{j=0}^{k-1} \beta_j' \geq \lceil M \rceil (\lceil M \rceil - 1) \quad \text{and}$$

$$\sum_{j=i}^{k-1} \beta_j' \leq (\beta_i' - \beta_{k+1}' + 1)\lceil M \rceil, \quad \text{for } i = k-1, k-2, \cdots, 1.$$

Rename by $\gamma_1, \gamma_2, \cdots, \gamma_p$ the decreasing series

$$\underbrace{\frac{1}{\beta_{k-1}'}, \cdots, \frac{1}{\beta_{k-1}'}}_{\beta_{k-1}'}, \underbrace{\frac{1}{\beta_{k-2}'}, \cdots, \frac{1}{\beta_{k-2}'}}_{\beta_{k-2}'}, \cdots, \underbrace{\frac{1}{\beta_1'}, \cdots, \frac{1}{\beta_1'}}_{\beta_1'},$$

where $p = \sum_{j=1}^{k-1} \beta_j'$, so that $\gamma_1 \geq \gamma_2 \geq \cdots \geq \gamma_p$. Let $q = \lceil p/\lceil M \rceil \rceil$, and for $i = 1, 2, \cdots, q-1$, define

$$\alpha_i = \sum_{j=(i-1)\lceil M \rceil + 1}^{i\lceil M \rceil} \gamma_j \quad \text{and} \quad \alpha_q = \sum_{j=(q-1)\lceil M \rceil + 1}^{p} \gamma_j.$$

It then follows that

$$q \leq \beta_1' - \beta_{k+1}' + 1 \quad \text{and} \quad \alpha_i \leq \frac{\lceil M \rceil}{\beta_{k+1}' + i - 1} \quad \text{for } i = 1, \cdots, q-1, q,$$

which implies

$$\sum_{i=1}^{q} \alpha_i \leq \lceil M \rceil \left(\frac{1}{\beta_1'} + \frac{1}{\beta_1' - 1} + \cdots + \frac{1}{\beta_{k+1}'} \right).$$

From inequality (8.13), we get

$$\frac{c}{c^*} \leq \frac{c'}{c^*} < \sum_{i=1}^{q} \alpha_i + \beta_k' \frac{1}{\beta_k'} + \lfloor M \rfloor \left(\frac{1}{\beta_{k+1}'} + \frac{1}{\beta_{k+1}' - 1} + \cdots + \frac{1}{2} + 1 \right) + \beta_{k+1}'$$

$$\leq \lceil M \rceil \left(\frac{1}{\beta_1'} + \frac{1}{\beta_1' - 1} + \cdots + \frac{1}{2} + 1 \right) + \frac{\lceil M \rceil}{\beta_{k+1}'} + \beta_{k+1}' + 1$$

$$\leq \lceil M \rceil (\log \beta_1' + 1) + \lceil M \rceil + 2 = \lceil M \rceil \left(\log \lceil \frac{g}{M} \rceil + 2 \right) + 2,$$

which proves the theorem. □

The approximation ratio in the above theorem is rather high and reflects again the hardness of the general problem. So in the next subsection we will study the minimum tree coloring problem in some special graphs.

8.2.3.2 *For Special graphs*

In this subsection we will discuss the impact of graph topology on the approximability of minimum tree coloring problem. We will study approximation algorithms with guaranteed performance ratios for three special graphs: trees, tori, and rings.

Recall that in trees, the minimum tree coloring problem is equivalent to the tree coloring problem: color all trees in $\mathcal{T}$ with a minimum number of colors, since the tree family $\mathcal{T}$ of minimal trees over a given set $\mathcal{S}$ of groups is unique. We refer to groups in $\mathcal{S}$ simply as trees in $\mathcal{T}$. Straightforwardly, the approximation algorithm for the vertex coloring problem [128] on $G_{\mathcal{T}}$ carries over to the minimum tree coloring problem on $\mathcal{T}$, and gives the following immediate upper bound [51], which is close to the lower bounds given in Theorem 8.8.

Theorem 8.13 *The minimum tree coloring problem in trees is approximable within a ratio* $O(g(\log \log g)^2/(\log g)^3)$.

When the size of each group is upper bounded by a constant k as in Steiner k-tree problem, i.e., $|S| \le k$ for every $S \in \mathcal{S}$, we have a tree family $\mathcal{T} = \{T_1, T_2, \cdots, T_g\}$, called a k-*tree family* such that the maximum degree of every T_i is no more than k. Notice that the minimum tree coloring problem on a k-tree family in a tree graphs is NP-hard even when $k = 2$ [116]. Fortunately, by the nice property of k-tree family and the acyclic structure of the underlying graph, this problem admits a k-approximation algorithm. Initially, it picks an arbitrary vertex (*root*) r of the tree graph, and reorders the trees $T_1, T_2, \cdots, T_g$ in the family as $T_1', T_2', \cdots, T_g'$ such that, for every $i = 1, 2, \cdots, g - 1$, the shortest path from r to a vertex of T_i' is not longer than the shortest path from r to every vertex of T_{i+1}'. Then it runs in g steps: in the i-th step, assign T_i' the first available color, i.e., the smallest positive integer that has not been assigned to any trees in $\{T_1', T_2', \cdots, T_{i-1}'\}$ sharing an edge with T_i'. Naturally, this is called *first fit algorithm*.

Theorem 8.14 *The minimum tree coloring problem is approximable within* $(k - \frac{k-1}{L})$ *for any given k-tree family in a tree graph of maximum load L.*

Proof. Let T be a k-tree family in a tree graph G, and let L be the maximum load of G. Since $c_{opt}(G, T) \geq L$, it suffices to show that the first fit algorithm requires at most $(kL - k + 1)$ colors. Suppose for a contradiction that the algorithm uses at least $(kL - k + 2)$ colors. Then for some i, at the beginning of the i-th step, T_i' is uncolored, and $(kL - k + 1)$ trees from $\{T_1', T_2', \cdots, T_{i-1}'\}$, say $T_1'', T_2'', \cdots, T_{kL-k+1}''$, have used up $(kL - k + 1)$ different colors, and every T_h'' ($1 \leq h \leq kL - k + 1$) shares a common edge e_h with T_i'. Let v be the vertex of T_i' nearest to the root r. Recall that, for every $h = 1, 2, \cdots, kL - k + 1$, the shortest path from r to a vertex of T_h'' is not longer than the path from r to v. The tree structure of G enables us to take the $(kL - k + 1)$ common edges e_h, $1 \leq h \leq kL - k + 1$, such that all of them are incident with v. Nevertheless, as the maximum degree of vertices in T_i, T_i' can share with at most $k(L-1)$ other trees edges incident with v, a contradiction. The proof is then finished. $\qquad\square$

We now make a brief discussion on graphs of torus. In view of the 4-regularity of tori, the comparison of Theorem 8.9 and Theorem 8.14 shows that neither low degree nor regularity necessarily implies improvements on the approximability of the minimum tree coloring problem. Consider an instance of the problem (G, S) where G is a 2-dimensional torus. Since G is 4-edge connected, there are two edge-disjoint spanning trees T and T' in G [217; 256]. It is easy to see that either $\{(T, i), (T', i) \mid 1 \leq i \leq g/2\}$ (when g is even) or $\{(T, i), (T', i) \mid 1 \leq i \leq (g-1)/2\} \cup \{(S, (g+1)/2)\}$ (when g is odd) is a solution to (G, S).

Theorem 8.15 *The minimum tree coloring problem in 2-dimensional tori is approximable within $\lceil g/2 \rceil$.*

Finally, we investigate the minimum tree coloring problem in ring graphs. A *ring graph* is a cycle with no chord. So in a ring, a tree is simply a path traversing all vertices in a group. Due to the simple topology, the minimum tree coloring problem in rings is relatively easy, and admits a 2-approximation algorithm using the same technique as for path routing and coloring in rings [231]: pick an arbitrary edge e of the ring G, and output an optimal solution $\{(T_i, c_i) \mid i = 1, 2, \cdots, g\}$ to the instance $(G \backslash \{e\}, S)$ of the minimum tree coloring problem in the chain $G \backslash \{e\}$ [124], where the solution uses $|\{c_1, c_2, \cdots, c_g\}| = L_{\{T_1, \cdots, T_g\}}$ distinct colors. To obtain the approximation ratio 2, consider T^* a tree family over S in an optimal solution to the instance (G, S) of minimum tree coloring problem. Let T_e^* consist of trees in T^* using e. Then both $L_{T_e^*}$ and $L_{T^* \backslash T_e^*}$ are at most L_{T^*}, and $L_{\{T_1, \cdots, T_g\}} \leq L_{T_e^*} + L_{T^* \backslash T_e^*} \leq 2L_{T^*}$. Now the performance guarantee 2 follows from the fact that the optimal solution to the instance $(G, \mathcal{S})$ must use at least L_{T^*} colors.

The following theorem [51] shows that further improvement can be made by investigating solutions for two chains and returning the better.

Theorem 8.16 *The minimum tree coloring problem in rings is approximable within* $(2 - 1/g)$.

Proof. We may assume that $|S_1| \geq 2$ (otherwise the problem is trivial). For every $v \in S_1$, let v', v'' be the two neighbors of v in the ring G. The approximation algorithm goes as follows. For $w = v', v''$, let e_w be the edge of G joining w and v, and let $S_w = (S_1 \setminus \{v\}) \cup \{w\}$ and $\mathcal{S}_w = \{S_w, S_2, \cdots, S_g\}$; find an optimal solution $\mathcal{C}_w = \{(T_{wi}, c_{wi}) | i = 1, 2, \cdots, g\}$ to the instance $(G \setminus \{e_w\}, \mathcal{S}_w)$ of minimum tree coloring problem in which T_{w1} is a minimal tree over S_{w1}. Switch v' and v'' if necessary so that $\mathcal{C}_{v'}$ uses no more colors than $\mathcal{C}_{v''}$. Take $u \in S_1$ with $|\mathcal{C}_{u'}| = \min\{|\mathcal{C}_{v'}| \,|\, v \in S_1\}$. Return tree routing and coloring $\{(T_i, c_{u'i}) | i = 1, 2, \cdots, g\}$ for the instance $(G, \mathcal{S})$ of minimum tree coloring problem, where T_1 is the tree obtained by connecting $T_{u'1}$ with u via edge $e_{u'}$, and $T_i = T_{u'i}$ for $i = 2, 3, \cdots, g$.

To analyze the algorithm, let $\mathcal{C}^* = \{(T_i^*, c_i^*) | i = 1, \cdots, g\}$ be an optimal solution to the instance $(G, \mathcal{S})$ of minimum tree coloring problem, and suppose that T_1^* is a minimal tree over S_1 and uses edge e_w with $w = v'$ or v'' to connect one of its leaves v. Let $\mathcal{T}_{e_w}^*$ consist of trees in $\mathcal{T}^* = \{T_1^*, T_2^*, \cdots, T_g^*\}$ using e_w, and $\mathcal{S}_{e_w}^*$ consist of groups in $\mathcal{S}$ such that $\mathcal{T}_{e_w}^*$ is over $\mathcal{S}_{e_w}^*$ and $\mathcal{T}^* \setminus \mathcal{T}_{e_w}^*$ is over $\mathcal{S} \setminus \mathcal{S}_{e_w}^*$. Observe that $L_{\mathcal{T}^*} \geq L_{\mathcal{T}_{e_w}^*} = |\mathcal{T}_{e_w}^*| = |\mathcal{S}_{e_w}^*|$ and $S_1 \in \mathcal{S}_{e_w}^*$. Moreover, $\mathcal{T}_1 = (\mathcal{T}^* \setminus \mathcal{T}_{e_w}^*) \cup \{T_1^* \setminus \{v\}\}$ is a tree family in the chain $G \setminus \{e_w\}$ over $(\mathcal{S} \setminus \mathcal{S}_{e_w}^*) \cup \{S_{w1}\}$, and straightforwardly $L_{\mathcal{T}_1} \leq L_{\mathcal{T}^*}$. Let $\mathcal{T}_\in$ be the family of the minimal trees in $G \setminus \{e_w\}$ over $\mathcal{S}_{e_w}^* \setminus \{S_1\}$. Then $L_{\mathcal{T}_2} \leq |\mathcal{S}_{e_w}^*| - 1 \leq L_{\mathcal{T}^*} - 1$, and $\mathcal{T}_1 \cup \mathcal{T}_2$ is a tree family in $G \setminus \{e_w\}$ over $((\mathcal{S} \setminus \mathcal{S}_{e_w}^*) \cup \{S_{w1}\}) \cup (\mathcal{S}_{e_w}^* \setminus \{S_1\}) = \mathcal{S}_w$ which has load $L_{\mathcal{T}_1 \cup \mathcal{T}_2} \leq L_{\mathcal{T}_1} + L_{\mathcal{T}_2} \leq 2L_{\mathcal{T}^*} - 1$ and can be colored with $L_{\mathcal{T}_1 \cup \mathcal{T}_2}$ colors. Recalling the optimality of $\mathcal{C}_w$, we have $|\mathcal{C}_w| = |\{c_{w1}, c_{w2}, \cdots, c_{wg}\}| \leq L_{\mathcal{T}_1 \cup \mathcal{T}_2} \leq 2L_{\mathcal{T}^*} - 1$.

Now consider the number of colors used by the solution $\{(T_i, c_{u'i}) | i = 1, \cdots, g\}$ for the instance $(G, \mathcal{S})$. We deduce from the choice of $\mathcal{C}_{u'}$ that $|\{c_{u'1}, c_{u'2}, \cdots, c_{u'g}\}| = |\mathcal{C}_{u'}| \leq |\mathcal{C}_w| \leq 2L_{\mathcal{T}^*} - 1 \leq 2c_{opt}(G, \mathcal{S}) - 1$. It follows from $c_{opt}(G, \mathcal{S}) \leq g$ that $|\{c_{u'1}, c_{u'2}, \cdots, c_{u'g}\}| / c_{opt}(G, \mathcal{S}) \leq 2 - 1/g$. The proof is then finished. $\square$

Recalling the general large gap between the minimum number of colors and the minimum maximum load stated in Theorem 8.11, we see that as a processing step of minimum tree coloring problem, the minimization of the maximum load works much better in rings than in trees and tori.

8.3   Discussions

In this chapter we have studied two versions of Steiner tree coloring problem, maximum tree coloring problem and minimum tree coloring problem. The former asks for the way for routing and coloring maximal number of multicast requests with given colors, while the latter asks for the way for routing and coloring all multicast requests with minimal number of colors. More sophisticated studies also put *Quality of Service* (QoS) constraints into consideration.

Jia et al. [154] consider QoS multicast in WDM networks. Given a set of QoS multicast requests, the objective is to find a set of cost sub-optimal QoS routing trees and assign wavelengths to them. The objective is to minimize the number of wavelengths in the system. This is a more general version of minimum tree coloring problem since it involves not only optimal QoS multicast routing, but also the optimal wavelength assignment. Usually setting up a channel in WDM networks is done in two separate steps: routing and wavelength assignment, which, however, has limited power in minimizing the number of wavelengths. They propose two methods for minimization of the number of wavelengths used that integrate routing and wavelength assignment via *light approximate shortest path tree* (refer to Section 5.3). One minimizes the number of wavelengths through reducing the maximal link load in the network; while the other does it by trying to free out the least used wavelengths. Simulation results demonstrate that the proposed methods can produce sub-optimal QoS routing trees while substantially saving the number of wavelengths used.

Jia et al. [155] study how to assign given wavelengths to (unicast) communication requests such that the *overall blocking* in the network is minimized. A communication request has to be *blocked* (also called *rejected*) if it can not be satisfied due to limited number of available wavelengths or load capacity. This problem can be considered as a general version of maximum tree coloring problem since the number of blocked requests is equal to the number of given requests minus that of the satisfied requests. They transform this problem into the *maximum weight k-cut problem*. Given an edge-weighted graph $G(V, E)$ and a k-partition $\{V_1, V_2, \cdots, V_k\}$ of V, a k-cut is a subset of E consisting of edges whose two endpoints are in two different sets V_i and V_j, for $i \neq j$. The weight of a k-cut is the sum of weights of all edges in the cut. The objective is to find the k-partition that produces a k-cut with maximum weight. They propose a local search algorithm with approximation ratio of $(1 - 1/k)$ for this NP-hard problem [163].

Chapter 9

Steiner Tree Scheduling Problem

Steiner tree scheduling problems arise from the applications of information dissemination in *Wireless Sensor Networks* (WSNs) [2]. A WSN usually consists of a large number of small-sized and low-powered sensor deployed over a geographical area and a sink node where the end user can access data. All nodes are equipped with capabilities of sensing, data processing, and communicating with each other by means of a wireless ad hoc network. A wide range of tasks can be performed by these tiny devices, such as condition-based maintenance and the monitoring of a large area with respect to some given physical quantity, e.g., temperature, humidity, gravity and seismic information.

The stringent resource constraint and the sheer number of sensor nodes in WSNs pose unique challenges on time-efficient information dissemination. First, the sensor nodes operate on batteries and employ low-power radio transceivers to enable communications. Data packet sent by a senor (sender) reaches all its neighbor nodes within the transmission range of the sender; Sensors far from the data sink have to use intermediate nodes to relay data transmission. Second, collision resulting from a large number of simultaneous sending creates response implosion [148]: when two or more sensors send data to a common neighbor at the same time, collision occurs at this node, which will not receive any of these data. Third, the data sent by a sender is received by any its neighbor (*receiver*) at which no collision occurs; the receiver fuses the data received with its own data (possibly null), and stores the fused data as its new data. In addition, the time consumed by a single sending-receiving-fusing-storing is typically normalized to one; parallel sending-receiving are desirable for reducing the network delay. Fourth, with the large population of sensor nodes, it may be impractical or energy consuming to pay attention to each individual

253

nodes in all situations; for instance, the user wound be more interested in querying "what is the highest temperature in some specified areas?"

Usually, it is assumed that each sensor node knows its geometric position in the network, which is considered as the unique ID of the sensor (the aggregated data may include some of these IDs). We further assume that the sink has global knowledge of IDs of all sensors in the WSN. When it needs some data of particular interests at some sensor nodes, it informs those nodes, by multicasting, of the transmission schedule which may be represented by IDs of senders and receivers. Upon receiving the request, sensor nodes will send their data or receive data from others as specified in the schedule. In such a way, the schedule guarantees collision-free data aggregation. It also enables significant energy savings since sensor nodes are in an energy conserving state when they do not participate in sending/receiving. Prior to the scheduled time for data aggregation, a node switched from the energy conserving state to the energy consuming state, transmits or receives data and then go back to the energy conserving state.

Motivated by various applications of time-efficient information dissemination for query-based monitoring WSNs (e.g. battlefield communications and rescue operations), we will study in this chapter two Steiner tree scheduling problems:

(1) *Minimum aggregation time problem.* It adopts a collision-free transmission model, which guarantees the energy-efficiency since no data need to be transmitted more than once. The objective is how to, given a WSN with a distinguished data sink d which is interested in data on a subset S of sensor nodes, determine a data transmission schedule such that all data on S are sent and aggregated to d in minimal time.
(2) *Minimum broadcast time problem* [110]. It is very similar to minimum aggregation time problem. Initially d has a message to be broadcasted to every node in the network; At each time round any node that has received the message is allowed to communicate the message to at most one of its neighbors. The objective is how to compute a broadcast schedule of minimal number of time rounds that is required for every node to receive the message.

Clearly, the minimum broadcast time problem schedules the data to flow from d to all nodes in S while the minimum aggregation time problem schedules data to flow from all nodes in S towards d.

The remainder of this chapter is organized as follows. In Sections 9.1 and 9.2 we will study minimum aggregation and broadcast time problems, respectively. For each problem, we will first give NP-hardness proofs, and then propose some approximation algorithms. In Section 9.3, we will conclude the chapter with some remarks.

9.1 Minimum Aggregation Time

9.1.1 *Problem Formulation*

In view of miniature design of sensor devices, we assume that all sensors in Wireless Sensor Networks (WSNs) are fixed and homogeneous. More specifically, the WSN under investigation consists of stationary nodes (sensor nodes and a sink node) distributed in the Euclidean plane. Assuming the transmission range of any sensor node is a unit disk (circular region with unit radius) centered at the sensor, we model a WSN as a *Unit Disk Graph* (UDG) $G = (V, E)$ in which two nodes $u, v \in V$ are considered neighbors,i.e., there is an edge $uv \in E$ joining u and v, if and only if the Euclidean distance $|uv|$ between u and v is at most one. Hereafter we reserve symbol G for UDGs modelling WSNs, and Δ for the maximum degree of G. It is always assumed that G is connected. We assume that communication is deterministic and proceeds in synchronous rounds controlled by a global clock. Moreover, we assume that in each time round,

(1) Each node can send data (be a sender) or receive data (be a receiver) but cannot do both;
(2) Each node can receive data from at most one of its neighbors;
(3) Data packet sent by any sender reaches simultaneously all its neighbors;
(4) All nodes can receive data only if exactly one of their neighbors sends data.

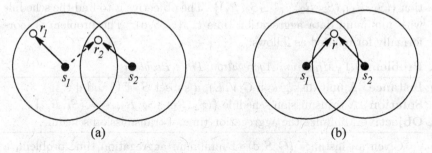

(a) (b)

Fig. 9.1 Description of network assumptions.

Note that the above assumptions guarantee collision-free data transmit/receiving between senders and receivers. In Fig.9.1(a), two time rounds are required if node s_i needs to send its data to node r_i for $i = 1, 2$ since when they send their data in the same time round r_2 will receive data from both s_1 and s_2 causing collision, which is not allowed (due to Conditions (3-4)). For the same reason, in Fig.9.1(b), two time rounds are required if s_1 and s_2 both need to send their data to r. Moreover, we assume that each receiver updates its data as the combination of all data received in different rounds, this enforces that each node needs to send data at most once.

In contrast to traditional networks (e.g., the Internet) which are address-centric, WSNs are intrinsically data-centric. In some applications of WSN, the end user needs to extract information from the sensor field with low latency. In this case, data sensed at some sensors related to the same physical phenomenon need to be aggregated and sent to the data sink efficiently. Real time data aggregation is a combination of data from different sensors according to a certain aggregation function, e.g., duplicate suppression, logical AND/OR, minima and maxima, and all requested data should be periodically delivered to sink node within a certain period of time from the moment they are requested (after that data may be useless).

An instance of minimum aggregation time problem is denoted by (G, S, d), where the set $S \subseteq V(G)$ consists of nodes whose data are requested by the sink node $\mathsf{d} \in V$. The solution of (G, S, d) is a *transmission schedule* $\{(S_1, R_1), (S_2, R_2), \cdots, (S_s, R_s)\}$ such that S_r (resp. R_r) is the set of senders (resp. receivers) in r-th round for $r = 1, 2, \cdots, s$, and all data on nodes in S must be aggregated to d within s rounds. Note that every (S_r, R_r) gives implicitly the 1-1 correspondence between S_r and R_r in a way that $v \in S_r$ corresponds to its receiver in R_r which is the only neighbor of v in R_r. The value s is called the *data aggregation time* of solution $\{(S_1, R_1), (S_2, R_2), \cdots, (S_s, R_s)\}$. The objective is to find the schedule with minimum data aggregation time $t_{opt}(G, S, \mathsf{d})$. The problem is more formally formulated as follows.

Problem 9.1 *Minimum Aggregation Time Problem*

Instance A unit disk graph $G(V, E)$, a subset S of V and $\mathsf{d} \in V \setminus S$.
Solution A transmission schedule $\{(S_1, R_1), (S_2, R_2), \cdots, (S_s, R_s)\}$.
Objective Minimize the aggregation time of transmission schedule.

Given an instance (G, S, d) of minimum aggregation time problem, a *shortest path tree* T of (G, S, d) is a tree in G consisting of shortest paths from d to nodes in S. The *height* of T, denoted by $h(G, S, \mathsf{d})$, equals to the length of the longest path in T from d to leaves of T. The following lower bound can be easily obtained by applying the same argument used in the estimation of multicasting time in telephone networks [23].

Lemma 9.1 *For any instance* (G, S, d) *of minimum aggregation time problem,* $t_{opt}(G, S, \mathsf{d}) \geq \max\{h(G, S, \mathsf{d}), \log_2 |S|\}$.

However, data aggregation in WSNs is not simply the reverse of broadcast/multicast in traditional telephone network. For example, Gandhi et al. [105] showed that the broadcast time in a WSN is at most 648 times the height of the shortest path tree. Note also when the underlying topology G of WSNs is a complete graph, we have $t_{opt}(G, S, \mathsf{d}) = n$ while the shortest path tree gives a multicast time equal to 1 and minimum broadcast time problem has minimum broadcast time of $log_2 |V|$.

9.1.2 *Complexity Analysis*

In order to prove the NP-hardness of minimum aggregation time problem, we will apply some known results on orthogonal planar drawing. An *orthogonal planar drawing* of a planar graph H is a planar embedding of H in the plane such that all edges are drawn as sequences of horizontal and vertical segments. A point where the drawing of an edge changes its direction is called a *bend* of this edge. All vertices and bends are drawn on integer points. If the drawing can be enclosed by a box of width g and height g, we call it an embedding with grid size $(g \times g)$. By a *plane graph* we mean a planar graph together with a planar embedding of it. Biedl and Kant [36] proved the following lemma.

Lemma 9.2 *Given a simple plane graph H on g vertices that is not an octahedron and has maximum degree at most 4, there is a linear algorithm which produces an orthogonal planar drawing of H with grid size $(g \times g)$ such that the number of bends along each edge is at most 2.*

Using the above lemma we can deduce the following lemma [53].

Lemma 9.3 *Let H be a plane graph on g vertices with maximum degree at most 4. Suppose that H is not an octahedron, and let H' be the graph obtained from H by replacing each edge in H with a path of length $120g^2$. Then H' is a unit disk graph and an orthogonal planar embedding of H' of grid size $(40g^2 + 40g) \times (40g^2 + 40g)$ can be computed in time polynomial in g.*

Proof.   Note first that the orthogonal planar embedding stated below is derived in time polynomial in g.

(C1) H has an orthogonal planar drawing D of grid size $(g \times g)$ in which every edge of H has at most 2 bends.

An orthogonal planar drawing D' of H with grid size $(40g^2 + 40g) \times (40g^2 + 40g)$ can be obtained from D in such a direct way that point (x, y) in D maps to point $(40nx, 40ny)$ in D', and every vertical (resp. horizontal) segment of unit length (*segment* for short) in D maps to a vertical (resp. horizontal) path of length $40g$ in D' between the images (in D') of the ends of the segment (in D). (The additional $40g$ is set for the further modification on the drawing.) It is straightforward from claim (C1) that

(C2) each edge of H has length at most $120g^2$ in D'.

We propose to modify D' into an orthogonal planar drawing of H' with grid size $(40g^2 + 40g) \times (40g^2 + 40g)$ in which every edge of H' is a vertical or horizontal segment and any two nonadjacent vertices of H' are drawn on two points with distance at least two. To this end, we consider a $18g \times 18g$ grid K and a, b the two corners of K on one side. We call a path from a to b an *a-b path*.

(C3) For every even integer j with $18g \le j \le 120g^2$, K contains an a-b path P_j of length j which is a unit disk graph.

Suppose, without loss of generality, that a and b are located on $(0,0)$ and $(18g, 0)$, respectively. Let $k \in \{18g, 18g - 2\}$ be such that $(k/2 + 1)$ is an even number (i.e. $\{0, 2, 4, \cdots, k\}$ is of even size). Denote by I_{2i} $(0 \le i \le k/2)$ the path consisting of points and segments on the $2i$-th (vertical) column of K. For each $0 \le i \le k - 6$ which is a fold of 4, let J_i (resp. J_{i+2}) be the shortest path connecting the top ends of I_i and I_{i+2} (resp. the bottom ends of of I_{i+2} and I_{i+4}) and let J_{k-2} (resp. J_k) be the shortest path connecting the top ends of I_{k-2} and I_k (resp. the bottom end of I_k and b). Thus $J_0, J_2, J_4, \cdots, J_{k-2}$ are all horizontal paths of length 2, J_k is horizontal paths of length 0 (when $k = 18g$) or 2 (when $k = 18g - 2$), and $J_0, J_4, \cdots, J_{k-2}$ (resp. $J_2, J_6, \cdots, J_{k-4}, J_k$) are contained in the top (resp. bottom) row of K. Now we get an a-b path $P_\ell := \bigcup_{i=0}^{k/2}(I_{2i} \cup J_{2i})$ of even length $\ell > (k/2 + 1)18g > 120g^2$. Clearly P_ℓ is induced, and so is a unit disk graph. Inductively, suppose that K contains an a-b path P_j of even length j $(\ge 18g + 2)$ such that P_j is a unit disk graph containing $J_2 \cup J_6 \cup \cdots \cup J_{k-4} \cup J_k$, and all vertical segments of P_j are contained in

$\cup_{i=0}^{k/2} I_{2i}$. Since $j \geq 18g + 2$, we see the statement holds for $(j - 2)$ in a way that P_{j-2} is obtained from P_j by replacing a subpath of P_j of length 4 consisting of two vertical segments and two horizontal segments with a horizontal path in K of length 2 between the same ends. Consequently, Claim (C3) is proved by induction.

We now turn back to the drawing D'. Consider an arbitrary edge e of H of length ℓ_e in D'. Let S_e be the subpath (in the $(40g^2 + 20g) \times (40g^2 + 40g)$ grid) of the drawing of e consisting of a maximal sequence of vertical (or horizontal) segments. Let u_e and v_e be the ends of S_e such that the summation of coordinates of u_e is less than the summation of coordinates of v_e. Then

(C4) In D', point u_e (resp. v_e) is a bend or an end of e, and S_e is a vertical or horizontal path of even length at least 40g.

Now we put a $(18g \times 18g)$-grid K_e such that $K_e \cap S_e$ is an a_e-b_e path Q_e which is a side of K_e, and $|a_e v_e| > |b_e v_e| = 2$. By (a) and (b), let P_e be the a_e-b_e path in K_e of length $120g^2 - \ell_e + 18n$ which is a unit disk graph. Construct drawing D'' from D' by replacing each Q_e in S_e with P_e in K_e. Let H' be the graph whose vertex (resp. edge) set consists of all points (resp. segments) in D''. Then D'' is an orthogonal planar drawing of G' of grid size $(40g^2 + 40g) \times (40n^2 + 40g)$ and G' is a graph obtained from G by replacing each edge of G with a path of length $120g^2$. Moreover,

(C5) In the embedding D'' of H' the distance between any point (vertex) in $P_e \backslash \{a_e, b_e\}$ and any point (vertex) in $H' \backslash V(K_e)$ is at least two.

Suppose the contrary that $x \in V(P_e) - \{a_e, b_e\}$ and $y \in V(H') - V(K_e)$ has distance one in D''. Since any two vertical (resp. horizontal) segments in D' have distance at least 40g, it follows from claim (C3) and $|a_e v_e| > |b_e v_e| = 2$ that $y \in P_f$ for some other edge f of H. From the positions of K_e on S_e and K_f on S_f, we deduce that S_e or S_f has length less than $40g$, a contradiction to claim (C4). Hence claim (C5) holds.

Combining claims (C3-4), we conclude that H' is a unit disk graph and establish the lemma. □

We now explain how to prove that minimum aggregation time problem by reducing the restricted planar 3-SAT problem to it. Let $\{x_1, x_2, \cdots, x_n\}$ and $\{c_1, \cdots, c_m\}$ denote, respectively, the sets of variables and clause in a Boolean formula φ in conjunctive normal form, where each clause has at most 3 literals. Associate with φ the formula graph $G_\varphi =$

$(\{x_1, x_2, \cdots, x_n\} \cup \{c_1, c_2, \cdots, c_m\}, E_1 \cup E_2)$, where $E_1 := \{x_i c_j \mid x_i \in c_j$ or $\bar{x}_i \in c_j\}$ and $E_2 := \{x_i x_{i+1} \mid 1 \le i \le n - 1\} \cup \{x_n x_1\}$. Boolean formula is called *planar* if G_φ is a planar graph. The *planar 3-SAT problem* is to decide if there is a truth assignment that satisfies all clauses in a planar boolean formula, where each clause has at most three literals. A planar 3-SAT problem is said to be *restricted* if

(a) each variable (unnegated or negated) appears at most three clauses,
(b) both unnegated and negated forms of each variable appear; and
(c) at every variable node in the planar embedding of G_φ, the edges in E_2 that are incident with node x separate the edges in E_1 incident to x such that all edges representing nonnegative appearance are incident to one side of x and all edges representing negative appearance are incident to the other side.

It is known that the restricted planar 3-SAT problem is NP-complete [186].

Theorem 9.1 *The decision version of minimum aggregation problem is NP-complete even when the underlying topology is a subgraph of a grid.*

Proof. The proof is based on a reduction from restricted planar 3-SAT. Given any restricted planar 3-SAT instance φ on n variable and m clauses, from its planar formula graph G_φ, we construct planar graphs G_k for positive integer k as follows.

To every variable x_i, $1 \le i \le n$, we associate a rectangle X_i and two node-disjoint paths $P_i, \bar{P}_i$ such that

(i) X_i has exactly $10k$ nodes among which equally spaced nodes $p_i, q_i, r_i, s_i,$ $t_i, s'_i, rr'_i, q'_i, p'_i, t'_i$ are located in cyclic order of X_i, and P_i (resp. P'_i) has ends o_i and p_i (resp. o'_i and p'_i) with $P_i \cap X_i = \{p_i\}$ (resp. $P'_i \cap X_i = \{p'_i\}$), and

(ii) both P_i and P'_i are of length $(6i - 5)k - 1$. To every clause c_j, we associate a path C_j with ends b_j, c_j and of length $k - 1$. All $X_i \cup P_i \cup P''_i$'s and C_j's are pairwise node-disjoint. For every edge $x_i c_j$ (resp. $x'_i c_j$) in G_φ, there is a path P_{ij} in G_k such that P_{ij} has one end c_j and the other end in $\{r_i, s_i\}$ (resp. $\{r'_i, s'_i\}$), and for all $1 \le i, i' \le n, 1 \le j,$ $j' \le m$, we have

(iii) $P_{i,j} \cap (\bigcup_{h=1}^n C_h) = \{c_j\}$, $P_{i,j} \cap (\bigcup_{h=1}^n (X_i \cup P_i \cup P'_i))$ consists of a node in $\{r_i, s_i, r'_i, s'_i\}$;

(iv) $P_{ij} \cap P_{i'j'} \ne \emptyset$ iff $P_{ij} = P_{i'j'}$ or $i \ne i'$ and $P_{ij} \cap P_{i'j'} = \{c_j\} = \{c_{j'}\}$;

(v) P_{ij} is of length $(6i - 4)k$ when it has end in $\{r_i, r'_i\}$ and of length $(6i - 3)k$ when it has end in $\{s_i, s'_i\}$. Finally we add $(n - 1)$ pairwise disjoint paths $W_1, W_2, \cdots, W_{n-1}$ such that

(vi) W_h has length k and $W_h \bigcap \left(\bigcup_{i,j} (X_i \cup P_i \cup P_i' \cup C_j \cup P_{ij}) \right) = \{t_h, t_{h+1}\}$ for $h = 1, 2, \cdots, n-1$. This completes the construction of $G_k :=$ $\bigcup_{i,j} \left(X_i \cup P_i \cup P_i' \cup W_i \cup C_j \cup P_{ij} \right)$.

Let G_k^+ be obtained from G_k by adding $(2n + m)$ pendant edges such that the $2n + m$ degree-one nodes $o_1, o_1', o_2, o_2', \cdots, o_n, o_n', b_1, b_2, \cdots, b_m$ in G_k have degree two in G_k^+. Denote by g the number of nodes in G_1^+ and set $\ell = 120g^2$. It is easy to see that $g < 36n^2 + m$ and that G_ℓ^+ is obtained from G_1^+ by replacing each edge of G_1^+ with a path of length $120g^2$. Since G_φ is a planar graph with maximum degree at least 3, so is G_1. Thus a planar embedding of G_1 might be computed in time polynomial in $(n + m)$ [25]. By the construction, this planar embedding might be extended to be a planar embedding for G_1^+ in time polynomial in g and hence polynomial in $(n + m)$. Notice that G_1^+ is a plane graph other than octahedron and that the maximum degree of G_1^+ is at most 4. It follows from Lemma 9.3 that G_ℓ^+ is a unit disk graph, so is G_ℓ. Moreover from Lemma 9.3, we deduce that G_ℓ is a subgraph of a grid, and that both the size of G_ℓ and the construction time of G_ℓ are polynomial in $(n + m)$.

Next we show that the restricted planar 3-SAT instance φ is satisfiable if and only if the minimum aggregation time problem on $(G_\ell, V(G_\ell), t_n)$ has a solution (schedule) which aggregates all data on $V(G_\ell)$ into the sink t_n within $(6n-1)\ell$ rounds. To this end, let us first make some observations. For notational convenience, we set $W_n = \emptyset$ and use X_i^+ (resp. X_i^-) to denote the shortest path from p_i to s_i (resp. p_i' to s_i'), $i = 1, 2, \cdots, n$. Clearly, $X_i^+ \cup X_i^- \subseteq X_i$, $X_i^+ \cap X_i^- = \emptyset$, $\{p_i, q_i, r_i, s_i\} \subseteq X_i^+$ and $\{p_i', q_i', r_i', s_i'\} \subseteq X_i^-$ for all $1 \leq i \leq n$.

(a) Every shortest path from t_1' to t_n is of length $(6n - 1)\ell$ and must be contained in $\bigcup_{i=1}^n (X_i \cup W_i)$.

Suppose the contrary that P is a shortest path from $\bar{t}_1$ to t_n violating (a). Then by (i) and (vi), we have $P \nsubseteq \bigcup_{i=1}^n (X_i \cup W_i)$. Then there exist $1 \leq h < i \leq n$ and $1 \leq j \leq m$ such that $P_{hj} \cup P_{ij}$ is a subpath of P. Let u_h (resp. u_i) denote the the end of P_{hj} (resp. P_{ij}) in X_h (resp. X_i). By symmetry suppose that the shortest path in X_1 from $\bar{t}_1$ to r_1 is a subpath of P. It is clear that $P \backslash ((P_{hj} \cup P_{ij}) \backslash \{u_h, u_i\})$ consists of two paths P_1 and P_2 with P_1 containing $\bar{t}_1, r_1, u_h$ and P_2 containing u_i, t_n. Note from (i), (v) and (vi) that $(\bigcup_{i=1}^{i-1} (X_i \cup W_i)) \cup X_i$ contains a path Q from r_1 to u_i of length $|E(Q)| = |E(P_{ij})| - 2\ell$. It is not hard to see that $P_1 \cup Q \cup P_2$

contains a path P' from $\bar{t}_1$ to t_n shorter than P, a contradiction. So (a) holds.

Now combining (a) with (ii) and (v), we have

(b) For $1 \leq i \leq n$, $(\bigcup_{i=i}^{n}(R_i \cup W_i))\backslash t'_i$ contains every shortest path from p_i (resp. p'_i) to t_n (which is of length $(6(n-i)+4)\ell$), every shortest path from r_i (resp. r'_i) to t_n (which is of length $(6(n-i)+2)\ell$), and every shortest path from s_i (resp. s'_i) to t_n (which is of length $(6(n-i)+1)\ell$).

(c) For $1 \leq i \leq n$, every shortest path from o_i (resp. $\bar{o}_i$) to t_n has length $(6n-1)\ell - 1$ and its intersection with $\bigcup_{i=1}^{n}(X_i \cup \Lambda_i)$ is a path from p_i (resp. $\bar{p}_i$) to t_n containing X_i^+ (resp. X_i^-) and avoiding X_i^- (resp. X_i^+); and no path from o_i (resp. $\bar{o}_i$) to t_n has length $(6n-1)\ell$.

(d) For $1 \leq j \leq m$, each shortest path from b_j to t_n has length $(6n-1)\ell - 1$ and its intersection with $\bigcup_{i=1}^{n}(X_i \cup W_i)$ is a path; and no path from b_j to t_n has length $(6n-1)\ell$.

Now we assume that all data on $V(G_\ell)$ are aggregated into t_n within $(6n-1)\ell$ rounds. It is immediate from (a) that the data on $\bar{t}_1$ is aggregated up to t_n along a shortest path P from $\bar{t}_1$ to t_n in $\bigcup_{i=1}^{n}(R_i \cup W_i)$ without any delay; particularly, we have

(e) for $1 \leq i \leq n$, $|P \cap \{X_i^+, X_i^-\}| = 1$.

For $i = 1, 2, \cdots, n$, let $\{w_i, w'_i\} = \{p_i, p'_i\}$, $\{x_i, x'_i\} = \{q_i, q'_i\}$, $\{y_i, y'_i\} = \{r_i, r'_i\}$ and $\{z_i, z'_i\} = \{s_i, s'_i\}$ be such that P contains w_i, x_i, y_i, z_i. Using inductive arguments we can show that

(f) for $1 \leq i \leq n$, the aggregated data from t'_1 is received by t'_i in round $(6i-6)\ell$, by w_i in round $(6i-5)\ell$, by y_i in round $(6i-3)\ell$, by z_i in round $(6i-2)\ell$, and by t_i in round $(6i-1)\ell$.

Note from (c) that the data on o_i (resp. o'_i) must be aggregated towards t_n along a shortest path from o_i (resp. o'_i) to t_n containing $P_i \cup X_i^+$ (resp. $P'_i \cup X_i^-$), and o_i (resp. o'_i) sends data in 1-th round or in 2-nd round. Since by (i) and (ii) the length of $P_i \cup X_i^+$ (resp. $P'_i \cup X_i^-$) is $(6i-2)\ell - 1$, we see that o_i (resp. o'_i) sends data in 1-th round as otherwise either the aggregated data from t'_1 and that from o_i (resp. o'_i) collide on w_i in round $(6i-5)\ell$ or collide on t_i in round $(6i-1)\ell$. Let $v_i = o_i$ if $w_i = p'_i$ and $v_i = o'_i$ if $w_i = p_i$. Then we have

(g) for $1 \leq i \leq n$, the aggregated data from v_i is received by w'_i in round $(6i-5)\ell - 1$, by y'_i in round $(6i-3)\ell - 1$, by z'_i in round $(6i-2)\ell - 1$, and by t_i in round $(6i-1)\ell-1$.

Let us consider an arbitrary $j \in \{1, 2, \cdots, m\}$. It can be seen from (d) that the data on b_j is aggregated to t_n along a shortest path from b_j to t_n without any delay. Obviously this shortest path contains $C_j \cup P_{ij}$ as a subpath for some $i \in \{1, 2, \cdots, n\}$. Recall from (v) that $C_j \cup P_{ij}$ is of length $(6i-3)\ell - 1$ when one end of P_{ij} is in $\{r_i, r_i'\}$, and of length $(6i-2)\ell - 1$ when one end of P_{ij} is in $\{s_i, s_i'\}$. If y_i' is an end of P_{ij}, then data from v_i and the data from b_j collide on y_i' in round $(6i-3)\ell-1$. So we have $y_i' \notin P_{ij}$, and similarly $z_i' \notin P_{ij}$. It follows that for every $1 \le j \le m$, there exists some P_{ij} on the aggregation path from b_j to t_n such that $P_{ij} \cap \{y_i, z_i\} \ne \emptyset$. This allows us to derive a truth assignment for φ by setting $x_i := true$ if and only if $X_i^+ \subseteq P$.

Conversely, we consider the case where the restricted planar 3-SAT instance φ has a true assignment $\{x_1^*, x_2^*, \cdots, x_n^*\}$. Notice that for each $1 \le j \le m$, there exists an index $i(j) \in \{1, 2, \cdots, n\}$ such that either $true = x_{i(j)}^* \in C_j$ and $P_{i(j)j}$ connects c_j with $r_{i(j)}$ or $s_{i(j)}$, or $true = \bar{x}_{i(j)}^* \in C_j$ and $P_{i(j)j}$ connects c_j with $r_{i(j)}'$ or $s_{i(j)}'$. To define a schedule for the minimum aggregation time problem on $(G_\ell, V(G_\ell), t_n)$, from G_ℓ we construct a spanning T of G_ℓ rooted at t_n by deleting some edges of G_ℓ as follows: For each $1 \le i \le n$, we delete the edge incident with t_i' contained in X_i^- (resp. X_i^+) when $x_i^* = true$ (resp. $\bar{x}_i^* = true$); for each $1 \le j \le m$, we delete all edges incident with c_j but the two in $C_j \cup P_{i(j)j}$. From (i-vi) and (a-d), it is easy to see that the height of T is $(6n-1)\ell$. Now for $i = 1, 2, \cdots, (6n-1)\ell$, let S_i consist of all nodes such that the paths in T from t_n to them have length $(6n-1)\ell - i + 1$, and let R_i consist of parents (in T) of nodes in S_i. Again (i-vi) and (a-d) assure that $\{(S_1, R_1), (S_2, R_2), \cdots, (S_{(6n-1)\ell}, R_{(6n-1)\ell})\}$ is a schedule with data aggregation time $(6n-1)\ell$ for the minimum aggregation time problem on $(G_\ell, V(G_\ell), t_n)$. The theorem is then proved. $\square$

9.1.3 *Greedy Algorithms*

In this subsection we will present four approximation algorithms for minimum aggregation time problem, A_{sda}, A_{lsda}, A_{psda} and A_{esda}, all adopt the *shortest data aggregation strategy* that aggregates data along shortest paths towards the sink. Among them A_{sda} is a basic algorithm for minimum aggregation problem, while others are its variations for different cases. We also present the theoretical analysis for the worst-case performance ratios of all these four algorithms.

9.1.3.1 *Basic Algorithm*

Algorithm A_{sda} proceeds by incrementally constructing smaller and smaller shortest path trees rooted at d that span all nodes in S. It, initially, sets T_1 to a shortest path tree of (G, S, d). A number of iterations is implemented by A_{sda} (refer to the pseudo-code below) and each iteration produces a schedule of a round. In the r-th iteration, T_r is a shortest path tree rooted at d spanning a set of nodes that possess all data aggregated from S till round $(r-1)$. A_{sda} selects from the leaves of T_i as the senders for round r. In Steps 4-9, the variable Z^r with initial value {leaves of T_r}\{d} is used for selection. The set Z^r maintains the property that every non-leaf neighbor of a leaf in T_r other than d has a neighbor in Z^r. The leaves of T_r other than d are examined in the decreasing order of the number of their neighbors in G that are non-leaf node in T_r. A leaf is eliminated from Z^r if and only if the elimination does not destroy the property of Z^r. When all leaves of T_r other than d are examined, the remaining nodes in Z^r form the set S_r of the senders in the r-th round. Subsequently, A_{sda} eliminates S_r from its consideration by setting $T_{r+1} := T_r \backslash S_r$ and ends the $(r+1)$-th iteration.

Algorithm 9.1 *Shortest Data Aggregation*

---

(1) $r \leftarrow 1$, $I_1 \leftarrow U$, $T_1 \leftarrow$ a shortest path tree T of (G, S, d)
(2) **while** $T_r \neq \{\mathsf{d}\}$ **do begin**
(3) $T_r \leftarrow T_r \backslash (\{\text{leaves of } T_r\} \backslash (\{\mathsf{d}\} \cup I_r))$
(4) $Z^r = Z_r \leftarrow \{\text{leaves of } T_r\} \backslash \{\mathsf{d}\}$;
         $Y_r \leftarrow N_G(Z_r) \cap V(T_r)$, $S_r \leftarrow \emptyset$, $R_r \leftarrow \emptyset$
(5)   **while** $Z^r \backslash S_r \neq \emptyset$ **do begin**
(6) $z \leftarrow$ a node in $Z^r \backslash S_r$ of the max number of neighbors in Y_r
(7)   **if** $Y_r \subseteq N_G((Z^r \cup S_r) \backslash \{z\})$ **then** $Z^r \leftarrow Z^r \backslash \{z\}$
(8)   **else** $y_z \leftarrow$ a node from $Y_r \backslash N_G((Z^r \cup S_r) \backslash \{z\})$;
                 $S_r \leftarrow S_r \cup \{z\}$, $R_r \leftarrow R_r \cup \{y_z\}$
(9)   **end-while**
(10) $I_{r+1} \leftarrow I_r \cup R_r$, $T_{r+1} \leftarrow T_r \backslash S_r$
(11) $r \leftarrow r + 1$
(12) **end-while**
(13) **return** $s \leftarrow r-1$ and $A_{sda}(G, S, \mathsf{d}) \leftarrow \{(S_1, R_1), (S_2, R_2), \cdots, (S_s, R_s)\}$

One of main ideas of the shortest data aggregation based algorithms is to apply degree sorting and assign parallel transmissions (e.g. Steps 3-9 of A_{sda}). Intuitively speaking, we prefer to assign nodes of small degrees to send data before those of large degrees, and to arrange nodes of similar degrees to send data simultaneously. Both preferences increase potentially the number of parallel sendings/receivings and therefore reduce potentially the data aggregation time. Next we analyze theoretically the correctness and the performance of algorithm A_{sda}. For a subset of $U \subset V$ in $G = (V, E)$, the notation $N_G(U)$ is a shorthand of $\{v \mid uv \in E, u \in U\}$, and $N_G(\{u\})$ is simply written as $N_G(u)$.

Lemma 9.4 *Let $R_0 = S_0 = Z_{s+1} = \emptyset$. Then for each r with $1 \leq r \leq s$*
(i) *I_r consists of all informed nodes at the beginning of the r-th round.*
(ii) *T_r is a subtree of $T \backslash (\bigcup_{i=0}^{r-1} S_i)$ with root d such that $|V(T_r)| \geq 2$ and all data on $S \backslash V(T_r)$ have been aggregated to nodes in $V(T_r)$ at the beginning of the r-th round (i.e. the end of the $(r-1)$-th round.*
(iii) *Y_r and Z_r are nonempty subsets of $V(T_r)$ such that $Y_r \subseteq N_G(Z_r)$.*
(iv) *$Y_r \subseteq N_G(S_r)$ in Step 10.*
(v) *There is an 1-1 mapping between S_r and R_r in such a way that every sender $z \in S_r$ corresponds to its receiver $y_z \in R_r$.*
(vi) *$\emptyset \neq R_r \subseteq I_{r+1} \cap Y_r \subseteq V(T_{r+1}) \subseteq V(T_r) \backslash S_r \subsetneq T_r$, $S_r \subseteq Z_r \subseteq I_r$, and $Z_r \backslash S_r \subseteq Z_{r+1}$.*

Proof. We apply inductive arguments on r. First we examine the base case of $r = 1$. Statements (i-iii) are trivially true since T is a shortest path tree whose leaves must be all in S. Using $Y_1 \subseteq N_G(Z_1)$, it is easily checked that in the $|Z_1|$ times implementations of the inter **while**-loop (Steps 3-9), $Y_1 \subseteq N_G(Z^1)$ always holds, and $z \in Z_1$ is put into S_1 and y_z is put into R_1 if and only if $y_z \in N_G(z)$ and S_1 *will* not contain any neighbor of y_z other than z. Statements (iv) and (v) follow (note that $Z^1 = S_1$ ultimately in Step 10. Step 5 guarantees $S_1 \subseteq Z_1 \subseteq I_1$, which in turn gives $Z_1 \backslash S_1 \subseteq Z_2$. Since $\emptyset \neq Y_1 \subseteq N_G(S_1)$ (by (iii) and (iv)), we deduce from (v) that $|R_1| = |S_1| > 0$. It is easily checked from Step 10 and Step 5 that (vi) holds.

Then we proceed to inductive steps. We verify statements (i-vi) one by one for $2 \leq r \leq s$ under the hypothesis that (i-vi) are true for $(r-1)$. For the simplicity of description, we use superscripts $(r-1)$ and r to distinguish the conclusions (i-vi) with respect to $(r-1)$ and r, respectively, i.e., (i)$^{r-1}$, (ii)$^{r-1}, \cdots$, (vi)$^{t-1}$ and (i)^r, (ii)$^r, \cdots$, (vi)^r. Statement (i)^r is true by (i)$^{r-1}$, (v)$^{r-1}$, (vi)$^{r-1}$, and $I_r = I_{r-1} \cup R_r$ in Step 10. Since S_{r-1} ($\subseteq Z_{r-1}$) consists of some leaves of T_{r-1} (by (vi)$^{r-1}$), $T_{r-1} \backslash S_{r-1}$ is a tree, so is T_r in Step 10 and in Step 4. Moreover $|V(T_r)| \geq 2$ in Step 4. If all nodes other than d are deleted from T_r in Step 5, then $Z_r = \emptyset$, and it follows from (vi)$^{r-1}$ that $R_{r-1} = \{d\}$ (by $R_{r-1} \subseteq V(T_r)$) and $\emptyset \neq V(T_{r-1}) \setminus (S_{r-1} \cup \{d\}) \subseteq Z_{r-1} \setminus S_{r-1} \subseteq Z_r$, a contradiction. Thus we have $|V(T_r)| \geq 2$ in Step 4. Now (ii)^r follows from (ii)$^{r-1}$, (vi)$^{r-1}$, Step 5 and (i)^r. Obviously, (iii)^r follows from $|V(T_r)| \geq 2$. Statements (iv)^r-(vi)^r can be justified by applying arguments similar to those used in the base case with script r in place of script 1. The lemma is then proved. $\square$

Corollary 9.1   (i) $S_1, S_2, \cdots, S_s$ *are pairwise disjoint,* (ii) $S_r \cap (R_r \cup R_{r+1} \cup \cdots \cup R_s) = \emptyset$ *for all* $1 \leq r \leq s$, *and* (iii) $T = T_1 \supsetneq T_2 \supsetneq \cdots \supsetneq T_s \supsetneq T_{s+1} = \{d\}$.

Theorem 9.2   *Given an instance* (G, S, d) *of minimum aggregation time problem, Algorithm* A_{sda} *produces a schedule in time of* $O(|V|^2 \log|V| + |V| \cdot |E|)$.

Proof.   The termination of algorithm A_{sda} is guaranteed by Corollary 9.1(iii). From $T_{s+1} = \{d\}$ and Lemma 9.4(v,vi), it can be verified that T_s is a 2-node tree on $R_s = \{d\}$ and it is the only sender in S_s. Since, by Lemma 9.4(ii), all data on S have been aggregated to $V(T_s)$ at the end of the $s - 1$-th round, the schedule $\{(S_1, R_1), \cdots, (S_{s-1}, R_{s-1}), (S_s, R_s)\}$ returned by algorithm A_{sda} aggregates all data on S to d within s rounds.

To estimate the running time of algorithm A_{sda}, note that the computation of a shortest path tree in Step 1 requires time $O(|V| + |E|)$ and A_{sda} executes the external **while**-loop (Steps 2-12) at most $|V|$ times, i.e., $s \leq |V|$. Since within the r-th iteration (of the external **while**-loop) sorting degree of nodes in Z^r and selecting nodes to form S_r can be accomplished in time $O(|V| \log|V|)$ and $O(|E|)$, respectively, we deduce that the time complexity of SDA is $O(|V|^2 \log|V| + |V| \cdot |E|)$. The proof is then completed. $\square$

We now study the approximation performance ratio of algorithm A_{sda}. Let $h = h(G, S, d)$, and $L_i = \{$nodes in T at i hops away from d$\}$ for every $0 \leq i \leq h + 1$; in particular, $L_0 = \{d\}$ and $L_{h+1} = \emptyset$. Set $T_i = \emptyset$ for all $i \geq s + 2$.

Lemma 9.5   *For every* $0 \leq i \leq h - 1$, $L_{h+1-i} \cap V(T_{(\Delta-1)i+1}) = \emptyset$, *where* Δ *is the maximum degree of given graph* G.

Proof.   We prove the lemma by mathematical induction on i. The base case where $i = 0$ is justified by $L_{h+1} = \emptyset$. Proceeding inductively, suppose that $j = (\Delta - 1)(i - 1) + 1$ and $L_{h+1-(i-1)} \cap V(T_j) = \emptyset$, which implies that any node in $L_{h+1-i} \cap V(T_j)$ is a leaf in T_j. We aim to show $L_{h+1-i} \cap V(T_{j+\Delta-1}) = \emptyset$.

By contradiction, let $v \in L_{h+1-i} \cap V(T_{j+\Delta-1})$. Note from Step 4 (the definition of Z_r) and Lemma 9.4(vi) that $v \in Z_{j+k}$ for every $0 \leq k \leq \Delta - 1$. Hence there exists $u \in L_{h-i}$ such that $u \in Y_{j+k}$ for every $0 \leq k \leq \Delta - 1$. It follows from Lemma 9.4(iv) and Corollary 9.1(i) that there exist Δ distinct nodes $v_0, v_1, \cdots, v_{\Delta-1}$ such that $v_k \in S_{j+k} \cap N_G(u)$ for $k = 0, 1, \cdots, \Delta - 1$. Recall from Lemma 9.4(ii-iii) that $S_{j+\Delta-1} \cap Y_{j+\Delta-1} = \emptyset$ and u is a node in tree $T_{j+\Delta-1} \backslash S_{j+\Delta-1}$. Observe that $\mathrm{d} \in V(T_{j+\Delta-1}) \backslash S_{j+\Delta-1}$ and $u \neq \mathrm{d}$ (since $i \leq h - 1$). Therefore, u has a neighbor w in $T_{j+\Delta-1} \backslash S_{j+\Delta-1}$. Now u has $(\Delta + 1)$ distinct neighbors $w, v_0, v_1, \cdots, v_{\Delta-1}$. The contradiction completes the proof. $\qquad\square$

Theorem 9.3 *Given any instance (G, S, d), Algorithm A_{sda} returns a schedule whose data aggregation time $t_{sda}(G, S, \mathrm{d}) \leq \min\{(\Delta-1)h+1, (\Delta-1)t_{opt}(G, S, \mathrm{d})\}$.*

Proof. To prove the theorem, it suffices to show (i) $|t_{sda}(G, S, \mathrm{d})| \leq (\Delta - 1)h + 1$ and (ii) $|t_{sda}(G, S, \mathrm{d})| \leq (\Delta - 1)T_{opt}(G, S, \mathrm{d})$. Recall from Lemma 9.1 that $t_{opt}(G, S, \mathrm{d}) \geq h$. If $s = |t_{sda}(G, S, \mathrm{d})| \leq (\Delta - 1)(h - 1) + 1$ then we are done. So we assume $s > (\Delta - 1)(h - 1) + 1$.

To justify (i), we deduce from Lemma 9.5 that $L_2 \cap V(T_{(\Delta-1)(h-1)+1}) = \emptyset$, and then from Lemma 9.4(iii) that $V(T_{(\Delta-1)(h-1)+1}) \subseteq L_1 \cup \{\mathrm{d}\}$. Note that $|L_1| \leq \Delta$ and $T_{s+1} = \{r\}$. Thus by Corollary 9.1(iii) we obtain $s + 1 \leq (\Delta - 1)(h - 1) + 1 + |L_1| \leq \Delta(h - 1) + 2$, which implies (i).

Next we prove (ii). In case of $|L_h| \geq 2$, we have $t_{opt}(G, S, \mathrm{d}) \geq h + 1$ and (i) implies $s \leq (\Delta - 1)t_{opt}(G, S, \mathrm{d})$. It remains to consider the case where $|L_h| = 1$. We may assume $\Delta \geq 3$ (since otherwise, G is a path or a cycle and $s = h = t_{opt}(G, S, \mathrm{d})$). It is obvious that $S_1 = L_h$. Let $G' = G \backslash L_h$ and S' consist of the nodes in $S \backslash L_h$ and the neighbor (parent) of L_h in T. Then $T \backslash L_h$ is a shortest path tree of (G', S', r) that has height $h - 1$; moreover, there is an implementation of algorithm A_{sda} on (G', S', d) which outputs $\{S'_1, S'_2, \cdots, S'_{s-1}\}$ with $S'_i = S_{i+1}$ for all $1 \leq i \leq s - 1$. Using (i), we have $s - 1 = |t_{sda}(G', S', \mathrm{d})| \leq (\Delta - 1)(h - 1) + 1$ since the maximum degree of G' is upper bounded by Δ. It follows from $\Delta \geq 3$ that $s \leq (\Delta - 1)h \leq (\Delta - 1)t_{opt}(G, S, \mathrm{d})$, and (ii) is proved. The proof is then complete. $\qquad\square$

9.1.3.2   *Algorithms for Special Cases*

In this subsection, we show that, when Algorithm A_{sda} is applied to some special instances of minimum aggregation time problem, it will return solutions with better theoretical guarantees. First, in view of Theorem 9.3(ii), A_{sda} returns a $(\Delta - 1)$-approximation for Unit Disk Graphs (UDGs) with maximum degree Δ. In the application of wireless sensor networks, sensor devices cannot be too close or overlapped; thus it is reasonable to assume that the distance between any two nodes is no less than a positive constant λ. The UDG modelling such a sensor network is called a λ-*precision unit disk graph* [141; 208]. Krumke et. al [181] showed that the maximum degree of a λ-precision UDG is at most $\lceil 2\pi/\lambda^2 \rceil$. Consequently, we have the following result.

Corollary 9.2   *Given any instance* (G, S, d) *with* λ-*precision* G, *Algorithm* A_{sda} *returns a schedule whose data aggregation time* $t_{sda}(G, S, \mathsf{d}) \leq \frac{2\pi}{\lambda^2} t_{opt}(G, S, \mathsf{d})$.

Next we exhibit some local properties of UDGs, which ensure that Algorithm A_{sda} has better performance guarantee in some other special cases. Let v be a node in a unit disk graph $G = (V, E)$ of degree d. Then all nodes in $N_G(v) = \{v_0, v_1, \cdots, v_{d-1}\}$ are located within a disk centered at v with radius 1 and boundary B of length 2π. Corresponding to every v_i $(0 \leq i \leq d - 1)$, let b_i be a point on B such that $|b_i v_i|$ is minimized. If $|v_i v_j| > 1$, then the angle between the ray originated at v through v_i, b_i and the ray originated at v through v_j, b_j is greater than $\pi/3$. This implies $|b_i b_j| > 1$. Thus for any $0 \leq i, j \leq d - 1$,

(C0) $l_{i,j} \leq \pi/3 \Rightarrow |b_i b_j| \leq 1 \Rightarrow |v_i v_j| \leq 1$,

where $l_{i,j}$ denotes the length of the clockwise arc in B from b_i to b_j.

Lemma 9.6   *Let* $G = (V, E)$ *be a planar unit disk graph and* $v \in V$ *be a node of degree* d. *Then* $d \leq 15$ *and*
(i) *every node* $w \in N_G(v)$ *has a neighbor in any subset* W *of* $N_G(v) \backslash \{w\}$ *with* $|W| \geq 13$.
(ii) *there exist distinct* y, z, y', z' *in* $N_G(v)$ *such that* $N_G(y) \cap \{z, z'\} = \{z\}$ *and* $N_G(y') \cap \{z, z'\} = \{z'\}$ *if* $d \geq 9$.

Proof.   Denote $N_G(v) = \{v_0, v_1, \cdots, v_{d-1}\}$ and let $b_0, b_1, \cdots, b_{d-1}$ be d points on the boundary B of the unit disk centered at v that minimize $|b_i v_i|$, $i = 0, 1, \cdots, d - 1$. As usual, K_k stands for a complete graph on k vertices, and a subdivision of K_5 is the graph obtained from a K_5 by replacing each edge e of the K_5 with a path between the ends of e whose internal nodes (if any) all have degree 2. The planarity of G implies that

(C1) The subgraph induced by $\{v\} \cup N_G(v)$ contains no subdivision of K_5; and in particular no four nodes in $N_G(v)$ can induce a K_4.

If an arc in B of length at most $\pi/3$ contains four distinct points b_i, b_j, b_k, b_ℓ, then the distance between every pair from $\{b_i, b_j, b_k, b_\ell\}$ is not greater than 1, so is the distance between every pair from $\{v_i, v_j, v_k, v_\ell\}$ by claim (C0); it follows that $\{v_i, v_j, v_k, v_\ell\}$ induces a K_4, that contradicts claim (C1). Hence

(C2) any arc in B of length $\pi/3$ can contain at most three points from $b_0, b_1, \cdots, b_{d-1}$.

To see (i), for every $0 \le i \le d-1$, let $V_i = N_G(v) \backslash (\{v_i\} \cup N_G(v_i))$, and let A_i be the arc in B consisting of points at distance at least 1 from b_i. Clearly, A_i has length $4\pi/3$, and by claim (C0), contains every b_j with $v_j \in V_i$. Note from pigeonhole principle and claim (C2) that A_i can contain at most 12 points from $b_0, b_1, \cdots, b_{d-1}$. It follows that $|V_i| \le 12$, which implies (i).

We then prove (ii). If the subgraph of G induced by $N_G(v)$ contains a triangle, say $v_0 v_1 v_2 v_0$, then considering the distributions of $b_3, b_4, \cdots, b_8$ on B, without loss of generality we may assume that $l_{3,4} \le \pi/3$, and therefore, by (C0), $v_3 v_4 \in E$. Recall from (C1) that $\{v_0, v_1, v_2, v_3, v_4\}$ induces a subgraph of G containing no K_4. It is routine to check that there exist $1 \le i \ne j \le 3$ such that $v_i v_3 \notin E$ and $v_j v_4 \notin E$, which gives $y = v_i$, $z = v_4$, $y' = v_j$, $z' = v_3$ as desired. So we consider the case in which the subgraph induced by $N_G(v)$ is triangle-free. By $d \ge 9$, permuting indices if necessary, we may assume that $\{v_0 v_1, v_2 v_3\} \subseteq E$. It is routine to check that either $y = v_0, z = v_1, y' = v_2, z' = v_3$ or $y = v_0, z = v_1, y' = v_3, z' = v_2$ satisfy (ii).

Finally we prove $d \le 15$ by contradiction argument. Assuming $d \ge 16$, we consider $v_i, b_i, i = 0, 1, \cdots, 15$, and do all additions involving subscripts in modulo 16. Without loss of generality suppose that $b_0, b_1, \cdots, b_{d-1}$ are on B in clockwise order. It follows from (C0) that

$$l_{i,i+3} > \pi/3 \text{ for every } 0 \le i \le 15. \qquad (9.1)$$

If $l_{i,i+1} > \pi/3$ for some i, then (9.1) implies a contradiction

$$2\pi \ge l_{i,i+1} + (l_{i+1,i+4} + l_{i+4,i+7} + l_{i+7,i+10} + i_{i+10,i+13} + l_{i+13,i})$$
$$> \pi/3 + 5(\pi/3).$$

So $l_{i,i+1} \leq \pi/3$, and therefore, (C0) implies that

$$G \text{ contains a cycle } C \text{ with } V(C) = \{v_0, v_1, \cdots, v_{15}\}. \qquad (9.2)$$

Similarly, if both $l_{i,i+2} > \pi/3$ and $l_{i+2,i+4} > \pi/3$ for some i, then a contradiction occurs

$$2\pi \geq l_{i,i+2} + l_{i+2,i+4} + (l_{i+4,i+7} + l_{i+7,i+10} + i_{i+10,i+13} + l_{i+13,i})$$
$$> \pi/3 + \pi/3 + 4(\pi/3),$$

which also follows from (9.1). Thus

$$\min\{l_{i,i+2}, l_{i+2,i+4}\} \leq \pi/3 \text{ for every } 0 \leq i \leq 15. \qquad (9.3)$$

If $l_{i,i+2} \leq \pi/3$ and $l_{i+1,i+3} \leq \pi/3$ for some i, then $C \cup \{v\} \cup \{vv_i, vv_{i+1}, vv_{i+2}, vv_{i+3}, v_iv_{i+2}, v_{i+1}v_{i+3}\}$ is a subdivision of K_5 in G, a contradiction to (C1). So

$$\max\{l_{i,i+2}, l_{i+1,i+3}\} > \pi/3 \text{ for every } 0 \leq i \leq 15. \qquad (9.4)$$

By (9.3), suppose, without loss of generality, that $l_{0,2} \leq \pi/3$. We then have the following implications $l_{0,2} \leq \frac{\pi}{3} \overset{(9.4)}{\Longrightarrow} l_{1,3} > \frac{\pi}{3} \overset{(9.3)}{\Longrightarrow} l_{3,5} \leq \frac{\pi}{3} \overset{(9.4)}{\Longrightarrow} l_{4,6} > \frac{\pi}{3} \overset{(9.3)}{\Longrightarrow} l_{6,8} \leq \frac{\pi}{3} \overset{(9.4)}{\Longrightarrow} l_{7,9} > \frac{\pi}{3} \overset{(9.3)}{\Longrightarrow} l_{9,11} \leq \frac{\pi}{3} \overset{(9.4)}{\Longrightarrow} l_{10,12} > \frac{\pi}{3} \overset{(9.3)}{\Longrightarrow} l_{12,14} \leq \frac{\pi}{3} \overset{(9.4)}{\Longrightarrow} l_{13,15} > \frac{\pi}{3} \overset{(9.3)}{\Longrightarrow} l_{15,1} \leq \frac{\pi}{3} \overset{(9.4)}{\Longrightarrow} l_{0,2} > \frac{\pi}{3}$. The contradiction establishes (ii), and the proof is then finished. $\qquad \square$

Corollary 9.3 *Algorithm A_{sda} returns a schedule to the minimum aggregation time problem within approximation ratios 3 for grid graphs and $\sqrt{12m}$ for UDGs with m edges.*

Proof. The first bound comes directly from Theorem 9.3(ii) and Lemma 9.6(iii) immediately. We prove the second bound by showing $\Delta - 1 \leq \sqrt{12m}$. To this end, consider a node v in unit disk graph G of maximum degree Δ and the boundary B of the unit disk centered at v. We may partition $N_G(v) = \{v_0, v_1, \cdots, v_{\Delta-1}\}$ into six disjoint subsets $V_1, V_2, \cdots, V_6$ such that $\{b_j \mid v_j \in V_i\}$ is contained by an arc in B of length $\pi/3$. By (C0), each $V_i \cup \{v\}$ $(1 \leq i \leq 6)$ induces a $K_{|V_i|+1}$ in G. It follows that $\sum_{i=1}^{6} |V_i| = \Delta$, and the number of edges in the subgraph of G induced by $\{v\} \cup N_G(v) =$

$\{v\} \bigcup (\bigcup_{i=1}^{6} V_i)$ is lower bounded by

$$\sum_{i=1}^{6} \frac{|V_i|(|V_i|+1)}{2} = \Big(\sum_{i=1}^{6} \frac{|V_i|^2}{2} \Big) + \frac{\Delta}{2}$$

$$\geq \frac{1}{12} \Big(\sum_{i=1}^{6} |V_i| \Big)^2 + \frac{\Delta}{2} = \frac{\Delta^2}{12} + \frac{\Delta}{2},$$

which implies $m > \Delta^2/12$. Hence $\Delta - 1 \leq \sqrt{12m}$, and then the proof is then finished. $\qquad\square$

9.1.4 Partition Algorithm

In this subsection we will present an approximation algorithm that uses a geometrical partition technique. It has two stages: first construct an aggregation tree, and then schedule data aggregation/transimssion of the tree. We will describe each of these two stages in the following subsections.

9.1.4.1 Aggregation Tree Construction

The algorithm A_{cat} for constructing aggregation tree consists of two main steps: first we partition the plane into small cells of hexagons, and then construct an auxiliary graph G'.

In the first step, given an instance (G, S, d) of minimum aggregation time problem, we obtain the Unit Disc Graph (UDG) $G = (V, E)$, the geometric representation of underlying wireless sensor network. We partition the plane into regular hexagons of size 0.5 in such a way that every node in V is in exactly one of the hexagons. Refer to Fig. 9.2, where H_{ij} denotes the small hexagon for $i = 1, 2, \cdots, m$ and $j = 1, 2, \cdots, n$.

Let V_{ij} denote the set of nodes in H_{ij}, which may be empty. Then we have $V = \bigcup_{i=1}^{m} \bigcup_{j=1}^{n} V_{ij}$. Now we construct an auxiliary graph $G' = (V', E')$ of G as follows: We associate every nonempty subset V_{ij} with a node v_{ij} in V'. Thus

$$V' = \{v_{ij} \mid V_{ij} \neq \emptyset, i = 1, 2, \cdots, m; j = 1, 2, \cdots, n\};$$

There is an edge between two different nodes v_{ij} and v_{kl}, if there exist $u \in V_{ij}$ and $v \in V_{kl}$ such that $(u, v) \in E$, that is,

$$E' = \{(v_{ij}, v_{kl}) \mid v_{ij} \neq v_{kl}, \exists u \in V_{ij}, v \in V_{kl}, s.t.(u, v) \in E\}.$$

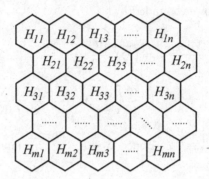

Fig. 9.2   The plane is partitioned into cells of hexagons.

With the auxiliary graph G' produced, we define the new data sink $d' \in V'$ in G' to be the node v_{ij} with $d \in V_{ij}$, and the new source field $S' = \{v_{ij} \in V' \mid V_{ij} \cap (\{S\} \cup d) \neq \emptyset\}$. In the end we can obtain a new instance (G', S', d') of minimum aggregation time problem, and find the shortest path tree T'_{spt} of (G', S', d').

In the second step, we construct an aggregation tree T_d for the original problem (G, S, d) with the help of T'. It is done as follows.

First, given T'_{spt} in G', let $V'(T'_{spt})$ (resp. $V(T)$) and $E'(T'_{spt})$ (resp. $E(T)$) be the node-set and edge-set of T'_{spt} (resp. T). For each edge $(v_{ij}, v_{kl}) \in E'(T'_{spt})$, by the definition of G', there must exist $u \in V_{ij}$ and $v \in V_{kl}$ such that $|uv| \leq 1$. There may be many such pairs of u and v. We just take any of them for each edge of $E'(T'_{spt})$, and put nodes u, v into $V(T)$, and edge (u, v) to $E(T)$. At the same time, we construct two node-sets V_{ij}^h and V^c which stand for the *head* set in V_{ij} and set of *clusters* of all V_{ij}, respectively. That means each V_{ij} may have many heads, but has at most one cluster. Suppose now that v_{ij} is the parent of v_{kl} in T'_{spt}, we put u into V_{ij}^h and v into V^c. Notice that each hexagon has at most one cluster. When all edge $(v_{ij}, v_{kl}) \in E'(T'_{spt})$ have been considered, we put d into both $V(T)$ and V^c. Now every hexagon has exactly one cluster. Next we connect these components in the current graph to form the tree T. For every $1 \leq k, l \leq m$, let v be the unique cluster in V_{kl}, then we connect v with all nodes in V_{kl}^h whenever $V_{kl}^h \neq \emptyset$. In particular, the cluster d is connected with all nodes in the V_{ij}^h with $d \in V_{ij}$. It is easy to see that T is a tree.

Secondly, let $\widetilde{V}_{ij} = (S \setminus V(T)) \cap V_{ij}$, then $\widetilde{V}_{ij}$ is the set of nodes in the intersection of S and the hexagon S_{ij} but outside T. Notice that each nonempty $\widetilde{V}_{ij}$, T contains at least one node in the hexagon S_{ij}, say u. According to the way of partitioning the plane, $\{u\} \cup \widetilde{V}_{ij}$ induces a complete subgraph. Now connect u with every node in $\widetilde{V}_{ij}$, and in the end delete all leaf nodes outside S recursively in the current tree constructed. The final tree is the data aggregation tree T_d as desired.

Algorithm 9.2 *Constructing Aggregation Tree*

---

(1) Embed $G = (V, E)$ into the plane, and partition it into $(m \times n)$ hexagons S_{ij} of size 0.5 for $i = 1, 2, \cdots, m, j = 1, 2, \cdots, n$.
$V_{ij} \leftarrow V \cap S_{ij}$

(2) Generate auxiliary graph $G' = (V', E')$ and instance (G', S', d')
$T' \leftarrow$ a shortest path tree of (G', S', d') and $E_0 \leftarrow E'(T')$

(3) $V_{ij}^h \leftarrow \emptyset$, $i = 1, 2, \cdots, m$, $j = 1, 2, \cdots, n$
$V^c \leftarrow \{\mathsf{d}\}$, $V(T) \leftarrow \{\mathsf{d}\}$, and $E(T) \leftarrow \emptyset$

(4) **while** $E_0 \neq \emptyset$ **do begin**

(5)   Take $(v_{ij}, v_{kl}) \in E_0$ and $(u, v) \in E$ such that
v_{ij} is the parent of v_{kl} in T' and $u \in V_{ij}$ and $v \in V_{kl}$

(6) $V(T) \leftarrow V(T) \cup \{u, v\}$ and $E(T) \leftarrow E(T) \cup \{(u, v)\}$

(7) $V_{ij}^h \leftarrow V_{ij}^h \cup \{u\}$ and $V^c \leftarrow V^c \cup \{v\}$

(8) $E_0 \leftarrow E_0 \setminus \{(v_{ij}, v_{kl})\}$

(9) **end-while**

(10) **for** every $v \in V^c$ **do begin**

(11) $V_{ij} \leftarrow$ the set containing v

(12)   Put an edge (u, v) into $E(T)$ for every $u \in V_{ij}^h$

(13) **end-for**

(14) $T_\mathsf{d} \leftarrow T$

(15) $\widetilde{V}_{ij} \leftarrow (S \setminus V(T_\mathsf{s})) \cap V_{ij}$, $i = 1, 2, \cdots, m, j = 1, 2, \cdots, n$

(16) **for** every $i = 1, \cdots, m, j = 1, \cdots, n$ with $\widetilde{V}_{ij} \neq \emptyset$ **do begin**

(17)   Take a node $u \in V_{ij} \cap V(T_\mathsf{d})$

(18)   Connect u with every node in $\widetilde{V}_{ij}$

(19) **end-for**

(20) $T_\mathsf{d} \leftarrow$ a shortest path tree of $(T_\mathsf{d}, S, \mathsf{d})$

(21) **return** T_d

9.1.4.2 *Aggregation Time Schedule*

The algorithm A_{sat} for scheduling aggregation time also consists of two steps. In the first step, schedule all nodes in $\widetilde{V}_{ij}$ as follows: consider every seven hexagons as a group (see Fig.9.3). In each group, the nodes coming from different hexagons are set to aggregate and transmit data to their parents. In the second step, schedule all nodes in T obtained at the end of Step 13 at Algorithm A_{cat} by running Algorithm A_{sda} with T as an input.

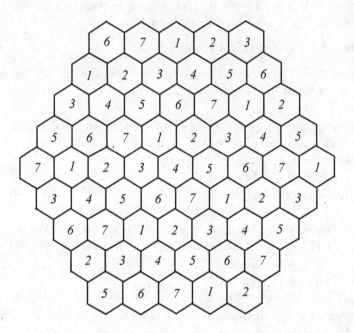

Fig. 9.3   Considering every seven hexagons as a group.

Algorithm 9.3 *Scheduling Aggregation Time*

---

(1) Merge all hexagons into groups of seven hexagons
(2) From hexagons labelled from 1 to 7, schedule all nodes in $\widetilde{V}_{ij}$ to their parents in T_{d} with aggregation time t_1
(3) Use T obtained at the end of Step 13 in Algorithm A_{cat} as the input in Step 1 of Algorithm A_{sda}, and then get the schedule $A_{sda}(T)$ with aggregation time t_2
(4) Return the combined schedule in above Steps 2 and 3 with final aggregation time $t = t_1 + t_2$

9.1.5 *Performance Analysis of Algorithm*

In this subsection, we will prove that integrating Algorithms A_{cat} with A_{sat} yields a new algorithm, denoted by A_{mat}, for minimum aggregation time problem that has guaranteed approximation performance ratio $(\frac{7\Delta}{\log_2 |S|} + c)$. For this purpose, we first show some lemmas in the following.

Lemma 9.7 *Each V_{ij} induces a complete subgraph of G.*

Proof. This is true since each hexagon H_{ij} has size of length 0.5 and there is an edge between two nodes in G if and only if the distance between them at most 1. □

Lemma 9.8 *Let t_1 be the aggregation time required for all nodes in $\widetilde{V}_{ij}$. Then $t_1 \leq \frac{7\Delta}{\log_2 |S|} t_{opt}(G, S, \mathsf{d})$.*

Proof. Let $\delta = max\{|\widetilde{V}_{ij}| \mid i = 1, 2, \cdots, m; j = 1, 2, \cdots, n\}$. Notice that in each group, the nodes located in different hexagons can send data to their parents at the same time without causing transmission conflict since the distance between these nodes are bigger than 1. Thus we have $t_1 \leq 7\delta \leq 7\Delta$. This, together with Lemma 9.1, leads to $t_1 \leq \frac{7\Delta}{\log_2 |S|} t_{opt}(G, S, \mathsf{d})$. The proof is then finished. □

Lemma 9.9 *Let Δ' be the maximum degree of nodes in auxiliary graph G'. Then $\Delta' \leq 18$.*

Proof. Recall that G is a unit disc graph and all hexagons v_{ij} have the same size of length 0.5. Since every node in V that is located in V_{ij} can reach only nodes located in those 18 hexagons as shown in Fig.9.4,

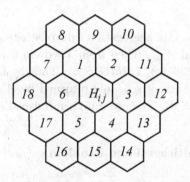

Fig. 9.4 The degree of v_{ij} is at most 18.

hence each $v_{ij} \in V'$ are adjacent to at most 18 other v_{ij}'s. The proof is then finished. □

Lemma 9.10 *Let $\Delta(T)$ be the maximum degree of T generated at the end of Step 13 in Algorithm A_{cat}, $\Delta(G(T))$ the maximum degree of induced graph $G(T)$, and T' the shortest path tree of the auxiliary graph G'. Then*

(i) *T is a tree rooted at d with $\Delta(T) \leq 18$ and $\Delta(G(T)) \leq 361$.*

(ii) *$h \leq 2h' \leq 2h(G, S, \mathsf{d})$, where h' and h are the heights of T' and T, respectively.*

Proof. (i) Recall that T is constructed in two steps. At first, all edges in T' are replaced by the corresponding edges in G. These edges may be separated since one node in T' may correspond to two or more different nodes in G. In the second step, we join every cluster with all heads in the same hexagon. Since T' is a tree, so is T. Notice that the maximum degree of T is no more than that of T'. Thus from Lemma 9.9 we deduce $\Delta(T) \leq \Delta' \leq 18$. By the way of constructing T, the number of nodes in each hexagon is at most 19.

(ii) By Lemma 9.7, in each hexagon the cluster can be joined with nodes in the head set directly. Then the length of each path in T is at most two times the length of the corresponding path in T'. Hence we have $h \leq 2h'$. By the way of constructing (G', S', d'), every path in G corresponds to a path in G' with equal or shorter length. Thus $h' \leq h(G, S, \mathsf{d})$. The proof is then finished. □

Theorem 9.4 *Given an instance (G, S, d) of minimum aggregation time problem, Algorithm A_{dat} returns an approximation solution whose data aggregation time is at most $(\frac{7\Delta}{\log_2 |S|} + c)$ times that of the optimal solution, where c is a constant.*

Proof. Let t_2 be the data aggregation time of schedule $A_{sda}(T)$ obtained in Step 3 of Algorithm A_{sat}. Then from Theorem 9.4 and Lemmas 9-10, we deduce $t_2 \leq \Delta h \leq 361h \leq 722h(G, S, \mathsf{d})$. Thus by Lemma 9.8, we have the data aggregation time of schedule returned by A_{dat} is $t = t_1 + t_2 \leq (\frac{7\Delta}{\log_2 |S|} + 722)t_{opt}(G, S, \mathsf{d})$. The proof is then finished. □

9.2   Minimum Multicast Time Problem

9.2.1   *Problem Formulation*

The models for data transmission and transmission schedule adopted for minimum multicast time problem [280] are almost the same as that of minimum aggregation time problem (Problem 9.1).

Now the source node s needs to send a query or message to some other nodes in the destination field D. It is assumed that communication is deterministic and proceeds in synchronous *rounds* controlled by a global clock. At each time round, any sensor that has received the message is allowed to communicate to at most one of its neighbors in the network; in other words, all other neighbors of this sensor hear nothing from this sensor. Due to collision-free constraint, two sensors can not send message to the same sensor simultaneously. The solution of (G, S, d) is a schedule $\{(S_1, R_1), (S_2, R_2), \cdots, (S_s, R_s)\}$ such that S_r (resp. R_r) is the set of senders (resp. receivers) in round r, $r = 1, 2, \cdots, s$, and all nodes in D receive the message of s within s rounds. Note that every (S_r, R_r) gives implicitly the 1-1 correspondence between S_r and R_r in a way that $v \in S_r$ corresponds to its receiver in R_r. The value s is called the *multicast time* of solution $\{(S_1, R_1), (S_2, R_2), \cdots, (S_s, R_s)\}$. The minimum multicast time problem is to find the optimal schedule with minimum multicast time $t_{opt}(G, D, \mathsf{d})$. The problem is more formally formulated as follows.

Problem 9.2 *Minimum Multicast Time Problem*

Instance A unit disk graph $G(V, E)$, a subset D of V and $\mathsf{s} \in V \setminus D$.
Solution A transmission schedule $\{(S_1, R_1), (S_2, R_2), \cdots, (S_s, R_s)\}$.
Objective Minimize the multicast time of transmission schedule.

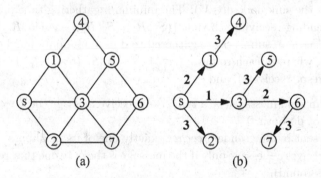

Fig. 9.5 An instance (G, D, s), where $D = \{v_i \mid i = 1, 2, \cdots, 7\}$.

In way of example, consider an instance (G, D, s) of minimum multicast time problem as shown in Fig.9.5(a), where the graph G contains the source node s and destination field $D = \{v_i \mid i = 1, 2, \cdots, 7\}$. In the first round, s sends message to v_3. In the second round, s sends message to v_1, and at the same time v_3 sends message to v_6. Although v_1 is located in the transmission range of v_3, no collision occurs at v_1 since the communication between s and v_1 and the communication between v_3 and v_6 use different frequencies. In the third round, s sends message to v_2, while v_1 sends message to v_4, v_3 sends message to v_5, and v_6 sends message to v_7. Fig.9.5(b) shows the schedule $\{(S_1, R_1), (S_2, R_2), (S_3, R_3)\} = \{(\{s\}, \{v_3\}), (\{s, v_3\}, \{v_1, v_6\}), (\{s, v_1, v_3, v_6\}, \{v_2, v_4, v_5, v_7\})\}$ with multicast time 3.

As in Section 9.1, we assume that each sensor knows its geometric position in the network, which is considered the unique ID of the sensor, and every sensor has global knowledge of all IDs in the WSN. The message will be sent from s to D following the schedule $\{(S_1, R_1), (S_2, R_2), \cdots, (S_s, R_s)\}$ determinately.

9.2.2 *Complexity Study*

It is known that minimum broadcast time problem is NP-hard [110] and has a 2Δ-approximation algorithm [246]. The NP-hardness proof for minimum multicast time problem is very similar to that for minimum aggregation time problem [53]. Let us consider a problem equivalent to the minimum broadcast time problem, the *minimum gathering time* problem [235]. In an instance (G, S, d) of minimum gathering time problem, the graph $G(V, E)$ models the WSN, the set § ($\subseteq V$) consists of nodes possessing messages requested by the sink node $\mathsf{d}(\in V)$. The minimum gathering time problem is to find a sending-receiving schedule $\{(S_1, R_1), (S_2, R_2), \cdots, (S_t, R_t)\}$ such that all messages on subset S are gathered to d in a minimum number t of time rounds, where in each round r ($r = 1, \cdots, t$), S_r (resp. R_r) is the set of senders (resp. receivers), and

(1) a node can send message (be a *sender*) or receive message (be a *receiver*) but cannot do both.
(2) message sent by any sender *reaches* exactly one of its neighbors.
(3) a node receives a message only if the message is the only one that reaches it (in this round).
(4) each receiver updates its message as the combination of its old message and the message received.

The following lemma is obvious from the definitions of the minimum multicast time problem and the minimum gathering time problem.

Lemma 9.11 *Given a schedule of minimum multicast tree, a schedule for minimum gathering time problem can be easily constructed just in the* *reverse order of the schedule of minimum gathering problem, and vice versa.*

Therefore, the minimum multicast time problem and the minimum gathering time problem are equivalent, and it suffices to prove that the minimum gathering time problem is NP-complete. The proof is very similar to the NP-hardness proof of the minimum aggregation time problem in Section 9.1.2 [53]. Applying results on orthogonal planar drawing and reduction from the NP-complete problem – restricted planar 3-SAT [186], we see that the minimum gathering time problem is NP-complete even when the underlying topology is a subgraph of a grid. Now the equivalence stated by Lemma 9.11 implies the following theorem.

Theorem 9.5 *The minimum multicast time problem is NP-complete even when the underlying graph is a subgraph of a grid.*

9.2.3 *Approximation Algorithm*

In this subsection we first exhibit some important properties of the minimum multicast time problem in wireless sensor networks. Then using these properties and a partition technique, we design a 15-approximation algorithm for the problem [281].

Given an instance (G, D, s) of minimum multicast time problem, a *shortest path tree* T of (G, D, s) is a tree in G consisting of shortest paths from s to nodes in D. The *height* of T, denoted by $h(G, D, \mathsf{s})$, equals to the length of the longest path in T from s to leaves of T. The following lower bound can be easily established by estimating multicasting time in a telephone network [23] (refer to Lemma 9.1).

Lemma 9.12 *Given an instance (G, D, s) of minimum multicast time problem, the minimal time of multicast schedule $t_{opt}(G, D, \mathsf{s})$ is at least* $\max\{h(G, D, \mathsf{s}), \lceil \log_2 |\{\mathsf{s}\} \cup D| \rceil\}$.

Lemma 9.13 *Suppose that H is a complete subgraph of G with $v \in V(H)$, then there exists a spanning tree T_H of H such that v can broadcast the message to all other nodes in H along the tree T_H within time of $\lceil \log_2 |V(H)| \rceil$ rounds.*

Proof. In fact, T_H can be constructed as follows: In each time round, every node that has received the message sends message to any one of its neighbor nodes which have not received the message yet. Then the whole process of multicast can be finished in time of $\lceil \log_2 |V(H)| \rceil$ rounds. In particular, if $\{\mathsf{s}\} \cup D'$ with $D' \subseteq D$ induces a complete subgraph of G, then

(G, D', s) is solvable in polynomial time, and the multicast time achieves its lower bound $\lceil \log_2 |\{\mathsf{s}\} \cup D'| \rceil$. The proof is then finished.   □

To make a multicast schedule we need to construct a multicast tree as shown in Fig.9.5(b). Sometimes the shortest path tree may be a good multicast tree. But the multicast schedule based on the shortest path tree can not be always a good approximate solution to the minimum multicast problem on G. The worst case occurs when $G = (V, E)$ is a complete graph with $\mathsf{s} \in V$ and $D = V \setminus \{\mathsf{s}\}$: the multicast schedule based on the shortest path tree, which is a star, needs time of $(|V| - 1)$ rounds while the optimal multicast time is $\lceil \log_2 |V| \rceil$. In the next subsection we will propose a new partition technique to construct a multicast tree which leads to an efficient multicast schedule.

The main idea of our algorithm is to construct a multicast tree T_s so that we make use of the optimal schedule on T_s to design an efficient multicast schedule for minimum multicast time problem. The algorithm consists of the following four basic steps.

Step 1. We firstly partition the plane into many small hexagons such that all nodes in each area induces a complete subgraph of G.

Step 2. With the obtained partition, we first construct an auxiliary graph G' with each node representing an area, and then define auxiliary instance $(G'; D', \mathsf{s}')$ and produce a shortest path tree T' of G'.

Step 3. Replacing each node in T' with some nodes in G, we obtain a Steiner tree T in G so that each nonempty area contains at least one node in T.

Step 4. Obtain T_s by joining the remaining nodes of D not on T with some nodes in T in such a way that we can apply Lemma 9.13 to assure an efficient schedule.

Since how to implement each of the above four stages has already been discussed in Section 9.1.4, we will just describe how to schedule the obtained multicast tree in the following.

Instead of using the simple implementation of an optimal broadcast algorithm on T_s, we present a different scheduling strategy that uses T_s just as a backbone tree (Messages are not multicasted strictly along the paths of the tree). This multicast schedule consists of two steps: Firstly, multicast on T generated at Step 13 of Algorithm A_{cat} using time t' rounds, and second multicast on $T_\mathsf{s} \setminus T$ by applying Lemma 9.13. The whole schedule algorithm A_{smt} on T_s is more formally presented below.

Algorithm 9.4 *Scheduling Multicast Tree*

(1) Label all nodes on T according to their hops from s
(2) $L(i) \leftarrow$ the set of nodes at the i-th level
(3) $L(0) \leftarrow \{s\}$
(4) **for** i from 0 to h **do begin**
(5)   apply Schedule(v) to every node $v \in L(i)$
(6)   Obtain the schedule and the time of t' rounds
(7) **end-for**
(8) **for** $u \in T_s \setminus T$ **do begin**
(9)   Apply Lemma 9.13 to node u
(10)   Obtain the schedule and the time of t'' rounds
(11) **end-for**
(12) **return** the final schedule with time of $t = t' + t''$ rounds

Let $C(v)$ denote the children of v on the tree T. Because of the property of unit disk graph, all nodes in $C(v)$ lie in the unit disk of v. From Lemma 9.9, we know the number of nodes in $C(v)$ is at most 18.

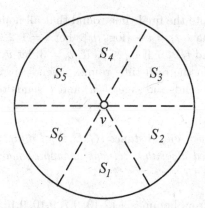

Fig. 9.6   Partition a disk into six equal sectors.

Now we design a multicast schedule for $\{v\} \cup C(v)$ which uses at most seven rounds. At first, we partition the unit disk into six equal sectors $s_1, s_2, \cdots, s_6$ as shown in Fig.9.6. Suppose that each sector s_i contains n_i number of nodes in $C(v)$, and assume, without loss of generality, that $n_1 \geq n_2 \geq \cdots \geq n_6$. Then since there are at most 18 nodes in the disk, we have $n_1 \leq 18$, $n_2 \leq 9, n_3 \leq 6, n_4 \leq 4, n_5 \leq 3, n_6 \leq 3$, and the nodes in each sector induce a complete graph. In the time round i, v sends the message to a node v_i in s_i. Then by Lemma 9.13, v_i finishes the multicast task in at most $\lceil \log_2 n_i \rceil$ rounds. This schedule denoted by Schedule(v) requires at most seven time rounds (see Lemma 9.15 in the next subsection).

9.2.4 *Performance Analysis of Algorithm*

In this subsection, we will prove that the proposed algorithm, which combines Algorithms A_{cat} with A_{smt}, is a 15-approximation algorithm for the minimum multicast tree problem. For this purpose, first, notice that Lemmas 9.9-10 also hold true for the proposed algorithm for minimum multicast time problem, and then we show two more technical lemmas.

Lemma 9.14   *If $D \cap V_{ij} \neq \emptyset$, then $V_{ij} \cap V(T) \neq \emptyset$.*

Proof.   The lemma is true since if $D \cap V_{ij} \neq \emptyset$, then $v_{ij} \in D'$, so T' must contain v_{ij}. $\square$

Lemma 9.15   *For any node $v \in T$, Schedule(v) finishes in seven rounds.*

Proof.   Let t_i denote the final time round that all nodes in s_i receive the message. Then we have $t_i = i + \lceil \log_2 n_i \rceil$. For $i = 1, 2, 3, 4, 5$, it is easy to verify that $t_i \leq 7$ and $t_6 \leq 7$ if $n_6 \leq 2$. If $n_6 = 3$, let v_6, u, w denote these there nodes in s_6. In the 6-th time round, v sends message to v_6. In the 7-th time round, v_6 sends message to u, and v sends to w. The proof is then finished. $\square$

Theorem 9.6   *Given any instance (G, D, s) of minimum multicast time problem, the proposed algorithm returns an approximation of the optimal solution within ratio 15.*

Proof.   It follows from Lemmas 9.12-13, 9.7, 9.10, 9.15, that the proposed algorithm returns a multicast schedule with time at most

$$
\begin{aligned}
t &= t' + t'' \\
&\leq 7h + \max\{\lceil \log_2(1 + |\tilde{V}_{ij}|) \rceil\} \\
&\leq 14h(G, D, \mathsf{s}) + \lceil \log_2 |\{\mathsf{s}\} \cup D| \rceil \\
&\leq 15 t_{opt}(G, D, \mathsf{s})
\end{aligned}
$$

The proof is then finished. $\square$

9.3   Discussions

In this chapter we have studied two versions of Steiner tree scheduling problem, minimum aggregating time problem and minimum multicast time problem. Both arise from the design and applications of Wireless Sensor Networks (WSNs) and aim at minimizing the transmission time of routing. It is worthy of pointing out that lots of work on routing problems in WSNs adopt the energy-efficiency as a metric to evaluate the performance of routing algorithms or protocols (refer to a recent survey [162] on the state-of-the-art routing techniques in WSNs). In the following we will discuss some of them as well as those which are closely related to minimum aggregation and multicast time problems.

9.3.1   *Convergecast*

There are two recent works [9; 166] on algorithm design of time efficient routing in WSNs. They study a special case of minimum aggregation time problem, called *convergecasting problem*, where $S = V \setminus d$ (i.e., data at all sensors are required to be sent and aggregated to the data sink). Annamalai et al. [9] propose a centralized heuristic that constructs a tree rooted at the sink node according to proximity criterion (a node is assigned as a child to the closest possible parent node) and to assign each node a code and a time slot to communicate with its parent node. More recently, Kesselman and Kowalski [166] propose a randomized distributed algorithm for convergecasting that has the expected running time $O(\log n)$ for any n-node network. They also prove that this bound is tight and any algorithm needs $\Omega(\log n)$ time steps. An assumption central to their model is that sensor nodes have capability of detecting collisions and adjusting transmission ranges, and that the maximum transmission range might be as large as the diameter of the network.

Kalpakis [161] considers the problem of *maximum lifetime data gathering* and aggregating in energy-constrained WSNs. They assume that each sensor periodically produces information as it monitors its vicinity. The basic operation in such a network is the systematic gathering and transmission of sensed data to a base station for further processing. During data gathering, sensors have the ability to perform in-network aggregation (fusion) of data packets enroute to the base station. The lifetime of such a sensor system is the time during which one can gather information from all the sensors to the base station. They study how to maximize the system lifetime, given the energy constraints of the sensors, and propose some heuristics for this problem.

Heinzelman et al. [133] propose an application-specific protocol architecture for data gathering in WSNs, which is known as LEACH (Low Energy Adaptive Clustering Hierarchy). It provides a solution where clusters are formed to fuse data before transmitting to the base station. By randomizing the cluster-heads chosen to transmit to the base station, LEACH achieves a factor of 8 improvement compared to direct transmissions in terms of when nodes die. Lots of work have been done to improve LEACH. For example, Lindsey et al. [190] propose an improved scheme, called PEGASIS (Power-Efficient GAthering in Sensor Information Systems), which is a near-optimal chain-based protocol that minimizes energy. In PEGASIS, each node communicates only with a close neighbor and takes turns transmitting to the base station, thus reducing the amount of energy spent per round.

Some works [133; 166] consider energy-efficiency and time-efficiency of data gathering at the same time. Heinzelman et al. [133] propose a scheme aiming at balancing the energy and delay cost for data gathering. Since most of the delay factor is in the transmission time, they measure delay in terms of number of transmissions to accomplish a round of data gathering. Therefore, delay can be reduced by allowing simultaneous transmissions when possible in the network. They introduce (*energy* × *delay*)-metric to evaluate the performances of proposed scheme. Kesselman and Kowalski [166] also study the trade-off between the energy and the latency of convergecast. They propose an algorithm that consumes at most $O(n \log n)$ times the minimum energy for n-node networks. They also demonstrate that for a line topology, the minimum energy convergecast takes n time steps while any algorithm performing convergecast within $O(\log n)$ time steps requires $\Omega(n/\log n)$ times the minimum energy.

9.3.2 *Multicast*

Lots of work have been done on energy-efficient multicast routing in WSNs, they adopt different network models. Wieselthier et al. [268; 269] assume that all sensors can adjust their transmission ranges continuously and the energy cost for data transmission between two sensors is measured in the square of Euclidean distance between them. They propose three greedy heuristics for minimum energy asymmetric broadcast routing: SPT (Shortest Path Tree), MST (Minimum Spanning Tree), and BIP

(Broadcasting Incremental Power). Based on the approach of pruning, they also propose three greedy heuristics for minimum energy asymmetric multicast routing: P-SPT (pruned shortest-path tree), P-MST (pruned minimum spanning tree), and P-BIP (pruned broadcasting incremental power). Wan et al. [261; 258] prove that SPT has an approximation ratio of at least $O(n^2)$ where n is the total number of nodes, and both MST and BIP have constant approximation ratios. They also prove that the approximation ratios of these three multicasting heuristics are at least $(n - 1/2)$, $n - 1$, and $n - 2 - o(1)$, respectively. Moreover, they show that any ρ-approximation Steiner tree algorithm gives rise to $c \cdot \rho$-approximation heuristic, where c is a constant between 6 and 12. In particular, Takahashi-Matsuyama Steiner tree algorithm [253] leads to an algorithm SPF (Shortest Path First) with approximation ratio at most $2c$. At the same time, they propose an algorithm MIPF (Minimum Incremental Path First) with approximation ratio between $(13/3)$ and $2c$. SPF and MIPF can be regarded as an adaptation of MST and BIP, respectively, but in a different manner from pruning method.

Li et al. [185] consider energy efficient broadcast routing in WSNs where each node is assumed to have a fixed level of transmission power. They prove that the problem is NP-hard and propose three centralized algorithms using shortest path tree, greedy strategy, and node weighted Steiner tree-based heuristic, respectively.

Papadimitriou and Georgiadis [223] study the *minimum energy broadcast problem* in WSNs. Their approach differs from the most commonly used one where the determination of the broadcast tree depends on the source node, thus resulting in different tree construction processes for different source nodes. They attempt to use a single broadcast tree to simplify the tree maintenance problem and allow scaling to larger networks. They show that, using the same broadcast tree, the total power consumed for broadcasting from a given source node is at most twice the total power consumed for broadcasting from any other source node. They also propose an algorithm that returns a tree with total power consumed for broadcasting from any source node within $2H(n-1)$ from the optimal, where n is the number of nodes in the network and $H(n)$ is the harmonic function. Moreover, they show that this approximation ratio is close to the best achievable bound in polynomial time.

9.3.3   *Convergecast and Multicast*

Multicast (broadcast) is to disseminate data/information from a central node to some (all other) nodes, while convergecast is to gather data/information from some (all) nodes towards a central node. They both are important communication paradigms in WSNs. Most of work focus on just one of these two. However, most sensor applications of WSNs involve both convergecasting and broadcasting, where the time taken to complete either of them has to be kept minimal. This can be accomplished by constructing an efficient tree for both broadcasting as well as convergecasting and allocating wireless communication channels to ensure collision-free communication. Note that a multicast/broadcast tree may not be time and energy efficient for convergecasting. For example, to minimize multicast time, sensors with long transmission rages are better. But they do not help for achieving time-efficient convergecast since collision will occur more likely (as a result, simultaneous transmissions can not be carried out).

Annamalai et al. [9] propose a heuristic algorithm (convergecasting tree construction and channel allocation algorithm) which constructs a tree with schedules assigned to nodes for collision free convergecasting. The algorithm is capable of code allocation, in case multiple codes are available, to minimize the total time required for convergecasting. They also show that the same tree can be used for broadcasting and is as efficient as a tree exclusively constructed for broadcasting.

Chapter 10

Survivable Steiner Network Problem

Survivable Steiner network problem arises from the survivable network design and applications. Given a set of terminals P, the network of minimum length that interconnects all terminals is Steiner Minimum Tree (SMT). Since such a network has a tree structure, it is vulnerable to the failures of links or nodes in networks. In fact, if any link or a certain node breaks down, then the resultant network will become disconnected, that is, there does not exist a path between some pairs of (good) nodes. Thus a *survivable network* must be multi-connected.

A graph $G(V, E)$ is called *k-edge connected* for any integer $k \geq 1$ if, after any set S of $(k-1)$ edges are removed from G, there still exits (at least) one path between any pair of vertices in $G(V, E \setminus S)$. Similarly, a graph $G(V, E)$ is called *k-vertex connected* if, after any set S of $(k-1)$ vertices (along with all edges incident to vertices in S) are removed from G, there still exists a path between any pair of vertices in $V \setminus S$. Clearly, k-vertex-connectivity is stronger than k-edge-connectivity in the sense that for any $k \geq 1$, a graph G is k-edge connected if it is k-vertex connected. In this chapter for the simplicity of presentation, we will use k-connectivity to denote either k-edge connectivity or k-vertex connectivity when a statement or claim is true for both. It is well known that a simple graph is k-edge (resp. k-vertex) connected if there exist k-edge (resp. k-vertex) disjoint paths between any pair of vertices.

For a set P of terminals, $N(V, E)$ is called a *Steiner network* on P if vertex-set V contains terminal-set P. It is called a *spanning network* on P if $V = P$. A Steiner network $N(V, E)$ on P is called a *weak k-edge* (resp. *k-vertex*) *connected* if after any $(k-1)$ edges (resp. vertices) are removed from the network, there still exists a path between any two vertices in $V \setminus P$. Similarly, a Steiner network $N(V, E)$ on P is called a *strong k-edge* (resp.

k-vertex) connected if after any $(k-1)$ edges (resp. vertices) are removed from the network, there still exists a path between any two vertices in V.

There are two versions of survivable Steiner network problem, which are formally defined as follows.

Problem 10.1 *Steiner Minimum k-Connected Network in Metric Spaces*

Instance A set $P = \{t_1, t_2, \cdots, t_n\}$ of terminals in a metric space M and an integer $k \geq 1$.

Solution A k-connected Steiner network $N(V, E)$ for P.

Objective Minimizing the total length of the edges in N, i.e.,
$$l(N) \equiv \sum_{e \in E} l(e).$$

Problem 10.2 *Steiner Minimum k-Connected Network in Graphs*

Instance A connected graph $G(V, E)$ with a cost[1] $l(e)$ on each edge $e \in E$, a terminal set $P \subset V$, and an integer $k \geq 1$.

Solution A k-connected Steiner network $N(V', E)$ for P with $V' \subset V$.

Objective Minimizing the total length of the edges in N, i.e.,
$$l(N) \equiv \sum_{e \in N} l(e).$$

Note that in the above formulations, we do not specify strong or weak connectivity. In fact, each has two versions associated with these two concepts. Moreover, we may also define a spanning network version of Problem 10.1 as follows.

Problem 10.3 *Minimum k-Connected Spanning Network in Metric Spaces*

Instance A set $P = \{t_1, t_2, \cdots, t_n\}$ of terminals in a metric space M and an integer $k \geq 1$.

Solution A k-connected spanning network $N(V, E)$ for P, that is, all edges have endpoints in P.

Objective Minimizing the total length of the edges in N, i.e.,
$$l(N) \equiv \sum_{e \in E} l(e).$$

Given an instance of Problem 10.1 and Problem 10.3, denote the lengths of minimum k-connected Steiner and spanning networks by $l_{smn-k}(P)$ and $l_{msn-k}(P)$, respectively. We can generalize Steiner ratio ρ_M in metric space M, refer to (1.1), to $\rho_M(k)$ as the infimum of the length of Steiner minimum k-connected network over that of minimum k-connected spanning network

---

[1]To keep consistence we will still use $l(\cdot)$ to represent the cost function on edges.

for all terminal-set P in M; More formally,

$$\rho_M(k) \equiv \inf\left\{\frac{l_{smn-k}(P)}{l_{msn-k}(P)} \,\Big|\, P \subset M\right\}. \tag{10.1}$$

For the simplicity of presentation, we will denote by $r_k(P)$ the ratio of $l_{smn-k}(P)/l_{msn-k}(P)$.

In this chapter, we will study some properties of k-connected Steiner and spanning networks and the generalized Steiner ratio $r_M(k)$ (and $r_k(P)$). We will also discuss some algorithms for Steiner minimum k-connected network problem. In Section 10.1, we consider the case of general $k \geq 2$, and in Sections 10.2 and 10.3 we focus on the cases of $k = 2$ and $k = 3$ (mainly in the Euclidean plane), respectively. In Section 10.4, we discuss some related problems.

10.1 Minimum k-Connected Steiner Networks

In this section, we will first present some general structural properties of minimum k-connected spanning networks and a useful technique for studying minimum k-connected Steiner networks.

Bienstock et al. [37] proved the following lemma that displays some structural properties of minimum k-edge-connected spanning networks.

Lemma 10.1 *Let H be the class of minimum k-edge-connected spanning subgraphs, with a nonnegative, symmetric weight function satisfying the triangle inequality. Then H can be restricted to those subgraphs which, in addition to the connectivity requirements, satisfy the following two conditions:*

(1) *Every vertex has degree k or $k + 1$.*

(2) *Removing any $1, 2, \ldots,$ or k edges does not leave the resulting connected components all k-edge-connected.*

As a Steiner network $N(V, E)$ of P can be regarded as a spanning network $N(V, E)$ of V, Lemma 10.1 is also true for Steiner networks.

Given P in a metric space M, let $G(V, E)$ be a k-edge-connected Steiner network on P. (G may contain multiple-edge.) Again points in $V \setminus P$ are called *Steiner points* while points in P are called *terminal points* (also called regular points in some references). Denote by $\lambda(x, y; G)$ the maximal number of edge-disjoint paths between x and y in G. It is well known that if G is k-edge-connected, then

$$\lambda(G) \equiv \min_{x,y \in V(G)} \lambda(x, y; G) \geq k. \tag{10.2}$$

Denote by $[x, y]_G$ the set of edges between the vertices x and y in G and by xy an arbitrary single edge in $[x, y]_G$.

The length of edge between x and y is defined as the distance between x and y in underlying metric space M and denoted by $l(x, y)$. Accordingly $l(\ ,\ )$ is a nonnegative, symmetric function satisfying the triangle inequality. The length of $G(V, E)$, denoted by $l(G)$, is defined as the total length of edges in G, i.e.,

$$l(G) \equiv \sum_{xy \in E} l(x, y).$$

Let $a = xz \in [z, x]_G$ and $b = yz \in [z, y]_G$ with $x \neq y$, and denote by $G(a; b)$ the multigraph which arises from G by removing edges a and b and adding exactly one new edge between x and y, i.e., $G(a; b) = (V, E \setminus \{a, b\} \cup \{xy\})$. It follows from the triangle inequality of metric space that $l(G) \geq l(G(a; b))$. The multigraph $G(a; b)$ is called a *lifting* of G at z, arising from the lifting of edges a and b at vertex z. See Fig.10.1.

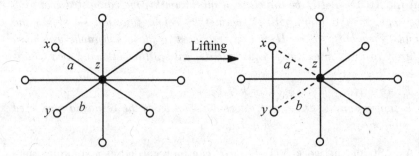

Fig. 10.1   Lifting at vertex z.

For a pair of vertices $x \neq y$ in $V(G) \setminus \{z\}$, it is obvious that $\lambda(x, y; G(a; b)) \leq \lambda(x, y; G)$. If for all such pairs $\lambda(x, y; G(a; b)) = \lambda(x, y; G)$ holds, we call the lifting *admissible*. See Fig.10.2.

For any vertex $x \in V$, denote the neighbor-set of x by $N(x; G) = \{y \in V \mid [x, y]_G \neq \emptyset\}$. Mader [200] proved the following basic result on admissible liftings, which will be frequently used in our study on the structure of k-edge-connected Steiner networks.

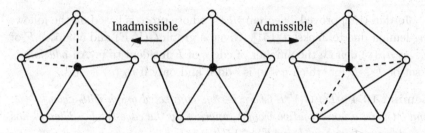

Fig. 10.2   Admissible lifting and inadmissible lifting.

Lemma 10.2   *Let z be a non-separating vertex whose degree is at least four, and $|N(z; G)| \geq 2$. Then there exists an admissible lifting of G at z.*

10.1.1   *Minimum Strong k-Connected Steiner Networks*

In this subsection we focus on minimum strong k-connected Steiner networks in metric spaces.

Lemma 10.3   *Given a set P of points and an integer $k \geq 2$, there exits a minimum strong k-edge connected Steiner network $G(V, E)$ on P such that any Steiner points in $V \setminus P$ has degree $(k + 1)$ for even k and degree k for odd k.*

Proof.   The theorem is obviously true for $k = 2$. We just consider the case of $k \geq 3$. By Lemma 10.1, we know that there exists a minimum-weight k-edge connected Steiner network $G(V, E)$ on P such that every vertex in G has degree k or $k + 1$. Let s be a Steiner point of G, then s must be a non-separating vertex of G with $|N(s, G)| \geq 2$. By Lemma 10.2, G has an admissible lifting at s if it has degree at least four. Moreover, we can find more admissible liftings at s until the degree of s is reduced to 3 or 2 since the degree of s is reduced by two each time of lifting while the degree of other vertices remain unchanged. If the final degree of s is two, then we can construct a new network $G'(V', E')$ by lifting at s and deleting s. Obviously, G' is still a k-edge connected Steiner network on P since $\lambda(x, y; G') = \lambda(x, y; G)$ for all pairs of distinct vertices x and y in $V \setminus \{s\} = V'$. In addition, by the triangle inequality we have $l(G') \leq l(G)$. Therefore, there exists a minimum-weight k-edge connected Steiner network on P such that each Steiner point has odd degree, which implies that it has degree $(k + 1)$ for even k and degree k for odd k. The lemma is then proved. □

Jordán [157] proved Theorem 10.1^2, whose proof is based on the following lemma due to Jackson [151]. Given a graph $G(V, E)$ and a subset F of E, let $n_F(v)$ denote the number of edges of F incident to v. An *odd join* is a set of $J \subseteq E$ for which $n_J(v)$ is odd if and only if $n_E(v)$ is odd.

Lemma 10.4   *Let $G(V, E)$ be a k-edge-connected graph with even $k \geq 2$, and let w be a non-negative weight function on the edge-set E. Then G has an odd join J with $w(J)/w(E) \leq 1/(k + 1)$.*

Proof.   The proof is by mathematical induction on $|E|$. Suppose that G has a vertex v with degree $d(v) \geq k + 2$. Since $k \geq 2$, we can apply Lemma 10.2 and split off two edges (v, u) and (v, w) in such a way that the resulting graph $G'(V, E')$ is k-edge-connected. Define a weight function w' on G' as follows: $w'(u, w) = w(v, u) + w(v, w)$ and $w'(x, y) = w(x, y)$ for all $(x, y) \neq (u, w)$. By the inductive hypothesis G' has an odd join J' with $w'(J)/w'(E') \leq 1/(k + 1)$. Since $w'(E') = w(E)$ and an odd join of G' yields an odd join of G of the same weight, the lemma follows.

Next suppose that G has vertex v of degree k. Since k is even, we can split off all the edges incident to v in pairs (and then delete v) such that the resulting graph G' remains k-edge-connected. As above, the lemma follows by induction.

In the end we may assume that G is $(k + 1)$-regular. By a result of Edmonds [93], G has a non-empty set S of perfect matchings such that each edge of G belongs to the same number of perfect matchings in S. Thus, for a minimum weight perfect matching $M \in S$ we have $w(M) \leq w(E)/(k+1)$. Hence M is the desired odd join of G. The proof is then finished. □

Theorem 10.1   *For a set of points P in metric space M, let $r_k(P)$ be the ratio of the length of minimum strong k-edge connected Steiner network of P over that of minimum k-edge connected spanning network of P. Then $r_k(P) \geq 1 - 1/(k + 2)$ for even $k \geq 2$ and $r_k(P) \geq 1 - 1/(k + 1)$ for odd $k \geq 3$.*

Proof.   Let $G^*(V, E)$ be a minimum strong k-edge connected Steiner network on P. By Lemma 10.3, we can assume that each vertex in $V \setminus P$ has degree $(k + 1)$ for even k or k for odd k.

---

^{2}There is a flaw in the proof given in [79].

Now let $G'(V', E')$ be the graph obtained from $G^*(P)$ by contracting the set P of terminals to a single vertex t, and let the length of an edge of G' be equal to the length of the corresponding edge in G^*. Note that G' is k-edge-connected and each vertex $v \neq t$ has odd degree in G'. By applying Lemma 10.4 to $G'(V', E')$, we deduce that there is an odd join J with

$$l(J) \leq \frac{1}{k+1} \, l(E') \text{ if } k \text{ is even, or } l(J) \leq \frac{1}{k} \, l(E') \text{ if } k \text{ is odd.}$$

Since $l(E') \leq l(E)$, adding the set J of edges to G results in a graph $G''(V'', E'')$ where each vertex in $V'' \setminus P$ has even degree and $l(E'') \leq \frac{k+2}{k+1} \, l(E)$ for even k while $l(E'') \leq \frac{k}{k+1} \, l(E)$ for odd k. By Lemma 10.2 we can iteratively eliminate the vertices in $V'' \setminus P$ by admissible splitting of pairs of edges so that the resulting graph G_S is a k-edge connected spanning subgraph on P. Since $l(G_S) \leq l(G'')$, the proof is then finished. $\square$

Edmonds [94] obtained many results on edge-disjoint branching of (unweighted) graphs and digraphs. One of them is given below, which finds an interesting application for studying the length of k-edge-connected Steiner networks.

Lemma 10.5 *Let $\overline{G}(\overline{V}, \overline{E})$ be a digraph and v a vertex in $\overline{V}$. Suppose that for any subset $S \subseteq \overline{V} \setminus \{v\}$, there are k arcs from $\overline{V} \setminus S$ to S in $\overline{G}$. Then $\overline{G}$ has k arc-disjoint spanning arborescences routed at v.*

Lemma 10.6 *For a set P of terminals in a metric space M, let $G_k^*(V, E)$ be a minimum k-edge connected Steiner network on P, and let T_{smt} and T_{mst} be a Steiner minimum tree and minimum spanning tree on P, respectively. Then $l(G_k^*) \geq \frac{k}{2} \, l(T_{smt}) \geq \frac{k}{2}\rho \cdot l(T_{mst})$, where ρ is the Steiner ratio on M.*

Proof. Let $\overline{G}_k^*$ be a digraph obtained from G_k^* by replacing each edge $(u, v) \in E$ with two oppositely oriented arcs (u, v) and (v, u). We may choose a vertex v_0 in V with maximum sum of lengths of edges incident to v_0, i.e.,

$$\sum_{(v, v_0) \in E} w(v, v_0) = \max \left\{ \sum_{(v, u) \in E} w(v, u) \; \middle| \; u \in V \right\}. \tag{10.3}$$

Now denote by G_k^0 the digraph obtained from $\overline{G}_k^*$ by deleting all arcs entering v_0. It is easy to see that $\overline{G}_k^*$ satisfies the condition of Lemma 10.5. Thus for any vertex $v \in V$, there are k arc-disjoint spanning arborescences rooted at v in $\overline{G}_k^*$, and each of them has weight greater than or equal to the length of Steiner minimum tree T_{smt} of P. Hence by equality (10.3) we can get

$$(1 - \frac{1}{n})\, l(\overline{G}_k^*) \geq l(G_k^0) \geq k \cdot l(T_{smt}),$$

where n is the number of vertices in $\overline{G}_k^*$. The above inequality proves the lemma since $2l(G_k^*) = l(\overline{G}_k^*)$.   □

Note that we have proved Lemma 10.6 without using the triangle inequality of metric functions, and it matches the result $|E| \geq \frac{k}{2}|V|$ of unweighted graphs.

Theorem 10.2   *There exists an α-algorithm for minimum Strong k-edge connected Steiner network problem with $\alpha = 3/2$ for even $k \geq 2$ and $\alpha = (3/2 + 1/(2k))$ for odd $k \geq 3$.*

Proof.   A simple approximation algorithm is as follows: First apply Christofides' algorithm [61] to produce a Hamilton cycle C_H of given terminal-set P, and then construct a minimum spanning tree T_{mst} of P, in the end construct a spanning network G of P that is composed of $\lfloor k/2 \rfloor$ duplications of C_H and $(\lceil k/2 \rceil - \lfloor k/2 \rfloor)$ duplications of T_{mst}.

It is easy to see that the obtained graph G is a k-edge connected spanning network on P. Let $G_k^*(V, E)$ be a minimum k-edge connected Steiner network on P. Then by Lemma 10.6, we have

$$\begin{aligned} l(G) &= \lfloor \frac{k}{2} \rfloor l(C_H) + \left(\lceil \frac{k}{2} \rceil - \lfloor \frac{k}{2} \rfloor \right) l(T_{mst}) \\ &\leq \lfloor \frac{k}{2} \rfloor l(T_{mst}) + \left(\lceil \frac{k}{2} \rceil - \lfloor \frac{k}{2} \rfloor \right) l(T_{mst}) \\ &= \left(\lceil \frac{k}{2} \rceil + \frac{1}{2} \lfloor \frac{k}{2} \rfloor \right) l(T_{mst}) \\ &\leq \frac{2}{k} \left(\lceil \frac{k}{2} \rceil + \frac{1}{2} \lfloor \frac{k}{2} \rfloor \right) l(G_k^*), \end{aligned}$$

which proves the theorem.   □

Observe that the approximation algorithm described in the above proof produces a k-edge connected spanning network, and by using exactly the same argument we can prove the following corollary.

Corollary 10.1   *There exists a polynomial-time algorithm that produces a k-edge connected spanning network for any given P whose weight is at most α times that of the minimum k-edge connected spanning network for P, where $\alpha = 3/2$ for even $k \geq 2$ and $\alpha = 3/2 + 1/(2k)$ for odd $k \geq 3$.*

10.1.2 *Minimum Weak k-Connected Steiner Networks*

In this subsection we turn our attention to minimum weak k-connected Steiner networks in metric spaces.

Lemma 10.7 *Given a set of points P in a metric space M and $k \geq 2$, there exists a minimum weak k-edge-connected Steiner network $G(V, E)$ on P such that every Steiner point in $V \setminus P$ has degree three.*

Proof. Let s be a Steiner point of G that has degree at least four. Then s is a non-separating vertex of G, and $|N(s; G)| \geq 2$. By Lemma 10.2, G has an admissible lifting at s if it has degree at least four. Thus we can apply admissible liftings at s until the degree of s is reduced to 3 or 2, since the degree of s decreases by two each time of lifting while the degree of other vertices remain unchanged. If the final degree of s is two, then we can construct a new network G' by lifting at s and removing s (along with two edges incident to s). Obviously, G' is still a k-edge-connected Steiner network on P, since $\lambda(x, y; G') = \lambda(x, y; G)$, for all pairs $x \neq y$ in $V(G) \setminus \{s\} = V(G')$. Moreover, it follows from the triangle inequality that $l(G') \leq l(G)$. The proof is then finished. $\square$

Theorem 10.3 *For any set P of terminals in a metric space M, let $r_k(P)$ denote the ratio of the length of minimum weak k-edge connected Steiner network on P over that of minimum k-edge connected spanning network on P. Then $r_k(P) \geq 3/4$ for $k \geq 2$.*

Proof. Let $G^*(V, E)$ be a minimum k-edge connected Steiner network on P. By Lemma 10.3, we can assume that each vertex in $V \setminus P$ has degree three.

Now let $G'(V', E')$ be the graph obtained from $G^*(P)$ by contracting the set P of terminals to a single vertex t, and let the length of an edge of G' be equal to the length of the corresponding edge in G^*. Note that G' is three-edge-connected and each vertex $v \neq t$ has odd degree in G'. By applying Lemma 10.4 to $G'(V', E')$, we deduce that there is an odd join J with $l(J) \leq \frac{1}{3} l(E')$.

Since $l(E') \leq l(E)$, adding the set J of edges to G results in a graph $G''(V'', E'')$ where each vertex in $V'' \setminus P$ has even degree and $l(E'') \leq \frac{4}{3} l(E)$. By Lemma 10.2 we can iteratively eliminate the vertices in $V'' \setminus P$ by admissible splitting of pairs of edges so that the resulting graph G_S is a three-edge connected spanning subgraph on P. Since $l(G_S) \leq l(G'')$, the proof is then finished. $\square$

Observe that the simple approximation algorithm described in the proof of Theorem 10.2 can also be used to produce an approximate solution to minimum weak k-edge connected spanning network problem.

Theorem 10.4   *There exists an α-approximation algorithm for minimum weak k-edge connected spanning network problem with $\alpha = 2$ for even $k \geq 2$ and $\alpha = (2 + 4/3k)$ for odd $k \geq 3$.*

10.2   Minimum Weak Two-Connected Steiner Networks

In this section we focus on the minimum weak two-connected Steiner network problem in the Euclidean plane (unless specified otherwise). For the simplicity of presentation, by "two-connectivity" we mean "weak two-connectivity", and an edge between vertices u and v will be denoted by uv or vu, which also represent its length defined as the Euclidean distance between the vertices/points u and v.

We assume throughout the section that there does not exist a straight line that contains every point in P since otherwise, minimum k-connected Steiner network problem (i.e., Problem 10.1) will become trivial.

Notice that for any three points a, b, and c, if $ab + bc = ac$, then these three points are on a straight line. By applying this property and the arguments similar to the proofs of [101; 215], we are able to show the following two lemmas.

Lemma 10.8   *The length of the minimum two-edge connected spanning network on P is equal to the length of the minimum two-vertex connected spanning network on P.*

Lemma 10.9   *Let $G(V, E)$ be a minimum two-edge or two-vertex connected spanning network on P. Then every vertex in V has degree 2 or 3; Moreover, deleting any edge or pair of edges in E leaves a bridge in one of the resulting connected components of G.*

Since a Steiner network $G(V, E)$ on P can be regarded as a spanning network $G(V, E)$ on V, so the above two lemmas are also valid for Steiner networks, which further deduce the following two lemmas.

Lemma 10.10   *If $G(V, E)$ is a minimum two-edge or two-vertex connected Steiner network on P, then it has no two edges that have the same pair of endpoints.*

Lemma 10.11 *Every minimum two-edge connected spanning network on P is also a minimum two-vertex connected spanning network on P, and vice versa.*

By the above lemma and as far as minimum two-connect Steiner network problem is concerned, we will use the term "two-connected" in the rest of this section without specifying "two-edge-connected" or "two-vertex-connected" unless the specification is needed. The following lemma can be deduced directly from Lemma 1.1.

Lemma 10.12 *Let $G(V, E)$ be a minimum two-connected Steiner network on P. Then every Steiner point in $V \setminus P$ is incident to exactly three edges and any two of them form an inner angle of 120°.*

Given a set of terminals P, let G^* denote a minimum two-connected Steiner network on P. As in the study of Steiner minimum trees, we can partition G^* uniquely into a set of full Steiner subnetworks $G_1, G_2, \cdots, G_m$ in such a way that each G_i is maximal with respect to the fact that every pair of edges in G_i can be connected by a path in G whose internal vertices are Steiner points. We call G_i a *full Steiner subnetwork* of G^*. See Fig.10.3, where G^* is partitioned into five full Steiner subnetworks. Note that two distinct G_i and G_j can intersect only at a set of terminals. Moreover, Luebke and Provan [197] proved the following two structural properties of minimum two-connected Steiner networks.

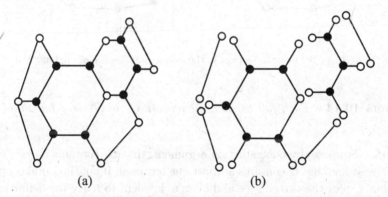

(a) (b)

Fig. 10.3 (a) A minimum two-connected Steiner network, and (b) it is partitioned into full Steiner subnetworks.

Lemma 10.13 *If two cycles C_1 and C_2 of G^* intersect in a simple path P_{ath}, then P_{ath} must have an internal vertex that is a terminal.*

Proof. Suppose, by contradiction argument, that all internal vertices in P_{ath} are Steiner points. Moreover, we assume that C_1 and C_2 are a pair of such cycles that P_{ath} has minimal number of edges. Suppose that they intersect at two points u and v. Let Steiner point s be an internal vertex in P_{ath}. Then there exists an edge e incident to s but not on P_{ath}, which is also not on either cycle. See Fig.10.4[3].

Note that edges e and e' must be on a cycle D since G^* is two-connected. Suppose that D intersects either C_1 or C_2, say C_1 at w as shown in Fig.10.4(a) (w may be on P_{ath}). Clearly, there exists a cycle C_3 that contains s, e, and e'. However, C_3 and C_2 intersect in a part of P_{ath}, which contradicts the choices of C_1 and C_2. Therefore, path P_{ath} must contain exactly a single edge (u, v). It can be verified that removing edge (u, v) from G^* will not spoil the two-connectivity of G^*, but this contradicts that G^* is a minimum two-connected Steiner network. The proof is finished. $\square$

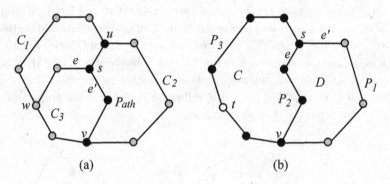

$$(a) \qquad\qquad (b)$$

Fig. 10.4   All internal vertices in the simple path are Steiner points.

Lemma 10.14   *Each full Steiner subnetwork G_i of G^* is a full Steiner subtree.*

Proof.   Suppose, by contradiction argument, that G_i contains a cycle C. We can assume that C contains at most one terminal; If not, it contains one terminal t, pick the two edges e and e' on C incident to t. By the definition of full Steiner subnetworks, there must be a cycle in G_i that contains e and e' but does not contain any other terminals except for t. Repeating this argument leads to a desired cycle.

---

[3] In this and other figures a point in G is marked by a circle with dots inside when its property is not clear, or this is not important with respect to our argument.

Let s be a Steiner point on C incident two edges e and e'. Suppose that edge e' is on C but e not. Since G^* is two-connected, there must exist a cycle D containing e and e' that intersects C at a Steiner point v. Now t and v partition the edges of C into two paths P_2 and P_3. Note that since C contains at most one terminal, at least one of P_2 or P_3, say P_2, has all of its internal vertices being Steiner points as shown in Fig.10.4(b). Thus the cycle formed by paths P_1 and P_2 intersects cycle C in a simple path P_2, violating the conditions of Lemma 10.13. Therefore, G_i is a tree. By the definition of full Steiner subnetworks, G_i can have terminals only at the end vertices, hence G_i must be a full Steiner tree. The proof is then finished. □

10.2.1   Complexity Study

In this subsection we will prove that minimum two-connected Steiner network problem (i.e., Problem 10.1 with $k = 2$) is NP-hard. The proof is given by Luebke and Provan [197], which is based on the reduction from a special version of the Euclidean Travelling Salesman Problem (TSP) to it. We first present some properties about the problem in a special case.

Lemma 10.15 *Suppose that all given points in P are lattice points of integral coordinates. Let G^* be a minimum two-connected Steiner network of P. Then*
(i) *If $l(G^*) = |P|$, then G^* is a TSP cycle such that every two successive points are adjacent in the lattice;*
(ii) *If $l(G^*) \neq |P|$, then $l(G^*) \geq |P| + \sqrt{2} - 1$.*

Proof. By Lemma 10.14, we can partition G^* into a set of edges $e_1, e_2, \cdots, e_m$ whose endpoints are terminals and a set of full Steiner trees $T_1, T_2, \cdots, T_s$. Let k_i be the number of terminals in T_i for $i = 1, 2, \cdots, s$. Note that the length of minimum spanning tree on these k_i terminals are at least $(k_i - 1)$ since the distance between any two points is at least one and T_i has $(k_i - 1)$ edges. Recall that the Steiner ratio in Euclidean plane is $\sqrt{3}/2$ [81], we know that the length of T_i is at least $\sqrt{3}(k_i - 1)/2 > k_i/2$. Hence the length of G is

$$l(G^*) = \sum_{i=1}^{m} l(e_i) + \sum_{j=1}^{s} l(T_j) \qquad (10.4)$$

$$\geq m + \sum_{j=1}^{s} \frac{3}{2}(k_i - 1) \qquad (10.5)$$

$$\geq m + \frac{1}{2} \sum_{j=1}^{s} k_i. \qquad (10.6)$$

Note that in the above inequality (10.4) is strict unless $l(e_i) = 1$ for all i, and inequality (10.5) is strict unless $s = 0$ (G^* has no Steiner points). Since each terminal in P must be contained in at least two distinct sets of $\{e_i \mid i = 1, 2, \cdots, m\}$ and $\{T_j \mid j = 1, 2, \cdots, s\}$, we have

$$2m + \sum_{j=1}^{s} k_i = 2|P|,$$

which, together with inequality (10.6), implies $l(G^*) \geq |P|$.

Note that $l(G^*) = |P|$ if and only if $s = 0$ and $l(e_i) = 1$ for each $i = 1, 2, \cdots, m$, this is the case that the minimum two-connected Steiner network is a Hamilton cycle through the points in P with each edge joining two points having distance 1 apart. This proves (i).

Now we prove (ii). Suppose that $l(G^*) > |P|$. We will consider the following four cases.

Case 1. G^* contains an edge between two non-adjacent lattice points p and q. In this case, $pq \geq \sqrt{2}$. So $l(G^*) \geq |P| + (\sqrt{2} - 1)$.

Case 2. G^* contains a full Steiner tree T_i interconnecting at least four terminals for some i. In this case, we use $l(T_i) \geq \sqrt{3}(4 - 1)/2$ instead of $l(T_i) \geq 4/2$ in inequalities (10.4-6), and then obtain

$$l(G^*) \geq |P| + \sqrt{3}(4 - 1)/2 - 4/2 > |P| + (\sqrt{2} - 1).$$

Case 3. G^* contains at least two full Steiner trees each interconnecting at least three terminals. As in case 2, we deduce

$$l(G^*) \geq |P| + 2(\sqrt{3}(3 - 1)/2 - 3/2) > |P| + (\sqrt{2} - 1).$$

Case 4. G^* contains exactly one full Steiner tree T_i interconnecting three terminals for some i. Since each terminal in P has degree at least three in G^*, then the number of edges in G^*, excluding those in T_i, must be at least $\lceil (2|P| - 3)/2 \rceil = |P| - 1$, each has length at least one. Using the bound $l(T_i) \geq \sqrt{3}(3 - 1)/2 = \sqrt{3}$, we have $l(G^*) \geq |P| - 1 = \sqrt{3}$.

As (ii) is proved and the proof of lemma is then complete. $\quad\square$

Theorem 10.5 *Minimum two-connected Steiner network problem is NP-hard.*

Proof. Recall that Itai et al. [150] show that the following problem is NP-complete: Given a set P of integral lattice points in the plane, decide if there exists a Hamilton cycle through all points in P such that the distance between each successive pair of points in the cycle is 1.

It follows immediately from Lemma 10.15 that for any given set P, minimum two-connected Steiner network problem has a solution of length k if and only if the above decision problem has a desired Hamilton cycle. The proof is then finished. $\quad\square$

10.2.2 Generalized Steiner Ratio

Let $G(V, E)$ be a Steiner network of terminal-set P in the Euclidean plane. Clearly, G consists of a cycle, denoted by $\mathcal{C}(G)$, and the union of some connected subnetworks inside $\mathcal{C}(G)$. We call G *basic* if $G \setminus \mathcal{C}(G)$ does not contain any cycle as shown in Fig.10.5(a), and *nonbasic* otherwise as shown in Fig.10.5(b). The following theorem shows some properties of basic networks.

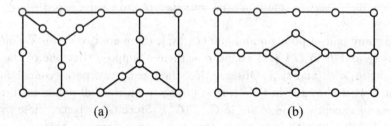

(a)   (b)

Fig. 10.5   (a) A basic Steiner network, and (b) a nonbasic Steiner network.

Theorem 10.6 *Suppose that $G^*(V, E)$ is a basic minimum two-connected Steiner network of P. Then*

(i) *There does not exist two Steiner points in $G \setminus \mathcal{C}(G^*)$ that are adjacent to each other;*

(ii) *$G^* \setminus \mathcal{C}(G^*)$ is a union of 3-Steiner minimum trees;*

(iii) *There does not exist any terminal point in $G \setminus \mathcal{C}(G^*)$ that is adjacent to two Steiner points.*

Proof.   (i) Suppose, by contradiction argument, that there exists an edge st in $G^* \setminus \mathcal{C}(G^*)$ such that both s and t are Steiner points. We will show that $G^* \setminus \{st\}$ is also a two-connected Steiner network on P contradicting that G has the minimum length. It is clear that there exist two paths from s to $\mathcal{C}(G^*)$ since G^* is two-connected; We denote one of them by $su_1u_2 \cdots u_k$ while the other by $sv_1v_2 \cdots v_m$. Similarly, there are two paths from t to $\mathcal{C}(G^*)$, $tw_1w_2 \cdots w_n$ and $tr_1r_2 \cdots r_i$, respectively. See Fig.10.6. Note that these four paths are all vertex-disjoint to one another and u_k, v_m, w_n, and r_i are four distinct points on $\mathcal{C}(G^*)$ since G^* is basic.

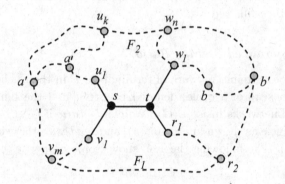

Fig. 10.6   Four vertex-disjoint paths connects $\mathcal{C}(G^*)$ to s and t, respectively.

For any pair of points a and b in $G \setminus \{st\}$, there are two vertex-disjoint paths Q and R in $G \setminus \{st\}$ connecting them. Suppose that one of them contains u_1s, st, and tw_1. Observe that there exist two paths connecting a and b with a' and b' in $\mathcal{C}(G^*)$, respectively, such that all points in these two paths except a' and b' are in $G \setminus \mathcal{C}(G^*)$. Since G^* is basic, these two paths are disjoint with $sv_1v_2 \cdots v_m$ and $tr_1r_2 \cdots r_i$. Moreover, there are two paths F_1 and F_2 on $\mathcal{C}(G^*)$ connecting a' with b' and v_m with r_i, respectively. Therefore, a and b are on the following common cycle:

$$\{a' \cdots a \cdots u_1s\} \cup \{sv_1 \cdots v_m\} \cup F_1 \cup \{r_i \cdots r_1t\} \cup \{tw_1 \cdots b \cdots b'\} \cup F_2.$$

When Q or R contains u_1s, st, and tr_1, or v_1s, st, and tr_1, by using the same argument we can arrive at the same conclusion that a and b are on a common cycle. Hence $G^* \setminus \{st\}$ is two-connected.

(ii) It follows directly from (i) and Lemma 10.14.

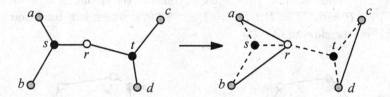

Fig. 10.7 A two-connected Steiner network with less length can be obtained.

(iii) Suppose, by contradiction argument again, that there exists a terminal point r incident to two Steiner points s and t. Now assume that s is adjacent to a and b, and t is adjacent to c and d as shown in Fig.10.7. Without loss of generality, we further assume that $|rt| \geq |rs|$. By the triangle inequality we obtain

$$|ar| < |as| + |sr| \leq |as| + |rt|, \ |br| < |bs| + |sr|, \text{ and } |cd| < |td| + |tc|.$$

Adding up the above three inequalities leads to

$$|ar| + |br| + |cd| < (|sa| + |sb| + |sr|) + (|tc| + |td| + |tc|).$$

By applying an argument similar to that used in (i), we can deduce that $G^* \setminus \{sa, sb, sr, tc, td, tr\} \cup \{ra, rb, cd\}$ is a two-connected Steiner network on P. This produces a contradiction since G^* has the minimum length. The proof is then finished. □

Observe that in the above proof of Theorem 10.6(i), we only use the assumption that s and t have degree three. Thus we can prove the following corollary using the same argument.

Corollary 10.2 *Suppose that G^* is a basic minimum two-connected Steiner network, then no edge in $G^* \setminus C(G^*)$ is incident to two points of degree three.*

Theorem 10.7 *Suppose that all terminal points in P are on the sides of the convex hull of P. Then the minimum two-connected Steiner network on P is the minimum two-connected spanning network on P, which is a Hamilton cycle of P on the sides of the convex hull of P.*

Proof. Let $G^*(V, E)$ be a minimum two-connected Steiner network on P, and let C_H be the convex hull of P. Since all points in P are on the sides of C_H, P is the set of extreme points of C_H. So we are able to label them by t_i in such a way that t_i is adjacent to t_{i-1} and t_{i+1} for $i = 1, 2, \cdots, |P|$, where $t_0 \equiv t_{|P|}$ and $t_{|P|+1} = t_1$. Now let us consider the spanning network $G'(V, E')$ on P with $E' = \{t_i t_{i+1} | i = 1, 2, \cdots, |P|\}$, which is a Hamilton cycle of P on the sides of C_H.

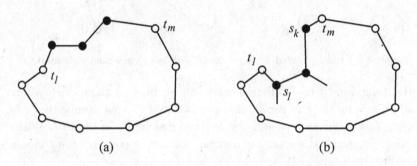

(a)   (b)

Fig. 10.8   Two possible cases when $G' \neq \mathcal{C}(G^*)$.

We will show, in fact, $G' = \mathcal{C}(G^*)$, which implies that $G' = \mathcal{C}(G^*) = G^*$ since G' is two-connected. Suppose, by contradiction, that there exists a path in G^* joining t_1 and t_m that consists of only Steiner points. We consider two possible cases.

Case 1. The path lies outside $\mathcal{C}(G^*)$ as shown in Fig.10.8(a). But this contradicts the geometrical structure of Steiner points specified in Lemma 1.1.

Case 2. The path lies inside $\mathcal{C}(G^*)$ as shown in Fig.10.8(b). In this case we can get a two-connected Steiner network of shorter length by removing all edges incident to Steiner points on the path and joining t_1 with t_m directly. This contradicts that G^* has the minimum length.

In each case, a contradiction will occur. The theorem is proved.   □

Theorem 10.8   *Suppose that every point in P except one is on the sides of the convex hull of P. Then the minimum two-connected Steiner network on P is the minimum two-connected spanning network on P, which is a Hamilton cycle of P.*

Proof.   By applying a similar analysis of Theorem 10.7, we deduce that the only possible way to cut the length short through introducing some Steiner points is to add two Steiner points adjacent to that unique terminal point lying inside C_H. However, using the same proof as for Theorem

10.6(iii), we can easily exclude this possibility. Hence the minimum two-connected Steiner network G^* on P is a spanning network on P.

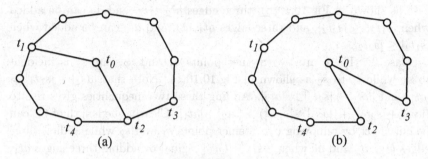

(a)   (b)

Fig. 10.9   The length of G^* could be shortened.

Suppose that $\mathcal{C}(G^*)$ is composed of the sides of C_H. Assume that the unique point inside $\mathcal{C}(G^*)$, denoted by t_0, is incident to two other terminal points t_1 and t_2 as shown in Fig.10.9(a). Without loss of generality, we assume that $|t_0t_1| \geq |t_0t_2|$. Denote two points adjacent to t_2 by t_3 and t_4, respectively. Note that

$$|t_0t_3| < |t_0t_2| + |t_2t_3| \leq |t_0t_1| + |t_2t_3|,$$

$$|t_0t_4| < |t_0t_2| + |t_2t_4| \leq |t_0t_1| + |t_2t_4|.$$

Thus the length of G^* can be shortened by replacing t_0t_1 and t_2t_3 (or t_2t_4) with t_0t_3 (or t_0t_4) as shown in Fig.10.9(b) while the resulting network remains two-connected. This is a contradiction since G has the shortest length. The proof is then finished.   □

Theorem 10.9   *If P contains at most five points, then the minimum two-connected Steiner network G^* is a spanning network on P, which is a Hamilton cycle of P.*

Proof.   The theorem is true for $|P| \leq 4$ due to Theorems 10.7-8. For the same reason, we only need to verify the case of $|P| = 5$ that there are two points, denoted by t_1 and t_2, which are inside the triangle formed by other three points, denoted by a, b, and c, respectively. In the following we will show that the length of G^* could be shortened if G^* contains some Steiner points by considering two cases.

Case 1. There are two Steiner points s_1 and s_2, and t_1 is incident to s_1 while t_2 is incident to s_1 and s_2. It can be verified that the length of G^*

can be shortened by deleting two Steiner points: In the case as shown in Fig.10.10(a), three edges at_2, t_1t_2, and bc can be added when $|s_1t_2| \le |s_2t_2|$, and three edges at_1, ct_2, and bt_2 can be added when $|s_1t_2| \le |s_2t_2|$; In the case as shown in Fig.10.10(b), three edges at_2, t_1t_2, and ac can be added when $|s_1t_2| \le |s_2t_2|$, and three edges at_1, ct_2, and at_2 can be added when $|s_1t_2| \le |s_2t_2|$.

Case 2. There are two Steiner points s_1 and s_2, and t_1 is incident to s_1 while t_2 to s_2 as shown in Fig.10.10(c). Note that $|as_1| + |s_1t_1| > |at_1|$ and $|as_2| + |s_2t_2| > |at_2|$. Adding these two inequalities gives rise to $(|as_1| + |s_2t_2|) + (|as_2| + |s_1t_1|) > |at_2| + |at_1|$. Clearly, the length of G^* can be cut short by removing two Steiner points s_1 and s_2 while adding three edges bt_1, at_2 and ac when $|as_1| + |t_2s_2| > |at_2|$ or adding three edges at_1, ct_2 and ab when $|as_2| + |s_1t_1| > |at_1|$ (since at least one of the inequalities must hold).

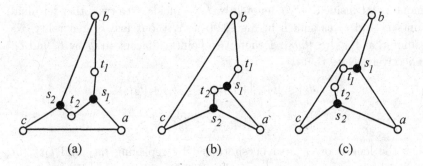

Fig. 10.10   Three possible two-connected Steiner networks when $|P| = 5$.

All other cases could be handled by using the length reducing technique developed in the proofs of Theorems 10.6-8. The proof is then finished. $\square$

No examples are discovered that show Theorem 10.9 is not true for $|P| = 6$ or 7. In fact, we believe that Theorem 10.9 remains true for $|P| \le 7$. In addition, Winter and Zachariasen [270] give an example, as shown in Fig.10.12(a), that demonstrates Theorem 10.9 is incorrect for six terminals in a graph of eight vertices with all edges having unit cost.

Moreover, Luebke and Provan [197] gave an example that demonstrates that Theorem 10.9 is not true for $|P| = 8$. As shown in Fig.10.11, there are eight terminals $\{t_i | i = 1, 2, \cdots, 8\}$ making a symmetric configuration with respect to left-right and up-down, four terminals $t_5, \cdots, t_8$ in the middle lying in a straight line, where $|t_5t_6| = |t_7t_8| = 1$ and $|t_6t_7| = \sqrt{3}/2$; Two

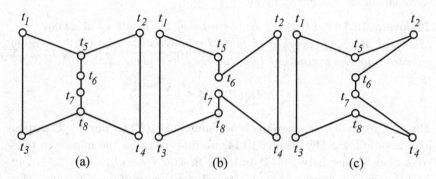

Fig. 10.11 Three two-connected spanning networks for an example of $|P| = 8$.

pairs of terminals t_1, t_3 and t_2, t_4 are on two sides of $t_5 t_8$, respectively, where $|t_1 t_3| = |t_2 t_4| = 4 + \sqrt{2}$, $t_1 t_3$ and $t_2 t_4$ both are parallel to $t_5 t_8$ with distance of 3. It can be verified that Fig.10.11(a) gives the minimum two-connected spanning networks with total length equal to 26.731, and two other solutions as shown in Fig.10.11(b) and (c) have lengths of 26.876 and 26.980, respectively. Clearly, the length of the optimal solution of Fig.10.11(a) can be shortened by replacing $t_5 t_2$ and $t_5 t_6$ with a Steiner minimum tree of t_2, t_5, and t_6, and replacing $t_8 t_3$ and $t_8 t_7$ with a Steiner minimum tree of t_3, t_7, and t_8.

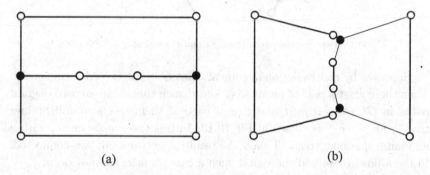

Fig. 10.12 Minimum two-connected Steiner networks: (a) in a graph and (b) in the Euclidean plane.

In addition, Winter and Zachariasen [270] give a similar example, as demonstrated in Fig.10.12(b), that shows the minimum two-connected Steiner network may not be a two-connected spanning network for eight terminals in the Euclidean plane.

Theorem 10.10   *Let $r_2(P)$ be the ratio of the length of minimum two-connected Steiner network over that of minimum two-connected spanning network both on terminal-set P in the Euclidean plane, then*

$$\frac{\sqrt{3}}{2} \leq \inf\{r_2(P) \mid P\} \leq 2\frac{\sqrt{3}+2}{\sqrt{3}+6}.$$

Proof.   We will first prove the lower bound of $r_2(P)$ for any P. The main idea is as follows: By Lemma 10.14, we first partition the minimum two-connected Steiner network G^* into full Steiner trees G_i, $i = 1, 2, \cdots, m$; And then we construct a spanning tree T_i for the set of terminals in G_i for each i; In the end, we construct a spanning network G consisting of all T_is. Recall that Steiner ratio in the Euclidean plane is $\sqrt{3}/2$ [81], we just need to show that G is a two-edge connected spanning network.

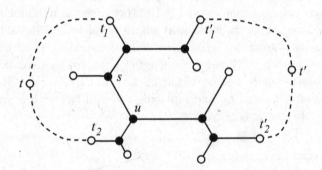

Fig. 10.13   Two disjoint paths joining t and t' meet at a full Steiner tree.

Suppose, by contradiction argument, that G is not two-edge connected. Then there exist a pair of terminals t and t' such that there are two disjoint paths in G^* connecting t and t' and each of them meets at full Steiner tree T_j for some j as shown in Fig.10.13. In this case, replacing T_j with a minimum spanning tree will make the resulting network not two-connected. In the following we will show that such a case, in fact, can not occur.

Let su be an edge on the path in T joining two paths which connects t_1 with t_1' and t_2 with t_2', respectively. Clearly, any path passing through edge su may go around via $t_1 \to t \to t_2$ or $t_1' \to t' \to t_2'$. Deleting edge su will yield a two-connected Steiner network with a shorter length, which contradicts that G^* has the shortest length. Hence there are two disjoint paths in G connecting t and t', each of them consisting of parts of different minimal spanning trees induced by their corresponding full Steiner trees. Thus G is a two-edge connected spanning network on P.

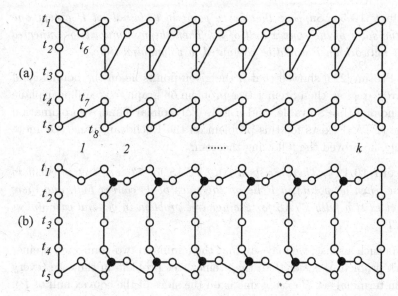

Fig. 10.14 (a) Minimum two-connected spanning network, and (b) minimum two-connected spanning network.

In the end, we will prove an upper bound of $r_2(P)$ by giving a series of terminal-sets P_k with $r_2(P_k) \le 2[(\sqrt{3}+2)/(\sqrt{3}+6)]$. Each P_k mainly consists of k identical units fastened one after another in a way specified in Fig.10.10, where $|t_it_{i+1}| = 1$ for $1 \le i \le 7$. It can be verified that the length of the minimum two-connected spanning network on P_k, as shown in Fig.10.14(a), is equal to $(\sqrt{3}+6)k + 6 - \sqrt{3}$, while the length of the minimum two-connected Steiner network on P_k, as shown in Fig.10.14(b), is $2(\sqrt{3}+2)k + 2(4 - \sqrt{3})$. It is easy to see that $r_2(P_k)$ tends to $2[(\sqrt{3}+2)/(\sqrt{3}+6)]$ as k approaches infinity. The proof is then finished. $\square$

The following corollaries follow directly from Theorem 10.10.

Corollary 10.3 *There exists a $\sqrt{3}$-approximation algorithm for the minimum two-connected Steiner network problem in the Euclidean plane.*

Corollary 10.4 *There exists a $\frac{3}{2}$-approximation algorithm for the minimum two-connected spanning network problem in the Euclidean plane.*

Moreover, from Theorems 10.7-8 and the well known fact [118] that the convex hull of any finite set could be computed in polynomial time, we deduce the following corollary.

Corollary 10.5   *Suppose that every point in terminal-set P except one is on the sides of the convex hull of P. Then the minimum two-connected Steiner network on P could be computed in polynomial time.*

As Provan [229] showed that if the given points lie on the boundary of a "convex" region, then Steiner tree problem on graphs or rectilinear plane is polynomial-time solvable, and a fully Polynomial Time Approximation Scheme (PTAS) exists for this problem on the Euclidean plane. Or more precisely, he proved the following theorem.

Theorem 10.11   *There exists a fully PTAS for Steiner tree problem in the case when all points in P lie on the sides of its convex hull, and there does not exist a fully PTAS for Steiner tree problem in general case unless $P = NP$.*

So in such a sense, we may say that the minimum two-connected Steiner network problem is easier than the Steiner tree problem in the case (every point in terminal-set P except one is on the sides of the convex hull of P).

Recall that Arora [10] proved that there exists a fully PTAS for the Steiner minimum tree problem in Euclidean plane (refer to Chapter 3). Luebke and Provan [197], however, proved the following negative result for the minimum two-connected Steiner network problem in the Euclidean plane.

Theorem 10.12   *There is no any fully PTAS for the minimum two-connected Steiner network problem in the Euclidean plane unless $P = NP$.*

Proof.   Consider the travelling salesman problem with all given terminals in P of lattice points. Recall Lemma 10.15, an optimal solution to the problem that is not a Hamilton cycle must have length at least $(|P|+\sqrt{2}-1)$. Now for $\epsilon = (\sqrt{2} - 1)/|P|$, an ϵ-approximate solution to the problem has length less than $(|P| + \sqrt{2} - 1)$ if and only if there exists a Hamilton cycle through the points of P such that the distance between each successive pair of points is 1. This means that if there exists such an ϵ-approximation algorithm, then it can be used to decide if P has a Hamilton cycle of length $|P|$ in polynomial time. Since this decision problem is known NP-complete, so there is no fully PTAS for the minimum two-connected Steiner network problem in the Euclidean plane unless $P = NP$. $\square$

10.2.3 *In the Rectilinear Plane*

By applying the same approach as used for the case of Euclidean plane, we can deduce a series of results for minimum two-connected Steiner networks on the *rectilinear plane* (refer to [138]) parallel to those obtained for the case of the Euclidean plane. In order to study and present these results easily, we need a few terminologies about the rectilinear geometry by referring to parallel terminologies adopted in the Euclidean geometry.

Given two points p and q in a plane, the *rectilinear edge* between p and q is defined as a set of all finite sequences of horizontal (vertical) segments alternating with vertical (horizontal) segments such that p and q are connected by them, and the total length of all segments in each sequence is equal to the rectilinear distance between p and q. In other words, the edge between p and q consists of infinite elements each has the same length. We, however, will denote the edge between p and q by pq or qp, which is used to denote one certain element in the edge between p and q (rather than the whole set) and its length as well as mathematical formulae.

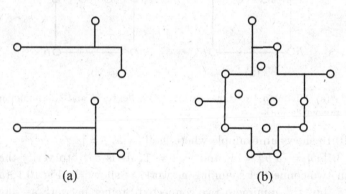

(a)   (b)

Fig. 10.15 (a) A convex hull of three points may not be a singleton, and (b) a convex hull is a singleton.

In addition, a set S of points in the plane is called a *rectilinear convex set* if for any two points p and q in S, there exists an edge pq such that every point on edge pq is in S. Note that if a set is a convex set in the Euclidean plane, then it is also a rectilinear convex set; but the reverse is not true. The *rectilinear convex hull* of a set S of points in the plane is a set of rectilinear convex sets containing S with minimal size. By contrast with corresponding concept in the Euclidean plane, the rectilinear convex hull of S is a collection of (maybe infinite number of) convex sets as shown in Fig.10.15 However, we can prove that the total length of sides in each convex set is equal. Hence, we denote the rectilinear convex hull of S by $C(S)$, which will be used to indicate a single convex set in the hull.

Adopting the above definitions and using the similar arguments applied to the case of the Euclidean plane, we can prove the following theorems.

Theorem 10.13 *Suppose that all terminals in P (maybe except one) are on the sides of the rectilinear convex hull of P. Then the minimum two-connected Steiner network on P in the rectilinear plane is a spanning network on P.*

Theorem 10.14 *In the rectilinear plane the minimum two-connected Steiner network for any P with $|P| \leq 5$ is a spanning network.*

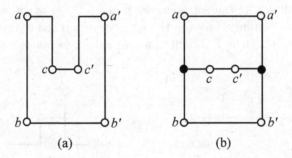

(a)   (b)

Fig. 10.16 (a) Minimum two-connected spanning network, and (b) minimum two-connected Steiner network.

Fig.10.16 shows an example where $|aa'| = |bb'| = |ac| = |a'c'| = |bc| = |b'c'| = 3$, $|ab| = |a'b'| = 4$, and $|cc'| = 1$. It is easy to verify that the minimum two-connected spanning network as shown in Fig.10.16(a) has length 18, but the minimum two-connected Steiner network as shown in Fig.10.16(b) has length 16.

Theorem 10.15 *Let $r_2(P)$ be the ratio of the length of minimum two-connected Steiner network over that of minimum two-connected spanning network both on terminal-set P in the rectilinear plane, then*

$$\frac{3}{4} \leq \inf\{r_2(P) \,|\, P\} \leq \frac{6}{7}.$$

Proof.   The lower bound of 3/4 follows from Theorem 10.1. The upper bound of 6/7 is achieved by a special class of set P_k as shown in Fig.10.17 with $k = 4$, where $|t_1 t_2| = |t_2 t_3| = |t_6 t_8| = 3$, $|t_1 t_4| = |t_2 t_5| = |t_2 t_6| =$

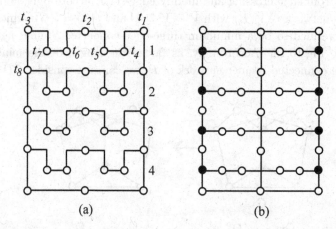

Fig. 10.17 (a) Minimum two-connected spanning network, and (b) minimum two-connected Steiner network.

$|t_3t_7| = |t_3t_8| = |t_7t_8| = 2$, and $|t_4t_5| = |t_6t_7| = 1$. It is easy to verify that the length of the minimum two-connected spanning network on P_k of Fig.10.17(a) is $2(7k + 3)$, and the length of the minimum two-connected Steiner network on P_k of Fig.10.17(a) is $12(k + 1)$, the ratio of the latter to the former achieves $6/7$ as k goes to the infinity. The proof is finished. $\square$

10.3 Minimum Weak Three-Edge-Connected Steiner Networks

In this section we study the minimum weak three-edge-connected Steiner network problem (i.e., Problem 10.1 with $k = 3$). Throughout the section, for the simplicity of presentation, by "three-connectivity" we mean "weak three-connectivity".

10.3.1 *In the Euclidean Plane*

In this subsection we focus on the case of terminal-set P on the Euclidean plane. To avoid triviality, we assume that there does not exist a straight line that contains every point in P. Let $G(V, E)$ be a minimum three-edge-connected Steiner network of P. In order to simplify our analysis and argument, we will get rid of crossing edges in $G(V, E)$ (if any) in the following way: We enlarge vertex-set V to include all points in V and all intersections produced from edge crossing and modify edge-set E accordingly, and then get a new network $\overline{G}(\overline{V}, \overline{E})$ with $\overline{V} \supseteq V \supseteq P$ and $l(\overline{G}) = l(G)$, which can be easily verified to be a minimum three-edge-connected Steiner network of P. See Fig.10.18. Accordingly, as far as the length of the minimum three-edge-connected Steiner network of P, we can just consider $\overline{G}(\overline{V}, \overline{E})$.

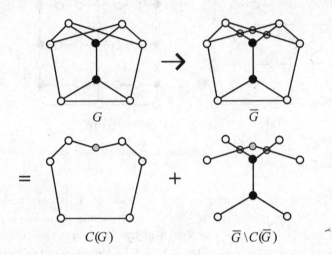

Fig. 10.18   Removing crossing edges from G produces $\overline{G} = C(\overline{G}) + \overline{G} \setminus C(\overline{G})$.

For the resulting network $\overline{G}(\overline{V}, \overline{E})$, we can describe it as a close trail $C(\overline{G})$ enclosing some subnetworks. See Fig.10.18. Let $V(C(\overline{G}))$ denote the set of points on $C(\overline{G})$. Observe that in Fig.10.18 all points in $V(C(\overline{G}))$ belong to V except one which is a crossing point of edges in E.

Lemma 10.16   *Let $G(V, E)$ be a minimum three-edge-connected Steiner network of P. Then $C(\overline{G})$ is a cycle, and $\overline{G} \setminus C(\overline{G})$ is a connected spanning network of $\overline{V}$ and a connected Steiner network of $V(C(\overline{G}))$.*

Proof.   Suppose, by contradiction argument, that $C(\overline{G})$ is not a cycle, then $C(\overline{G})$ is a union of at least two cycles, thus $C(\overline{G})$ contains a cut vertex c which is the joint of two cycles C_1 and C_2 in $C(\overline{G})$ and has degree four because of Lemma 10.1. Notice that c is incident to two edges in C_1 (and C_2, respectively). Clearly removing these two edges will cause $\overline{G}$ disconnected, this contradicts that $\overline{G}$ is three-edge-connected.

By contradiction argument again, suppose that $\overline{G} \setminus C(\overline{G})$ is not a connected spanning network of $\overline{V}$, then there exist two separated subnetworks G_1 and G_2 of $\overline{G} \setminus C(\overline{G})$, i.e., there is no path in $\overline{G} \setminus C(\overline{G})$ connecting G_1

and G_2. Since $C(\overline{G})$ is a cycle and $\overline{G} \setminus C(\overline{G})$ is inside $C(\overline{G})$, there exist two edges on $C(\overline{G})$ which connects G_1 and G_2 while deleting them will cause $\overline{G}$ to be disconnected, this contradicts that $\overline{G}$ is three-edge-connected.

By using the same method we can show that $\overline{G} \setminus C(\overline{G})$ is a connected Steiner network of $V(C(\overline{G}))$. The proof is then finished. ☐

In general, Lemma 10.16 is not true for original network $G(V, E)$, even if it is a minimum three-edge-connected spanning network of P. See Fig.10.19 for a simple example, where P has four points that produce a square.

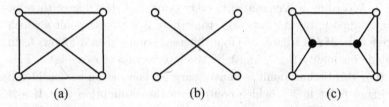

Fig. 10.19 (a) Minimum 3-edge-connected spanning network G, (b) $G \setminus C(G)$, and (c) minimum 3-edge-connected Steiner network.

In order to simplify our notations, we will keep using $G(V, E)$ instead of $\overline{G}(\overline{V}, \overline{E})$, and assume

(1) $G(V, E)$ is a minimum three-edge-connected Steiner network of P;
(2) $G(V, E)$ has minimal number of Steiner points in V, which consists of terminals, Steiner points, and crossing points.

From assumption (1) we know that for any pair of points u and v in V, there are three edge-disjoint paths in G connecting them. Denote them by $p_i(u, v), i = 1, 2, 3$.

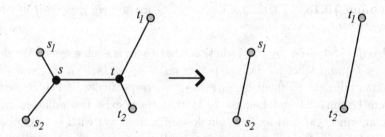

Fig. 10.20 Steiner lifting at st.

Let s and t be two adjacent Steiner points, where s and t are adjacent to two points s_i and t_i, $i = 1, 2$, respectively. The process of deleting s and t together with all edges incident to them and adding two edges $s_1 s_2$ and $t_1 t_2$ is called *Steiner lifting* of G at st. See Fig.10.20. It is further called *admissible* if the resulting network remains three-edge-connected.

From this definition and assumption (1) we know that G has no admissible Steiner lifting.

Lemma 10.17   *Each Steiner point in G is adjacent to exactly three edges meeting at angles of* $120°$.

Proof.   According to Lemma 10.1, every point in G has degree three or four. Now suppose, by contradiction, that there is a Steiner point s which has degree four. If s is adjacent to four different points, then it follows from the triangle inequality that s must be the intersection of two edges. This implies that this Steiner point is unnecessary (in fact, it can be considered as a crossing point in V), which contradicts the assumption (2). If s is adjacent to three different points a, b, and c, this means that one of these points, say a, is connected with s by multiple edges. Now construct a new network G' by removing Steiner point s from G (together with those four edges incident to it) and adding edges ab and bc. It is easy to verify that G' is a three-edge-connected Steiner network of P with $l(G') < l(G)$, this contradicts assumption (1).

Now suppose, by contradiction again, that a Steiner point is incident to three edges and two of them do not meet at an angle of $120°$. Then we can relocate this Steiner point in such a way that it is incident to these three edges while any two of them meet at an angle of $120°$. Clearly, the modified network is still a three-edge-connected Steiner network of P, while its length, according to Lemma 1.1, is shorter than that of G, this contradicts the assumption (1). The proof is then finished.   □

Lemma 10.18   *There is no cycle in G exclusively composed of Steiner points.*

Proof.   Suppose, by contradiction, that there is such a cycle. We denote it by C, and label all Steiner points on C by s_0, s_1, s_2, ..., and denote points adjacent to them by r_0, r_1, r_2, ..., respectively. It can be deduced from Lemma 1.1 and Lemma 10.17 that there exist two adjacent Steiner points on C, s_1 and s_2 (without lose of generality) with r_1 and r_2 being outside of C. We will produce a contradiction by showing that there is an admissible Steiner lifting of G at edge $s_1 s_2$ on C, or equivalently we will show that for any pair of points u and v in $V \setminus \{s_1, s_2\}$, there are three edge-disjoint paths connecting them in $G'(V', E')$ obtained from $G(V, E)$ by applying Steiner lifting at $s_1 s_2$.

Case 1. There exists an edge xy such that $\{xy, s_0 s_1, s_2 s_3\}$ is a *cut set* [4] of G. In this case, xy is not incident to s_1 or s_2 (otherwise there will exist a smaller cut set of G), and $G \setminus \{xy, s_0 s_1, s_2 s_3\}$ consists of two separate subnetworks G_1 and G_2. See Fig.10.21.

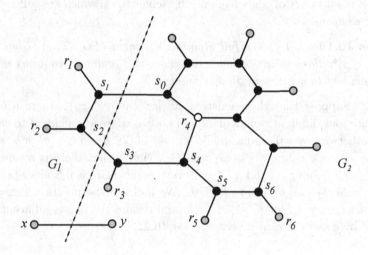

Fig. 10.21   Case 1 in the proof of Lemma 7.

Now we will use three paths connecting r_1 with r_2, i.e., $p_i(r_1, r_2)$, $i = 1, 2, 3$, as auxiliary paths to reconstruct $\{p_i(u, v) \mid i = 1, 2, 3\}$ in G to make them to be three edge-disjoint paths in G' connecting u and v. It can be easily verified that all possible cases associated with $p_i(r_1, r_2)$, $i = 1, 2, 3$ can be reduced to the case that $\{p_i(r_1, r_2) \mid i = 1, 2, 3\}$ are in G_1, and $p_1(r_1 r_2)$ is $r_1 s_1 s_2 r_2$. Notice that when no path in $\{p_i(u, v) \mid i = 1, 2, 3\}$ includes $s_1 s_2$, then we are done, as they, after some minor modification (if necessary, replacing $r_1 s_1 s_0$ and $r_2 s_2 s_3$ in G with $r_1 s_0$ and $r_2 s_3$ in G', respectively), are three edge-disjoint paths in G' connecting u and v. Suppose that one of these three paths, say $p_1(u, v)$, includes $s_1 s_2$. Then $p_2(u, v)$ and $p_3(u, v)$ can not include any edge in $\{r_1 s_1, r_2 s_2, s_1 s_2, s_1 s_0, s_2 s_3\}$. Thus u and v are either in G_1 or G_2 (otherwise both $p_2(u, v)$ and $p_3(u, v)$ use xy). In the former case, $p_2(u, v)$ and $p_3(u, v)$ are both in G_1, and $p_1(u, v)$ can go around

---

[4]Removing all edges in the set from E will make resultant graph G disconnected.

s_1s_2 via a path of C in G_2. In the latter case, $p_2(u, v)$ and $p_3(u, v)$ are both in G_2, and $p_1(u, v)$ can go around s_1s_2 via path $p_2(r_1, r_2)$ in G_1.

Case 2. There exist k (> 1) edges, $x_1y_1, x_2y_2, \ldots x_ky_k$, such that no point in V is incident to two of them and $\{x_iy_i \mid i = 1, 2, \ldots k\} \cup \{s_0s_1, s_2s_3\}$ makes a cut set of G. The above argument for Case 1 can be used to deal with this case in a similar way. The proof is then finished. ☐

Now according to Lemma 10.18, splitting G at every terminal will decompose G into a set of edge-disjoint full Steiner trees, which are called *full Steiner components* of G.

Lemma 10.19 *Let T be a full Steiner component of G. Then G has no cut set of size three which includes two edges in T unless it contains three edges incident to a common Steiner point in T.*

Proof. Suppose that there exists a cut set $\{ab, a'b', xy\}$, where a, b, a' and b' are four different points of T, and no two edges are incident to each other (otherwise we will find a smaller cut set of G). Then $G \setminus \{ab, a'b', xy\}$ has two separate subnetworks G_1 and G_2. Notice that there is a unique path T' on T joining ab and $a'b'$, and every point on T' is a Steiner point. Let $T' = bs_1s_2 \cdots s_kb'$, where $k \geq 0$. We indicate the points which are adjacent to s_i by x_i, for $i = 1, 2, \ldots, k$, and denote the points adjacent to b and b' by c and c', respectively. See Fig.10.22.

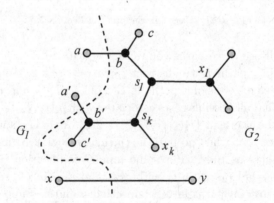

Fig. 10.22   Case 1 in the proof of Lemma 10.19.

First consider the case of $k \geq 1$. We will produce a contradiction by showing that G has an admissible Steiner lifting at bs_1, or equivalently, we will show that for any pair of points u and v in $V \setminus \{b, s_1\}$, there are three edge-disjoint paths connecting them in $G'(V', E')$ obtained from $G(V, E)$ by applying Steiner lifting at bs_1. This time we use three paths connecting c with x_1, i.e., $p_i(c, x_1)$, $i = 1, 2, 3$, as auxiliary paths to reconstruct $\{p_i(u, v) \mid i = 1, 2, 3\}$ in G to make them to be three edge-disjoint paths in G' connecting u and v. It is easy to verify that all possible cases associated with $p_i(c, x_1)$, $i = 1, 2, 3$ can be reduced to two cases: Case 1, $p_1(c, x_1) = cbs_1x_1$, $p_2(c, x_1)$ includes edges $a'b'$ and xy, and $p_3(c, x_1)$ is in G_2; Case 2, $p_1(c, x_1) = cbs_1x_1$, $p_2(c, x_1)$ and $p_3(c, x_1)$ are in G_2. In both cases, we can apply the same case-study as we have done in the proof of Lemma 10.18. The detailed analysis is omitted.

In the end, consider the case of $k = 0$, we can produce a contradiction by showing that G has an admissible Steiner lifting at bb' in a similar way. The proof is then finished. $\square$

Theorem 10.16 *For a terminal-set P in the Euclidean plane, let $r_k(P)$ be the ratio of the length of minimum three-edge connected Steiner network of P over that of minimum k-edge connected spanning network of P. Then*

$$\frac{\sqrt{3}}{2} \leq \inf\{r_3(P) \mid P\} \leq \frac{2 + \sqrt{3}}{4}.$$

Proof. First we will show that for any P, $\sqrt{3}/2 \leq r_3(P)$. In order to do this, we will, 1) decompose G into a set of full Steiner components of G, 2) replace each full Steiner component of G with its corresponding minimum spanning tree, and 3) prove that the resulting network G' is three-edge-connected. The whole process is demonstrated by a simple example in Fig.10.23. (Notice that G' may not be a shortest three-edge-connected spanning network of P.) In the end, recall that the Steiner ratio in the Euclidean plane is $\sqrt{3}/2$ [81], we have $l(G) \geq \frac{\sqrt{3}}{2} l(G')$, this implies the inequality, since $l(G')$ is greater than or equal to the length of the minimum three-edge-connected spanning network of P.

To show that G' is a three-edge-connected spanning network of P, consider any pair of terminals u and u in P. Given a full Steiner component T of G, from Lemma 10.19 we know that at most one path in $\{p_i(u, v) \mid i = 1, 2, 3\}$ includes some edges in T. If no path in $\{p_i(u, v) \mid i = 1, 2, 3\}$ includes edges in T, then after substitution of T they remain three edge-disjoint paths connecting u with v. Now suppose that $p_1(u, v)$ include some edges in T. When T is substituted by its corresponding minimum spanning tree T', we can reconstruct $p_1(u, v)$ by replacing these edges in T with a path in T' so that the modified path $p_1'(u, v)$ is still edge-disjoint with other two paths $p_2(u, v)$ and $p_3(u, v)$. Thus there always exist three edge-disjoint paths connecting u with v during the substituting process. This means that G' is a three-edge-connected spanning network of P.

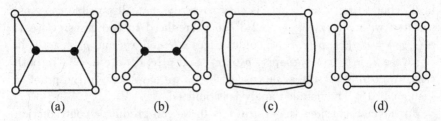

Fig. 10.23   (a) A Steiner network, (b) Split G at every terminal, (c) replace every full Steiner component with its corresponding minimum spanning tree, and then (d) produce a spanning network G'.

In the end we consider a set P_{12} that consists of twelve terminals; they form two regular hexagons with radius of one and $a > 1$, respectively as shown in Fig.10.24(b), where $b^2 = 1/4 + (a - \sqrt{3}/2)^2)$. It is not very difficulty to verify that the length of the minimum 3-edge-connected Steiner and spanning networks are $3(\sqrt{3} + 2)a + 6$ and $6(1 + a + b)$, respectively as shown in Fig.10.24(c,a). Now let a go to infinity, we obtain $r_3 \leq (\sqrt{3} + 2)/4$. The proof is then finished. $\square$

The following corollary follows directly from Theorem 10.16.

Corollary 10.6   *There is a polynomial time $\frac{5}{\sqrt{3}}$-approximation algorithm for the minimum three-edge-connected spanning network in the Euclidean plane.*

In the following study we restrict our attention on those P that all points in P lie on the sides of its convex hull. For this special case, we are able to label each terminal point in P by t_i, $i = 0, 1, \ldots, |P| - 1$, in clockwise order, and denote by $C(P)$ the cycle consisting of $(t_0 t_1 \cdots t_{|P|-1} t_0)$.

Lemma 10.20   *Suppose that all terminal points in P lie on the sides of the convex hull of P. Then $C(G) = C(P)$, and $G \setminus C(G)$ is a Steiner minimum tree of P.*

Proof.   (1) Suppose, by contradiction, that $C(G) \neq C(P)$, and $t_0 t_1 \notin C(G)$. Then there exists a path in $C(G)$ connecting t_0 with t_1. Denote this path by $t_0 q_1 q_2 \cdots q_k t_1$, where $k \geq 1$. Now we consider the following three cases.

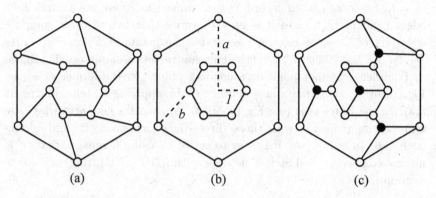

Fig. 10.24 (a) The minimum 3-edge-connected spanning network, (b) twelve terminals form two regular hexagons, and (c) the minimum 3-edge-connected Steiner network.

Case 1. q_1 is a terminal. Let $q_1 = t_i$. See Fig.10.25(a). According to Lemma 10.1, there are at most four edges incident to t_i. It is easy to check that two of them compose a cut set of G, since edge $t_0 t_i$ separates the points in P into two parts $\{t_1, t_2, \ldots, t_{i-1}\}$ and $\{t_i, \ldots, t_0\}$. This contradicts assumption (1).

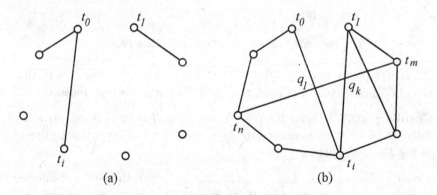

Fig. 10.25 For the proof of Lemma 10.20: (a) Case 1 and (b) Case 2.

Case 2. q_1 is an intersection of two edges $t_0 a$ and $q_2 b$. See Fig.10.25(b). Now produce a new network G' by replacing $t_0 a$ and $q_2 b$ in G with $t_0 q_2$ and ab. Clearly, $l(G') < l(G)$. According to Lemma 1.1 we can deduce that G' is three-edge-connected. This contradicts assumption (1).

Case 3. q_1 is a Steiner point, where q_1 is adjacent to s_1 and q_2. If

$s_1 = t_i$, for some i, then according to Lemma 10.1, there are at most four edges incident to t_i, and it is easy to verify that two of them compose a cut set of G, since edge $q_1 t_i$ separates the points in P into two parts $\{t_1, t_2, \cdots, t_{i-1}\}$ and $\{t_i, \cdots, t_0\}$. This contradicts assumption (1). Hence s_1 is either a Steiner point or a crossing point. In the former case (see Fig.10.26(a)), construct a new network G' by applying a Steiner lifting at $q_1 s_1$; In the latter case (see Fig.10.26(b)), construct a new network G' by removing q_1 (together with those three edges incident to it) and adding edge $t_0 q_2$. In both cases, it is easy to verify, by using Lemma 5, that G' is a three-edge-connected Steiner network with $l(G') < l(G)$, this contradicts assumption (1).

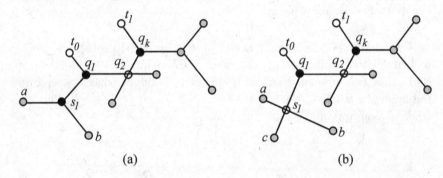

Fig. 10.26   Case 3 in the proof of Lemma 10.20.

Therefore, we have $C(G) = C(P)$. It follows from Lemma 10.16 that $G \setminus C(G)$ is a Steiner minimum tree of P. The proof is then finished.   □

Corollary 10.7   *Under the same assumption of Lemma 10.20, there is a fully PTAS for the minimum three-edge-connected Steiner network problem in the Euclidean plane.*

Proof.   According to Lemma 10.20, the minimum three-edge-connected Steiner network of terminal-set P consists of $C(P)$ and the Steiner minimum tree $T_{smt}(P)$ of P. As there is a polynomial time algorithm to construct $C(P)$ [118] and a fully PTAS for the Steiner tree problem in this case [229], integrating these two algorithms can make a fully PTAS for the minimum three-edge-connected Steiner network problem.   □

Theorem 10.17   *Under the same assumption of Lemma 10.20, $r_3(P) > (2 + \sqrt{3})/4$.*

Proof. Let G^* be a minimum three-edge-connected spanning network of P. Recall that it is proved in Chapter 1 [81] that Steiner ratio in the Euclidean plane is $\sqrt{3}/2$, by Lemma 10.20 we have

$$l(G) = l\big(C(P)\big) + l\big(T_{smt}(P)\big) \geq l\big(C(P)\big) + \frac{\sqrt{3}}{2}l\big(T_{mst}(P)\big).$$

Since a Hamiltonian cycle of P and a minimum spanning tree $T_{mst}(P)$ of P can compose a three-edge-connected spanning network of P, we have $l(G^*) \leq l\big(C(P)\big) + l\big(T_{mst}(P)\big)$. In addition, it is obvious that $l\big(T_{mst}(P)\big) < l\big(C(P)\big)$. Therefore, we have

$$r_3(P) = \frac{l(G)}{l(G^*)} \geq \frac{l\big(C(P)\big) + \frac{\sqrt{3}}{2}l\big(T_{mst}(P)\big)}{l\big(C(P)\big) + l\big(T_{mst}(P)\big)} > \frac{2 + \sqrt{3}}{4},$$

and then the proof is finished. $\square$

Let G^* be a minimum three-edge-connected spanning network of P. Then we can show the following result by applying the same approach that we have used to study the minimum three-edge-connected Steiner network.

Lemma 10.21 *Under the same assumption of Lemma 10.20, $C(G^*) = C(P)$.*

Note, however, that although $C(P)$ and a minimum spanning tree of P can be constructed in polynomial time, Lemma 10.21 does not imply that the minimum three-edge-connected spanning network of P can be produced in polynomial time, since $G^* \setminus C(P) = G^* \setminus C(G^*)$ may not be a minimum spanning tree of P. See Fig.10.19 for a simple example.

10.3.2 *In the Rectilinear Plane*

By applying the same approach as used for the case of Euclidean plane, we can deduce a series of results [138] for minimum three-connected Steiner networks on the rectilinear plane parallel to those obtained for the case of the Euclidean plane.

Theorem 10.18 *Suppose that all terminals in P (maybe except one) are on the sides of the rectilinear convex hull of P. Then the minimum two-connected Steiner network on P in the rectilinear plane is a spanning network on P.*

Theorem 10.19 *For a terminal-set P in the rectilinear plane, let $r_k(P)$ be the ratio of the length of minimum three-edge connected Steiner network of P over that of minimum three-edge connected spanning network of P. Then*

$$\frac{3}{4} \leq \inf\{r_3(P)\,|\,P\} \leq \frac{7}{8}.$$

Proof. Given a set P of terminals in the rectilinear plane, let G^* be the minimum three-edge connected Steiner network on P. We can construct a three-edge connected spanning network G' on P by replacing each full Steiner component in G^* with its corresponding minimum spanning tree. Recall that the Steiner ratio in the rectilinear plane is $2/3$ [142], we obtain that $l(G') \leq \frac{3}{2}l(G^*)$, which yields the lower bound.

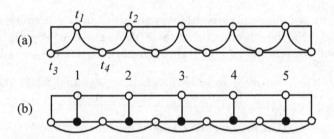

Fig. 10.27 (a) Minimum three-connected spanning network and (b) minimum three-connected Steiner network.

The upper bound of $7/8$ is achieved by a special class of set P_k as shown in Fig.10.27 with $k = 5$, where $|t_1t_2| = |t_3t_4| = |t_1t_3| = |t_2t_4| = |t_1t_4| = 2$. It is easy to verify that the length of the minimum two-connected spanning network on P_k of Fig.10.27(a) is $(8k + 2)$, and the length of the minimum two-connected Steiner network on P_n of Fig.10.27(a) is $(7k + 2)$, the ratio of the latter to the former achieves $7/8$ as k goes to the infinity. The proof is then finished. □

Theorem 10.20 *Suppose that all terminals in P are on the sides of the convex hull of P. Then there exists a minimum three-edge connected Steiner network on P that consists of a Hamilton cycle of P and a Steiner minimal tree of P.*

10.4　Discussions

We have obtained some lower bounds of generalized Steiner ratio $r_M(k)$ in metric space M for $k \geq 2$, and some estimations of the ratios in the Euclidean plane for $k = 2, 3$. It is a challenge to determine the exact value of $r_M(k)$, or narrow the gap between its lower and upper bounds, especially for small k in the Euclidean plane, which is more important in practice. However, the currently used arguments for this case heavily rely on the structural properties of Steiner minimum trees in the Euclidean plane. Thus acquiring similar results for minimum k-edge-connected Steiner networks with $k \geq 4$ demands new techniques, since every Steiner point is incident to at least four edges.

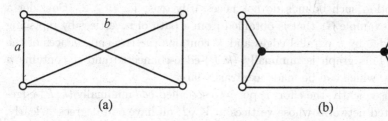

(a)　　　　　　　　　　　　　　　　(b)

Fig. 10.28　(a) The minimum 3-vertex-connected spanning network and (b) The minimum 3-vertex-connected Steiner network.

There are two other interesting cases of generalized Steiner ratio problem, which are also of great interest for further study. One is addressed under k-vertex-connectivity, and the other disallows multiple edges between any pair of points. Observe in these two cases Theorem 10.16 does not hold any more. For example, in Fig.10.28 four points in the Euclidean plane form a rectangle of size $a \times b$. It is easy to verify that when a is sufficiently smaller than b, the length of the shortest 3-vertex-connected Steiner and spanning networks are $(2 + \sqrt{3})a + 3b$ and $2(a + b + \sqrt{a^2 + b^2})$, respectively. Let a/b approach to zero, we get $r_3 \leq 3/4 < \sqrt{3}/2$. These two versions demand a new approach, since the technique of admissible lifting can not be applied.

10.4.1　*Minimally k-Edge Connected Networks*

Jordán [157] studied minimally k-edge-connected graphs, which are very similar to minimum weak k-connected Steiner networks. A graph $G(V, E)$

is called (k, P)-*edge-connected* for $P \subseteq V$ if there exist k-edge disjoint paths in G between each pair of vertices of P, and G is called *minimally* (k, P)-*edge-connected* if it is (k, P)-edge-connected but $G \setminus \{e\}$ is no longer (k, P)-edge-connected for any $e \in E$. Clearly, minimum weak k-connected Steiner networks are minimally (k, P)-edge-connected.

The focus of Jordán's work [157] is on determining the upper bounds on the number of vertices (and edges) of minimally (k, P)-edge-connected graphs in terms of k and $|P|$. It is easy to see that given a *minimally* (k, P)-edge-connected network $G(V, E)$, replacing any edge in E by a path between its two endpoints will produce a *minimally* (k, P)-edge-connected network with arbitrarily larger number of edges. Thus in general there does not exist such a bound. Based on this simple fact, it is assumed that every vertex of $V \setminus P$ has degree at least three. Unfortunately, even with this assumption such bounds do not necessarily exist for $k \geq 2$. Consider a simple example G_k that is obtained from a path of n vertices by replacing every edge by k parallel edges and P contains the two end-vertices of the path. This graph is minimally (k, P)-edge-connected and it contains n vertices, which can be made arbitrarily large.

Consequently, the effort is put into so-called *odd* minimally (k, P)-edge-connected networks whose vertices in $V \setminus P$ all have odd degrees. Clearly, minimum weak k-connected Steiner networks are odd minimally (k, P)-edge-connected. Jordán proved the following theorem among some other results.

Theorem 10.21   *Let $G(V, E)$ be an odd minimally (k, P)-edge-connected graph for some $P \subseteq V$. Then $|E| \leq 2k|P| - 3k$ and $|V| \leq (k + 1)|P| - 2k$.*

10.4.2   *Minimum k-Edge Connected Spanning Networks*

The minimum k-edge connected Steiner network problem will be reduced to *minimum k-edge connected spanning network problem* [169] if $P = V$. This problem is equivalent to the *source based minimum k-edge connected spanning network problem*: given an edge-weighted k-connected graph $G(V, E)$ and a vertex $r \in V$, find a minimum weight subgraph G' of G such that for every vertex $v \in V$, there are k edge-disjoint paths from r to v in G'. Any solution to this problem is also a solution to the minimum k-edge connected spanning network problem [257]. Using Edmonds' results [95], Khuller and Vishkin [169] proved the following theorem.

Theorem 10.22   *There is a 2-approximation algorithm for the minimum* *k-edge connected spanning network problem.*

Given an edge-weighted directed graph $G(V, E)$ and a vertex $r \in V$. A set $E' \subseteq E$ is called an *r-arborescence* if every vertex except r has in-degree 1. In fact, an r-arborescence is a directed spanning tree rooted at r. The *r-connectivity* of G is defined as the maximum k such that there are k edge-disjoint paths from r to v for any $v \in V \setminus \{r\}$. Edmonds [95] proved that maximum number of edge-disjoint r-arborescences in G is equal to the r-connectivity of G. As a result, Khuller and Vishkin [169] showed that to find a minimum weight subgraph $G' \subseteq G$ of r-connectivity is the same as to find a minimum weight subgraph $G' \subseteq G$ that has k edge-disjoint r-arborescences.

Moreover, Edmonds [95] proved that the edges of a directed $G(V, E)$ can be partitioned into k edge-disjoint r-arborescences if and only if, when directions of edges are ignored, E can be partitioned into k spanning trees, and the in-degree of every vertex except r is exactly k. From this result Khuller and Vishkin [169] deduced that the subgraph $G' \subseteq G$ of minimum weight that has k edge-disjoint arborescences can be computed in polynomial time.

Now given an edge-weighted undirected graph $G(V, E)$ and a vertex $r \in V$. We can obtain a weighted directed graph H by simply replacing each edge $(u, v) \in E$ with two edges of opposite directions each having the same weight as (u, v). Then we can compute, in polynomial time, a minimum weight subgraph of H that can be partitioned into k r-arborescences $A_k(H)$ and satisfies the following inequalities

$$l(G_{opt}) \leq l(A_k(H)) \leq 2 \cdot l(G_{opt}),$$

where $l(G_{opt})$ is the length of the minimum k-edge connected spanning network of G. Now ignoring the directions of edges in $A_k(H)$, we obtain a 2-approximation of minimum k-edge connected spanning network.

For the *unweighted version* of minimum k-edge connected spanning networks in graphs, that is, how to, given a k-edge connected graph $G(V, E)$, find a subgraph $G(V, E')$ of G that is k-edge connected and the number of edges in E' is minimized. Gabow et al. [104] proposed a LP-rounding based algorithm for the problem, and obtained a tight bound on the approximation ratio of $(1 + 3/k)$ and $(1 + 2/k)$ for undirected graphs with odd $k > 1$ and even k, respectively, and $(1 + 2/k)$ for directed graphs with arbitrary k. The first bound can be reduced to $(1 + 2/k)$ by using iterated rounding. In addition, Jothi et al. [159] proposed a 5/4-approximation algorithm for the case of $k = 2$ in undirected graphs.

For the case of multigraphs, Gabow [103] proved that the algorithm proposed by Khuller and Raghavachari [167] has approximation ration at most $(1 + \sqrt{1/e})$.

10.4.3   *Minimum k-Vertex Connected Spanning Networks*

As a contrast to the case of $k = 2$, minimum k-edge and minimum k-vertex connected spanning networks may have different weights for $k \geq 3$. Fig.10.29 demonstrates such an example of six terminal points in the Euclidean plane that make a regular hexagon of one unit. Fig.10.29(a) is a minimum 3-edge connected spanning network of length 11, Fig.10.29(b) is a minimum 3-vertex connected spanning network of length $(8 + 2\sqrt{3})$, and Fig.10.29(c) is a minimum 3-connected Steiner network of length $(6 + 3\sqrt{3})$.

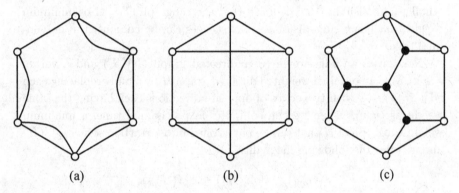

Fig. 10.29   Minimum 3-connected Networks of six terminals: (a) 3-edge-connected spanning network, (b) 3-vertex-connected spanning network, and (c) 3-vertex-connected Steiner network.

Bienstock et al.   [37] described a polynomial time algorithm that produces a k-vertex connected spanning network with weight at most $3k(k + 1)/4$ times that of the minimum 2-connected spanning network. This algorithm can be improved as follows: Produce a Hamilton cycle first and then connect vertices on the cycle in a systematic way as shown in Fig.10.30 The construction for even k is a little bit different from that for odd k, which is a more difficult case. Fig.10.30 show three k-vertex connected spanning networks of twelve terminals. Note that Fig.10.30(a-b) are for $k = 4$ and $k = 6$, respectively, both have the topologies of *double-loop networks* [139], while Fig.10.30(c) is for $k = 3$, which have the topology of *chordal ring* networks [140].

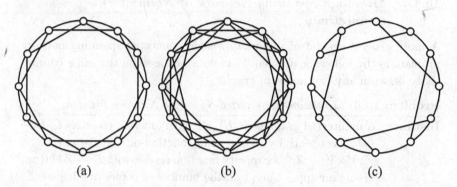

(a)   (b)   (c)

Fig. 10.30   Spanning networks: (a) 3-vertex-connected, (b) 6-vertex-connected, and (c) a 3-vertex-connected.

Theorem 10.23   *There exists an approximation algorithm for the minimum k-vertex connected spanning network problem in metric spaces with performance ratio of $3(k + 2)/8$ for even $k \geq 2$ and $3(k + 4 + 3/k)/8$ for odd $k \geq 3$.*

For the *minimum k-vertex connected spanning networks* in graphs, Aulettaa et al. [20] proposed a polynomial time algorithm that finds a ($\lceil k/2 \rceil + 1$)-connected spanning subgraph whose weight is at most twice the optimum of the problem. In particular, they obtained a 2-approximation algorithm for the case of $k = 3$. Recently, Kortsarz and Nutov [176] proved that for arbitrary weights on edges, there are a k-approximation algorithm for undirected graphs and a ($k + 1$)-approximation algorithm for directed graphs; For metric weights, the ratios could be reduced to ($2 + \frac{k-1}{n}$) and ($2 + \frac{k}{n}$), respectively.

More recently, Kortsarz and Nutov [177] proposed two algorithms for directed or undirected case. The first one has approximation ratio of $O(\frac{n}{n-k} \ln^2 k)$, which is based on properties of ℓ-connected p-critical graphs; The second one using the primal-dual method has approximation ratio of $O(\sqrt{n} \ln k)$.

For the *unweighted version* of minimum k-vertex connected spanning networks in graphs, that is, how to, given a k-vertex connected graph $G(V, E)$, find a subgraph $G(V, E')$ of G that is k-vertex connected and the number of edges in E' is minimized. Cheriyan and Thurimella [58] proposed a $(1 + 2/k)$-approximation algorithm for this problem.

10.4.4 *Minimum Spanning Network of Nonuniform Connectivity*

A more general version of minimum k-edge connected spanning network problem is the following problem that does not assume the same connectivity between any two nodes in graphs.

Problem 10.4 *Minimum Generalized Spanning Network Problem*

Instance   A undirected graph $G(V, E)$, a length function on edges $l : E \to$ Q^+, a connectivity requirement function on vertex-pairs $r : V \times V \to Z^+$, a capacity function on edges $c : E \to Z^+ \cup \{\infty\}$ giving an upper bound on the number of copies of edge $e \in E$ that are allowed to use; When $c(e) = \infty$, there is no upper bound on edge e.
Solution   A multigraph H on vertex-set V that has $r(u, v)$ edge-disjoint paths for each pair of vertices $u, v \in V$.
Objective Minimizing the total length of the edges in H, $$l(H) \equiv \sum_{e \in H} l(e), \text{ where each copy of edge } e \text{ in } H \text{ costs } l(e).$$

In the above formulation, the length function not necessarily satisfies the triangle inequality. Jain [152] proposed an *iterated rounding* algorithm for the above problem and proved the following theorem.

Theorem 10.24   *There is a 2-approximation algorithm for the minimum generalized Steiner network problem in graphs.*

Appendix A

More Steiner Tree Related Problems

In Chapters 1-10, we have studied the classical Steiner tree problem and some their generalizations or variants arising from the design of computer communication networks. In fact, there are many more Steiner tree related problems proposed in the past twenty years, which are not addressed in Chapters 1-10. In the following, we give a list for some of them, which is certainly not complete.

Problem A.1 *Steiner Forest Problem*

Instance A graph $G(V, E)$ with a cost $l(e)$ on each edge $e \in E$, a family of k disjoint subsets of V, $P_1, P_2, \cdots, P_k$.

Solution A subgraph G' in which each pair of vertices in P_i is connected for $i = 1, 2, \cdots, k$.

Objective Minimizing the total length of edges in G',
$$l(G') \equiv \sum_{e \in G'} l(e).$$

Positive result It has a 2-approximation algorithm [1]. There is an $O(\log^2 n)$-competitive algorithm for the online version of the *generalized Steiner tree problem* where each P_i contains exactly two vertices [28].

Problem A.2 *Acyclic Directed Steiner Tree Problem*[1]

Instance An acyclic directed graph $G(V, E)$ with a cost $l(e)$ on each arc $e \in E$, a terminal-set $P \subset V$, and a root r.

Solution A subgraph G' containing a path from r to each terminal in P.

Objective Minimizing the total length of edges in G',
$$l(G') \equiv \sum_{e \in G'} l(e).$$

---

[1] Also known as the *Steiner arborescence problem* in acyclic networks [144].

Positive results It admits an $O(|P|^\epsilon)$-approximation algorithm for any $\epsilon > 0$ for any ϵ [278]. For the online version for the case of bounded *edge-asymmetry* β, which equals the maximum ratio of the costs of antiparallel arcs in the graph, it has a greedy algorithm with competitive ratio of $O(\min\{|P|, \beta \log |P| / \log \log \beta\})$ [8].

Negative results Unless $NP \subseteq DTIME[|V|^{poly \log |V|}]$ no polynomial time algorithm can guarantee better then $(\ln |P|)/4$-approximation [199]. For the online version specified above, every deterministic algorithm has competitive ratio at least $\Sigma(\min\{k^{1-\epsilon}, \beta \log k / \log \log k\})$ for any arbitrarily small ϵ [8].

Problem A.3 *Power-p Steiner Tree Problem in Graphs*

Instance A graph $G(V, E)$ with a length $l(e)$ on each edge $e \in E$,
a terminal-set $P \subset V$ and an integer $p > 1$.
Solution A Steiner tree T that spans P in the distance graph $G_D(P)$ and
each Steiner point has degree at least 3.
Objective Minimizing the total power p length of edges in T, $\sum_{e \in T} l^p(e)$.

Negative result It is MAX SNP-hard [30]. It is conjectured that the problem is not finitely solvable for $p \geq 5$ [107].
Positive result The power-p Steiner ratio for $p = 2$ is between $[1/17.2, 1/23.3)$ [107; 30].

Problem A.4 *Polymatroid Steiner Tree Problem*

Instance A graph $G = (V, E)$ with a cost $l(e)$ on each edge $e \in E$, a
polymatroid $P(V)$ defined on V, and a terminal-set $P \subset V$.
Solution A Steiner tree T spanning at least one base of $P(V)$.
Objective Minimizing the total cost of edges in T, $\sum_{e \in T} l(e)$.

Positive Result It admits an approximation algorithm with polylogarithmic ratio [43]. It includes the group and covering Steiner tree problems as special cases, and it is reduced to Steiner tree problem when there is only one base $P \subset V$.

Problem A.4' *Polymatroid Directed Steiner Tree Problem*

Instance A directed graph $G(V, E)$ with a cost $l(e)$ on each arc $e \in E$,
a terminal-set $P \subset V$, a root $r \in V$, and a polymatroid $M(P)$
defined on P.
Solution A Steiner arborescence T connecting root r to a base of
polymatroid $M(P)$.

Objective Minimizing the total cost of arcs of T, $\sum_{e \in T} l(e)$.

Positive Result It has an approximation algorithm with polylogarithmic ratio [43].

Problem A.5 *Prize Collecting Steiner Tree Problems*

Instance A graph $G(V, E)$ with a prize $p(v)$ for each $v \in V$ and a cost $c(e)$ for each $e \in E$; A *quota bound* Q and a *budget bound* B.

Solution A Steiner tree T (or a Steiner tree T contain a specified root vertex $v_0 \in V$).

Objective (1) Minimizing the total cost of edges in T plus the prizes of vertices not in T, $\sum_{e \in T} c(e) + \sum_{v \notin T} p(v)$.

Objective (2) Maximizing the *net worth* of T, $\sum_{v \in T} p(v) - \sum_{e \in T} c(e)$.

Objective (3) Minimizing the total cost of edges in T under the quota constraint, $\min\{\sum_{e \in T} c(e) \mid \sum_{v \in T} p(v) \geq Q\}$.

Objective (4) Maximizing the total prize of vertices in T under the budget constraint, $\max\{\sum_{v \in T} p(v) \mid \sum_{e \in T} c(e) \leq B\}$.

Positive results It has a 2-approximation algorithm for objective (1) [115]. It has approximation algorithms for objective (3) with the same performance ratios as for *k-MST problem*[2] [156]. It has $(5 + \epsilon)$-approximation algorithm for objective (4) if there is a 3-approximation algorithm for k-MST problem [156]. Refer to [46] for some recent results on the variations of the problem.

Negative result It is NP-hard to approximate the problem for objective (2) within any constant factor [100].

Problem A.6 *Group Steiner Tree Problem*[3]

Instance A graph $G = (V, E)$ with a length $l(e)$ on each edge $e \in E$, a set of k disjoint groups $\{V_1, V_2, \cdots, V_k\}$ with $V_i \subset V$ for each i.

Solution A Steiner tree T that spans at least one node in V_i for each i.

Objective Minimizing the total cost of edges in T, $\sum_{e \in T} l(e)$.

Positive results It admits an $O(k^{\epsilon})$-approximation algorithm for any $\epsilon > 0$ [134], and an $O(\log^2 |V| \log k)$-competitive randomized algorithm for the online version in trees and an $O(\log^3 |V| \log k)$-competitive randomized algorithm in general graphs [4; 113].

---

[2]The problem for objective (3) is reduced to k-MST problem if all vertices have prize 1 and $Q = k$. k-MST problem has a 2-approximation algorithm [112].

[3]It is also called the *class Steiner tree problem*.

Negative result It cannot be efficiently approximated with a performance ratio less than $\ln k$ times the optimum [147].

Problem A.7 *Node-Weighted Steiner Tree Problem*

Instance A graph $G = (V, E)$ with two cost functions, $c_v(\cdot)$ defined on V and $c_e(\cdot)$ defined on E, and a terminal-set $P \subset V$.

Solution A Steiner tree T interconnecting all terminals in P.

Objective Minimizing the edge-cost and vertex-cost of vertices and edges in T, $\sum_{v \in T} c_v(v) + \sum_{e \in T} c_e(e)$.

Positive result It can be approximated within $2 \ln |P|$ [171].

Negative result It cannot be approximated to within less than a logarithmic factor unless $DTIME[n^{poly \log n}] \supseteq NP$ [1; 199].

Problem A.8 *Node-Weighted Geometric Steiner Tree Problem*

Instance A set $P = \{t_1, t_2, \cdots, t_k\}$ of terminal points in a metric space each with a penalty $p(t_i) > 0$, and a positive constant $c > 0$.

Solution A Steiner tree T that interconnects some terminals in P using Steiner points $S = \{s_1, s_2, \cdots, s_m\}$.

Objective Minimizing the total length of T plus the total penalty of T, $\sum_{e \in T} l(e) + \sum_{t_i \notin T} p(t_i) + \sum_{s_j \in S} c \cdot s_j$.

Negative result It admits a polynomial time approximation scheme in the Euclidean plane [236].

Problem A.9 *Prize Collecting Node-Weighted Steiner Tree Problems*

Instance A graph $G = (V, E)$, a profit $p(v)$ and a cost $c(v)$ for each $v \in V$, a terminal-set $P \subset V$. A *budget bound B* and a *quota bound Q*.

Solution A Steiner tree T interconnecting all points in P.

Objective (1) Maximizing the profit of T under budget constraint on edge cost of T, $\max\{\sum_{v \in T} p(v) \mid \sum_{v \in T} c(v) \leq B\}$.

Objective (2) Minimizing the cost of T under quota constraint on the prize of T, $\min\{\sum_{v \in T} c(v) \mid \sum_{v \in T} p(v) \geq Q\}$.

Objective (3) Minimizing the cost of T minus the profit of T (profit loss), $\sum_{v \in T} c(v) - \sum_{v \notin T} p(v)$.

Positive result It has an $O(\log |V|)$-approximation algorithm for the problem with all objectives [216].

Negative result It is at least as hard to approximate the problem for objectives (2-3) as the set cover problem [216].

Problem A.10 *Terminal Steiner Tree Problem*

Instance A terminal-set P in a metric space, or P in a graph $G(V, E)$ with $P \subset V$ and cost $l(e)$ on each edge $e \in E$.

Solution A terminal Steiner tree T that all vertices in P are leaves of T.

Objective Minimizing the total length of T, $\sum_{e \in T} l(e)$.

Positive Result It has an approximation algorithm with performance ratio $2\rho - \rho/(3\rho - 2)$ [203].

Negative Result It is MAX SNP-hard in metric spaces [189]. It has no polynomial time approximation algorithm with a performance ratio less than $(1 - o(1)) \ln n$ unless $NP = DTIME(n^{O(\log \log n)})$ [73].

Problem A.11 *Diameter Bounded Steiner Tree Problem*

Instance A graph $G(V, E)$ with a set of Steiner nodes S and two functions, length $l(\cdot)$ and cost $c(\cdot)$, defined on E. A diameter bound d^4.

Solution A Steiner tree T such that all the Steiner nodes belongs to S and the diameter of T (in terms of length $l(\cdot)$ is no more than d.

Objective Minimizing the total costs of T, $\sum_{e \in T} c(e)$.

Positive result It has $O(\log |V|)$-approximation algorithms for both the cost and diameter [202; 207].

Problem A.12 *Path-Length Bounded Steiner Tree Problem*

Instance A graph $G(V, E)$ with two functions, length $l(\cdot)$ and cost $c(\cdot)$, defined on E. A source vertex $s \in V$ and a set of terminal vertices in V, $t_1, t_2, \ldots, t_k$, and k delay bounds $d_1, d_2, \ldots, d_k$.

Solution A Steiner tree T such that the length of path from s to t_i in T is no more than d_i for each i.

Objective Minimizing the total cost of edges in T, $\sum_{e \in T} c(e)$.

Positive result It has a fully polynomial time approximation scheme when the Steiner topology is given [273].

Negative result It is NP-hard even when the topology of the Steiner tree is fixed [273].

---

[4] When the diameter bound d is set to be infinitely large, it becomes equivalent to Steiner tree problem.

Problem A.13 *Bounded-Skew Steiner Tree Problem*

Instance A terminal-set P in a metric space M and a bound $b \geq 0$.

Solution A *bounded-skew Steiner tree* T that is a rooted tree with two mappings $\pi : V(T) \to M$ and $c : E(T) \to R_+$ such that

     1) π is an 1-1 mapping between leaves of T and P,

     2) $c(u, v) \geq d(\pi(u), \pi(v))$ for each edge $e = (u, v) \in E(T)$, and

     3) length difference between any two root-leaf paths is at most b.

Objective Minimizing the total cost of edges in T, $\sum_{e \in T} c(e)$.

Positive result It has approximation algorithms with performance ratios 4 for bound $b = 0$ (zero-skew) and 14 for $b > 0$ in any metric space, 3 for $b = 0$ and 9 for $b > 0$ in the rectilinear plane, respectively [279].

Negative result It is NP-hard [47].

Problem A.14 *Multi-Weighted Steiner Tree Problem*

Instance A graph $G(V, E)$ with k costs $l_1(e), \cdots, l_k(e)$ with $l_i(e) \geq l_{i+1}(e)$ for each $1 \leq i \leq k - 1$ on each edge $e \in E$, and a hierarchical partition $P_1, \cdots, P_k$ of V.

Solution A spanning tree T of V with a subtree T_i spanning $P_1 \cup \cdots \cup P_i$.

Objective Minimizing the bottleneck length of edges in T,

$$\sum_{i=1}^{k} \sum_{e \in T_i \setminus T_{i-1}} l_i(e), \text{ where } T_0 = \emptyset^5.$$

Positive result It has $O(|P| \cdot |V|^2)$-time heuristics, which are conjectured to have the same performance ratio as any approximation algorithm for the Steiner tree problem [92].

---

[5]When $k = 1$, it is the minimum spanning tree problem; When $l_i(e) = 0$ for each $e \in E$ and all $i = 2, \cdots, k$, it is the Steiner tree problem.

Bibliography

[1] A. Agrawal, P. Klein, and R. Ravi, When trees collide: An approximaa-
tion algorithm for the generalized Steiner tree problem on networks, *SIAM
Journal on Computing*, 24(3)(1995), 440-456.

[2] I. F. Akyildiz, W. Su, Y. Sankarasubramaniam, and E. Cayirci, Wireless
sensor networks: a survey, *Computer Networks*, 38(4)(2002), 393-422.

[3] M. Alfaro, M. Conger, K. Hodges, A. Levy, R. Kochar, L. Kuklinski,
Z. Mahmood, and K. von Haam, The structure of singularities of Φ-
minimizing networks in R^2, *Pacific Journal of Mathematics*, 149(2)(1991),
201-210.

[4] N. Alon, B. Awerbuch, Y. Azar, N. Buchbinder, and J. Naor, A general
approach to online network optimization problems, *ACM Transactions on
Algorithms*, 2(4)(2006), 640-660.

[5] M. Ali and J. S. Deogun, Power-efficient design of multicast wavelength-
routed networks, *IEEE Journal on Selected Areas in Communications*,
18(10)(2000), 1852-1862.

[6] K. D. Andersen, An efficient Netwon barrier method for minimizing a sum
of Euclidean norms, *SIAM Journal on Optimization*, 6(1)(1996), 74-95.

[7] K. D. Andersen, E. Christiansen, A. R. Conn, and M. L. Overton, An efficient primal-dual interior-point method for minimizing a sum of Euclidean
norms, *SIAM Journal on Scientific Computing*, 22(1)(2000), 243-262.

[8] S. Angelopoulos, Improved bounds for the online Steiner tree problem in
graphs of bounded edge-asymmetry, *Technical Report* CS-2006-36, David
R. Cheriton School of Computer Science, University of Waterloo, 2006.

[9] V. Annamalai, S. K. S. Gupta, and L. Schwiebert, On tree-based convergecasting in wireless sensor networks, *IEEE Wireless Communication and
Networking Conference* (WCNC), 2003, vol. 4, pp. 1942-1947.

[10] S. Arora, Polynomial-time approximation schemes for Euclidean TSP and
other geometric problems, *Proceedings of the 37th IEEE Symposium on
Foundations of Computer Science* (FOCS), 1996, 2-12. (Also appear in
Journal of the ACM, 45(5)(1998), 753-782.)

[11] S. Arora, Nearly linear time approximation schemes for Euclidean TSP
and other geometric problems. *Proceedings of the 38th IEEE Symposium*

on Foundations of Computer Science (FOCS), 1997, pp. 554-563.

[12] S. Arora, Approximation schemes for NP-hard geometric optimization problems: a survey, *Mathematical Programming*, 97(1-2)(2003), 43-69.

[13] S. Arora, M. Grigni, D. Karger, P. Klein, and A. Woloszyn, Polynomial time approximation scheme for weighted planar graph TSP, *Proceedings of the Ninth Annual ACM-SIAM Symposium on Discrete Algorithms* (SODA), 1998, pp. 33-41.

[14] S. Arora and G. Karakostas, A $2 + \epsilon$ approximation algorithm for the k-MST problem, *Proceedings of the 11th annual ACM-SIAM symposium on Discrete algorithms* (SODA), 2000, pp. 754-759.

[15] S. Arora and C. Lund, Hardness of approximation, in *Algorithms for NP-Hard Problems*, D. Hochbaum (ed.), PWS Publishing Company, 1996, pp. 399-466.

[16] S. Arora, C. Lund, R. Motwani, M. Sudan, and M. Szegedy, Proof verification and hardness of approximation problems, *Journal of the ACM*, 45(3)(1998), 501-555. (Preliminary version appeared in *Proceedings of the 33rd IEEE Symposium on Foundations of Computer Science* (FOCS), 1992, pp. 14-23.)

[17] S. Arora, P. Raghavan, and S. Rao, Polynomial time approximation scheme for Euclidean k-medians and related problems, *Proceedings of the 30th Annual ACM Symposium on Theory of Computing* (STOC), 1998, pp. 106-113.

[18] S. Arora and M. Safra, Probabilistic checking of proofs: a new characterization of NP, *Journal of the ACM*, 45(1)(1998), 70-122. (Preliminary version appeared in *Proceedings of the 33rd IEEE Symposium on Foundations of Computer Science* (FOCS), 1992, pp. 2-12.)

[19] V. Arora, S. Vempala, H. Saran, and V. V. Vazirani, A Limited-backtrack greedy schema for approximation algorithms, *Lecture Notes in Computer Science*, 1994, Vol. 880, pp. 318-329.

[20] V. Aulettaa, Y. Dinitzb, Z. Nutovc, and D. Parented, A 2-approximation algorithm for finding an optimum 3-vertex-connected spanning subgraph, *Journal of Algorithms*, 32(1)(1999), 21-30.

[21] Y. Aumann and Y. Rabani, Improved bounds for all-optical routing, *Proceedings of the 6th Annual ACM-SIAM Symposium on Discrete Algorithms* (SODA), 1995, pp. 567-576.

[22] D. Banerjee and B. Mukherjee, A practical approach for routing and wavelength assignment in large wavelength-routed optical networks, *IEEE Journal on Selected Areas in Communications*, 14(5)(1996), 903-908.

[23] A. Bar-Noy, S. Guha, J. Naor, and B. Schieber, Message multicasting in heterogeneous networks, *SIAM Journal on Computing*, 30(2)(2000), 347-358.

[24] S. Baroni and P. Bayvel, Wavelength requirements in arbitrarily connected wavelength-routed optical networks, *IEEE Journal of Lightwave Technology*, 15(2)(1997), 242-251.

[25] G. Di Battista, G. Liotta, and F. Vargiu, Spirality of orthogonal representations and optimal drawings of series-parallel graphs and 3-planar graphs, *Lecture Notes in Computer Science*, 1993, vol. 709, pp. 151-162.

[26] J. E. Beasley, A heuristic for the Euclidean and rectilinear Steiner problems, *European Journal of Operational Research*, 58(2)(1992), 284-292.

[27] M. Bellare, O. Goldreich, and M. Sudan, Free bits and non-approximability-towards tight results, *SIAM Journal on Computing*, 27(3)(1998), 804-915.

[28] P. Berman and C. Coulston, On-line algorithms for Steiner tree problems, *Proceedings of the Twenty-Ninth Annual ACM Symposium on Theory of Computing* (STOC), 1997, 344-353.

[29] P. Berman and V. Ramaiyer, Improved approximation algorithms for the Steiner tree problem, *Journal of Algorithms*, 17(3)(1994), 381-408. (Preliminary version appeared in *Proceedings of the 3rd ACM-SIAM Symposium on Discrete Algorithms* (SODA), 1992, pp. 325-334.)

[30] P. Berman and A. Zelikovsky, On the approximation of power-p and bottleneck Steiner trees, in *Advances in Steiner Trees*, D.-Z. Du, J. H. Rubinstein, and J. M. Smith (eds.), Kluwer Academic Publishers, 2000, pp. 117-135.

[31] M. Bern, Two probabilistic results on rectilinear Steiner trees, *Proceedings of the 18th Annual ACM Symposium on Theory of Computing* (STOC), 1986, pp. 433-441.

[32] M. Bern and D. Eppstein, Approximation algorithms for geometric problems, *Approximation Algorithms for NP-Hard Problems*, D. Hochbaum (ed.), PWS Publishing Company, Boston, 1997, 296-345.

[33] M. W. Bern and R. L. Graham, The shortest-network problem, *Scientific American*, January, 1988, 84-89.

[34] M. Bern and P. Plassmann, The Steiner problem with edge length 1 and 2, *Information Processing Letters*, 32(3)(1989), 109-120.

[35] K. Bharath-Kumar and J. M. Jaffe, Routing to multiple destinations in computer network, *IEEE Transactions on Communications*, 31(3)(1983), 343-351.

[36] T. Biedl and G. Kant, A better heuristic for orthogoanl graph drawings, *Computational Geomety*, 9(3)(1998), 159-180.

[37] D. Bienstock, E. F. Brickell, and C. L. Monma, On the Structure of Minimum-Weight k-Connected Spanning Networks, *SIAM Journal on Discrete Mathematics*, 3(3)(1990), 320-329.

[38] R. S. Booth, Analytic formulas for full Steiner trees, *Discrete Computational Geometry*, 6(1)(1991), 69-82.

[39] A. Borchers and D.-Z. Du, The k-Steiner ratio in graphs, *SIAM Journal on Computing*, 26(3)(1997), 857-869.

[40] A. Borchers, D.-Z. Du, B. Gao, and P.-J. Wan, k-Steiner ratio in the rectilinear plane, *Journal of Algorithms*, 29(1998), 1-17.

[41] O. Bottems, *Geometric Inequalities*, 1968.

[42] Z.-P. Cai, G.-H. Lin, and G.-L. Xue, Improved approximation algorithms for the capacitated multicast routing problem, *Lecture Notes in Computer Science*, 2005, vol. 3595, pp. 136-145.

[43] G. Calinescu and A. Zelikovsky, The polymatroid Steiner problems, *Journal of Combinatorial Optimization*, 9(3)(2005), 281-294.

[44] S.-K. Chang, The generation of minimal trees with a Steiner topology, *Journal of the ACM*, 19(1972), 699-711.

[45] S.-K. Chang, The design of network configurations with linear or piecewise linear cost functions. *Symposium on Computer-Communications, Networks, and Teletraffic*, 1972, pp. 363-369.

[46] O. Chapovska and A. P. Punnen, Variations of the prize-collecting Steiner tree problem, *Networks*, 47(4)(2006), 199-205.

[47] M. Charikar, J. Kleinberg, R. Kumar, S. Rajagopalan, A. Sahai, and A. Tomkins, Minimizing wirelength in zero and bounded skew clock trees, *SIAM Journal on Discrete Mathematics*, 17(4)(2004), 582-595.

[48] C. Chiang, M. Sarrafzadeh, and C. K. Wong, A powerful global router: based on Steiner min-max tree, *IEEE Transactions on Computer-Aided Design*, 19(1990), 1318-1325.

[49] M. Charikar, J. Naor, and B. Schieber, Resource optimization in QoS multicast routing of real-time multimedia, *IEEE/ACM Transactions on Networking*, 12(2)(2004), 340-348.

[50] D. Chen, D.-Z. Du, X.-D. Hu, G.-H. Lin, L. Wang, and G. Xue, Approximations for Steiner trees with minimum number of Steiner points, *Journal of Global Optimization*, 18(1)(2000), 17-33.

[51] X.-J. Chen, X.-D. Hu, X.-H. Jia, Inapproximability and approximability of minimal tree routing and coloring, *Journal of Discrete Algorithms*, to appear in 2007.

[52] X.-J. Chen, X.-D. Hu, and T.-P. Shuai, Inapproximability and approximability of maximal tree routing and coloring, *Jouranl of Combinatorial Optimization*, 11(2)(2006), 219-229.

[53] X.-J. Chen, X.-D. Hu, and J.-M. Zhu, Minimum data aggregation time problem in wireless sensor networks, *Lecture Notes in Computer Sciences*, 2005, vol. 3794, pp. 133-142.

[54] X.-Z. Cheng, B. DasGupta, and B. Lu, A polynomial time approximation scheme for the symmetric rectilinear Steiner arborescence problem, *Journal of Global Optimization*, 21(4)(2001), 385-396.

[55] X.-Z. Cheng and D.-Z. Du, *Steiner trees in Industry*, Kluwer Academic Publishers, Dordrecht, 2001.

[56] X.-Z. Cheng, D.-Z. Du, J.-M. Kim, and H. Q. Ngo, Guillotine cut in approximation algorithms, in *Cooperative Control and Optimization*, R. Murphey and P. M. Pardalos (eds.), Springer US, 2002, pp. 21-34.

[57] X. Cheng, X. Hang, D. Li, W. Wu, and D.-Z. Du, A polynomial time approximation scheme for minimum connected dominating set in ad hoc wireless networks, *Networks*, 42(4)(2003), 202-208.

[58] J. Cheriyan and R. Thurimella, Approximating minimum-size k-connected spanning subgraphs via matching, *SIAM Journal on Computing*, 30(2)(2000), 528-560.

[59] S. Cheung and A. Kumar, Efficient quorumcast routing algorithms, *Proceedings of the 13th Annual Joint Conference of the IEEE Computer and Communications Societies* (INFOCOM), 1994, pp. 840-855.

[60] E. A. Choukhmane, Une heuristique pour le probleme de l'arbre de Steiner, *RAIRO Recherche Operationnelle*, 12(1978), 207-212.

[61] N. Christonfides, Worst-case analysis of a new heuristic for the traveling

salesman problem, *Technical Report*, GSIA, Carnegie-Mellon University, 1976.

[62] F. R. K. Chung and E. N. Gilbert, Steiner trees for the regular simplex, *Bulletin of Institute of Mathematics Academia Sinica*, 4(1976), 313-325.

[63] F. R. K. Chung and R. L. Graham, A new bound for Euclidean Steiner minimum trees, *Annals of the New York Academy of Sciences*, 440(1985), 328-346.

[64] F. R. K. Chung and F. K. Hwang, A lower bound for the Steiner tree problem, *SIAM Journal of Applied Mathematics*, 34(1)(1978), 27-36.

[65] V. Chvátal, A greedy heuristic for the set-covering problem, *Mathematics of Operations Research*, 4(3)(1979), 233-235.

[66] D. Cieslik, The Steiner ratio of Banach-Minkowski planes, In *Contemporary Methods in Graph Theory*, R. Bodendiek (ed.), BI-Wissenschatteverlag, Mannheim, 1991, pp. 231-248.

[67] D. Cieslik, The Steiner Ratio of Banach-Minkowski spaces - A Survey, In *Handbook of Combinatorial Optimization*, D.-Z. Du and P. M. Pardalos (eds.), 2005, pp. 55-81.

[68] D. Cieslik, *Steiner Minimal Trees*, Kluwer Academic Publishers, The Netherlands, 1998.

[69] D. Cieslik, *The Steiner Ratio*, Kluwer Academic Publishers, The Netherlands, 2001.

[70] R. Crourant and H. Robbins, *What is Mathematics?*, Oxford University Press, New York, 1941.

[71] J. R. Current, C. S. Revelle, and J. L. Cohon, The hierarchical network design problem, *European Journal of Operations Research*, 27(1)(1986), 57-66.

[72] V. F. Dem'yanov and V. N. Malozemov, *Introduction to Minimax*, Dover Publications, Inc., New York, 1974.

[73] D. E. Drake and S. Hougardy, On approximation algorithms for the terminal Steiner tree problem, *Information Processing Letters*, 89(1)(2004), 15-18.

[74] D. R. Dreyer and M. L. Overton, Two heuristics for the Steiner tree problem, *Journall of Global Optimization*, 13(1)(1998), 95-106.

[75] D.-Z. Du, Minimax and its applications, *Handbook of Global Optimization*, R. Horst and P. M. Pardalos (eds.), Kluwer Academic Publishers, 1995, pp. 339-367.

[76] D.-Z. Du, On component size bounded Steiner trees, *Discrete Applied Mathematics*, 60(1-3)(1995), 131-140.

[77] D.-Z. Du, B. Gao, R. L. Graham, Z.-C. Liu, and P.-J. Wan, Minimum Steiner trees in normed planes, *Discrete and Computational Geometry*, 9(1)(1993), 351-370.

[78] D.-Z. Du, D. F. Hsu, and K.-J. Xu, Bounds on guillotine ratio, *Congressus Nemerantium*, 58(1987), 313-318.

[79] X.-F. Du, X.-D. Hu, and X.-H. Jia, On shortest k-edge-connected Steiner networks in metric spaces, *Journal of Combinatorial Optimization*, 4(1)(2000), 99-107.

[80] D.-Z. Du and F. K. Hwang, A new bound for the Steiner ratio, *Transactions of American Mathematical Society*, 278(1)(1983), 137-148.

[81] D.-Z. Du and F. K. Hwang, A proof of Gilbert-Pollak's conjecture on the Steiner ratio, *Algorithmica*, 7(1)(1992), 121-135. (Preliminary version appeared in the proceedings of *Proceedings of the 31st IEEE Foundations of Computer Science* (FOCS), 1990, pp. 76-85.)

[82] D.-Z. Du and F. K. Hwang, Reducing the Steiner problem in a normed space with a d-dimensional polytope as its unit sphere, *SIAM Journal on Computing*, 21(6)(1992), 1001-1007.

[83] D.-Z. Du and F. K. Hwang, The State of art on Steiner ratio problems, *Lectures Notes Series on Computing* (2nd Edition), D.-Z Du and F. K. Hwang (eds.), World Scientific Publishing Cl. Pte. Ltd., 1995, vol. 4, pp. 195-222.

[84] D.-Z. Du, F. K. Hwang, M. T. Shing, and T. Witbold, Optimal routing trees, *IEEE Transactions on Circuits*, 35(10)(1988), 1335-1337.

[85] D.-Z. Du, F. K. Hwang, and E.-N. Yao, The Steiner ratio conjecture is true for five points, *Journal of Combinatorial Theorey* (Ser. A), 32(3)(1982), 396-400.

[86] D.-Z. Du, L.-Q. Pan, and M.-T. Shing, Minimum edge length guillotine rectangular partition, *Technical Report 0241886*, Mathematical Sciences Research Institute, University of California, Berkeley, 1986.

[87] D.-Z. Du, J. M. Smith and J. H. Rubinstein, *Advances in Steiner Trees*, Kluwer Academic Publishers, Dordrecht, 2000.

[88] D.-Z. Du, L.-S. Wang, and B.-G. Xu, The Euclidean bottleneck Steiner tree and Steiner tree with minimum number of Steiner points, *Lecture Notes in Computer Science*, 2001, vol. 2108, pp. 509-518.

[89] D.-Z. Du and Y.-J. Zhang, On heuristics for minimum length rectilinear partitions, *Algorithmica*, 5(1)(1990), 111-128.

[90] D.-Z. Du and Y.-J. Zhang, On better heuristic for Steiner minimum trees, *Mathematical Programming*, 57(2)(1992), 193-202.

[91] D.-Z. Du, Y.-J. Zhang, and Q. Feng, On better heuristic for Euclidean Steiner minimum trees, *Proceedings of the 32nd Annual IEEE Symposium on Foundations of Computer Science* (FOCS), 1991, pp. 431-439.

[92] C. Duin and A. Volgenant, The multi-weighted Steiner tree problem, *Annals of Operations Research*, 33(1-4)(1991), 451-469.

[93] J. Edmonds, Maximum matching and a polyhedron with 0,1-vertices, *Journal of Research of the National Bureau of Standards*, B 69(1965), 125-130.

[94] J. Edmonds, Edge-disjoint branchings, In *Combinatorial Algorithms*, R. Rustin (ed.), Academic Press: New York, USA, 1973, pp. 91-96.

[95] J. Edmonds, Matroid intersection. *Annals of Discrete Mathematics*, 4(1979), 185-204.

[96] T. Erlebach and K. Jansen, The complexity of path coloring and call scheduling, *Theorectical Computer Science*, 255(1-2)(2001) 33-50.

[97] C. El-Arbi, Une heuristique pour le probleme de l'Arbre de Steiner, *RAIRO Recherche Operationnelle*, 12(1978), 207-212.

[98] U. Feige, A threshold of $ln(n)$ for approximating set cover, *Journal of the*

ACM, 45(4)(1998), 634-652.

[99] U. Feige and J. Kilian, Zero knowledge and the chromatic number, *Journal of Computer System Sciences*, 57(2)(1998) 187-199.

[100] J. Feigenbaum, C. H. Papadimitriou, and S. Shenker, Sharing the cost of multicast transmissions, *Journal of Computer and System Sciences*, 63(1)(2001), 21-41.

[101] G. N. Fredrickson and J. Ja'Ja, On the relationship between the biconnectivity augmentation and traveling salesman problem, *Theoretical Computer Science*, 19(2)(1982), 189-201.

[102] J. Friedel and P. Widmayer, A simple proof of the Steiner ratio conjecture for five points, *SIAM Journal of Applied Mathematics*, 49(3)(1989), 960-967.

[103] H. N. Gabow, Better performance bounds for finding the smallest k-edge connected spanning subgraph of a multigraph, *Proceedings of the 14th Annual ACM-SIAM Symposium on Discrete Algorithms*, 2003, pp. 460-469.

[104] H. N. Gabow, M. X. Goemans, E. Tardos, and D. P. Williamson, Approximatin the smallest k-edge connected spanning subgraph by LP-rounding, *Proceedings of the 16th Annual ACM-SIAM Symposium on Discrete Algorithms*, 2005, pp. 562-571.

[105] R. Gandhi, S. Parthasarathy, and A. Mishra, Minimizing broadcasting latency and redundancy in ad hoc networks, *Proceedings of the 4th ACM international Symposium on Mobile ad hoc Networking and Computing* (MobiHoc), 2003, pp. 222-231.

[106] J. L. Ganley and J. S. Salowe, Optimal and approximate bottleneck Steiner trees, *Operations Research Letters*, 19(5)(1996), 217-224.

[107] J. L. Ganley and J. S. Salowe, The power-p Steiner tree problem, *Nordic Journal of Computing*, 5(2)(1998), 115-127.

[108] B. Gao, D.-Z. Du, and R. L. Graham, A tight lower bound for the Steiner ratio in Minkowski planes. *Discrete Mathematics*, 142(1)(1995), 49-63.

[109] M. R. Garey, R. L. Graham, and D. S. Johnson, The complexity of computing Steiner minimal trees, *SIAM Journal on Applied Mathematics*, 32(4)(1977), 835-859.

[110] M. R. Garey and D. S. Johnson, *Computers and Intractability: A Guide to the Theory of NP-completeness*, W. H. Freeman and Company, San Francisco, 1979.

[111] M. R. Garey and D. S. Johnson, The rectilinear Steiner tree problem is NP-complete, *SIAM Journal on Applied Mathematics*, 32(4)(1977), 826-834.

[112] N. Garg, Saving an epsilon: a 2-approximation for the k-MST problem in graphs, *Proceedings of the 37th Annual ACM Symposium on Theory of Computing* (STOC), 2005, pp. 396-402.

[113] N. Garg, G. Konjevod, and R. Ravi, A polylogarithmic approximation algorithm for the group Steiner tree problem, *Journal of Algorithms*, 37(1)(2000), 66-84.

[114] E. N. Gilbert and H. O. Pollak, Steiner minimal trees, *SIAM Journal on Applied Mathematics*, 16(1)(1968), 1-29.

[115] M. X. Goemans and D. P. Williamson, The primal-dual method for ap-

proximation algorithms and its application to network design problems. In D. S. Hochbaum, editor, *Approximation Algorithms for NP-Hard Problems*, PWS Publishing Company, Boston, 1997, pp. 144-191.

[116] M. C. Golumbic and R. E. Jamison, The edge intersection graphs of paths in a tree, *Journal of Combinatorial Theory* (Series B), 38(1)(1985), 8-22.

[117] T. Gonzalez and S. Q. Zheng, Bounds for partitioning rectilinear polygons, *Proceedings of the 1st Annual Symposium on Computational Geometry*, 1985, pp. 281-287.

[118] R. L. Graham, An efficient algorithm for determining the convex hull of a finite set of points, *Information Processing Letters*, 1(4)(1972), 132-133.

[119] R. L. Graham and F. K. Hwang, Remarks on Steiner minimal trees, *Bulletin of Institute of Mathematics Academia Sinica*, 4(1976), 177-182.

[120] C. Gröepl and S. Hougardy, Approximation algorithms for the Stiner tree problem in graphs, in *Steiner Trees in Industries*, D.-Z. Du and X. Cheng (eds.), Kluwer Academic Publishers, 2001, pp. 235-280.

[121] C. Gröepl, S. Hougardy, T. Nierhoff, and H. J. Pröemel, Steiner trees in uniformly quasi-bipartite graphs, *Information Processing Letters*, 83(4)(2002), 195-200.

[122] J. Gu, X.-D. Hu, X. Jia, and M.-H. Zhang, Routing algorithm for multicast under multi-tree model in optical networks, *Theoretical Computer Science*, 314(1-2)(2004), 293-301.

[123] J. Gu, X.-D. Hu, and M.-H. Zhang, Algorithms for multicast connection under multi-path routing model, *Information Processing Letters*, 84 (1)(2002), 31-39.

[124] U. I. Gupta, D. T. Lee, and Y.-T. Leung, Efficient algorithms for interval graphs and circular-arc graphs, *Networks*, 12(4)(1982) 459-467.

[125] R. L. Hadas, Efficient collective communication in WDM networks with a power budget, in proceedings of *IEEE International Conference on Computer Communication Network* (ICCCN), 2000, pp. 612-616.

[126] S. L. Hakimi, Steiner's problem in graphs and its implications, *Networks*, 1(2)(1971), 113-133.

[127] P. Hall, On representatives of subsets, *Journal of the London Mathematical Society*, 10(1935), 26-30.

[128] M. M. Halldórsson, A still better performance guarantee for approximate graph coloring, *Information Processing Letters*, 45(1)(1993), 19-23.

[129] M. M. Halldórsson, Approximations of weighted independent set and hereditary subset problems, *Journal of Graph Algorithms and Applications*, 4(1)(2000), 1-16.

[130] M. Hanan, On Steiner's problem with rectilinear distance, *SIAM Journal on Applied Mathematics*, 14(2)(1966), 255-265.

[131] H. A. Harutyunyan and A. L. Liestman, Improved upper and lower bounds for k-broadcasting, *Networks*, 37(2)(2001), 94-101.

[132] J. Hastad, Clique is hard to approximate within $n^{1-\epsilon}$, *Acta Mathematica*, 182(1)(1999), 105-142.

[133] W. Heinzelman, A. Chandrakasan, and H. Balakrishnan, An application-specific protocol architecture for wireless microsensor networks, *IEEE*

Transactions on Wireless Communications, 1(4)(2002), 660-670.

[134] C. H. Helvig, G. Robins, and A. Zelikovsky, Improved approximation scheme for the group Steiner problem, *Networks*, 37(1)(2001), 8-20.

[135] D. S. Hochbaum and W. Maass, Approximation schemes for covering and packing problems in image processing and VLSI, *Journal of ACM*, 32(1)(1985), 130-136.

[136] S. Hougardy and H. J. Prömmel, A 1.598-approximation algorithm for the Steiner problem in graphs, *Proceedings of ACM/SIAM Symposium on Discrete Algorithms* (SODA), 1999, pp. 448-453.

[137] D. F. Hsu and X.-D. Hu, On shortest two-edge-connected Steiner networks with Euclidean distance, *Networks*, 32(2)(1998), 133-140.

[138] D. F. Hsu, X.-D. Hu, and Y. Kajitani, On shortest k-edge-connected Steiner networks with rectilinear distance, *Nonconvex Optimization and its Application*, Kluwer Academic Publisher, Dordrecht, 1995, vol. 4, pp. 119-127.

[139] D. F. Hsu, X.-D. Hu, and G.-H. Lin, On minimum-weight k-edge connected Steiner networks on metric spaces, *Graphs and Combinatorics*, 16(3)(2000), 275-284.

[140] X.-D. Hu and F. K. Hwang, Reliability of chordal rings, *Networks*, 22(5)(1992), 487-501.

[141] H. B. Hunt III, M. V. Marathe, V. Radhakrishnan, S. S. Ravi, D. J. Rosenkrantz, and R. E. Stearns, NC-approximation schemes for NP- and PSPACE-hard problems for geometric graphs, *Journal of Algorithms*, 26(2)(1998), 238-274.

[142] F. K. Hwang, On Steiner minimal trees with rectilinear distance, *SIAM Journal on Applied Mathematics*, 30(1)(1976), 104-114.

[143] F. K. Hwang and D. S. Richards, Steiner tree problems, *Networks*, 22(1)(1992), 55-89.

[144] F. K. Hwang, D. S. Richards, and P. Winter, *The Steiner minimum tree problems*, North-Holland, Amsterdam, 1992.

[145] F. K. Hwang and J. F. Weng, Hexagonal coordinate systems and Steiner minimal trees, *Discrete Mathematics*, 62(1)(1986), 49-57.

[146] F. K. Hwang and Y. C. Yao, Comments on Bern's probabilistic results on rectilinear Steiner trees, *Algorithmica*, 5(1)(1990), 591-598.

[147] E. Ihler, The complexity of approximating the class Steiner tree problem, *Lecture Notes in Computer Science*, 1992, vol. 570, pp. 85-96.

[148] T. Imielinski and S. Goel, DataSpace: querying and monitoring deeply networked collections in physical space, *IEEE Personal Communication*, 7(5)(2000), 4-9.

[149] A. O. Ivanov and A. A. Tuzhilin, *Minimal Networks: The Steiner Problem and Its Generalizations*, Boca Raton, FL: CRC Press, 1994.

[150] A. Itai, C. H. Papadimitriou, and J. Szwarcfiter, Hamiltonian paths in grid graphs, *SIAM Journal on Computing*, 11(4)(1982), 676-686.

[151] B. Jackson, Shortest circuit covers and postman tours in graphs with a nowhere zero 4-flow, *SIAM Journal on Computing*, 19(4)(1990), 659-665.

[152] K. Jain, Factor 2 approximation algorithm for the generalized Steiner network problem, *Combinatorica*, 1 (2001), 39-60. (Preliminary version ap-

peared in *Proceedings of the 39th Annual Symposium on Foundations of Computer Science* (FOCS), 1998, pp. 448-457.)

[153] X.-H. Jia, D.-Z. Du, and X.-D. Hu, Integrated algorithms for delay bounded multicast routing and wavelength assignment in all optical networks, *Computer Communications*, 24(14)(2001), 1390-1399.

[154] X.-H. Jia, D.-Z. Du, X.-D. Hu, M. K. Lee, and J. Gu, Optimization of wavelength assignment for QoS multicast WDM networks, *IEEE Transaction on Communications*, 49(2)(2001), 341-350.

[155] X.-H. Jia, X.-D. Hu, and D.-Z. Du, *Multiwavelength Optical Networks*, Kluwer Academic Publishers, The Netherlands, 2002.

[156] D. S. Johnson, M. Minkoff, and S. Phillips, The prize collecting Steiner tree problem: theory and practice. *Proceedings of the 11th Annual ACM-SIAM Symposium on Discrete Algorithms* (SODA), 2000, pp. 760-769.

[157] T. Jordán, On minimally k-edge-connected graphs and shortest k-edge-connected Steiner networks, *Discrete Applied Mathematics*, 131(2)(2003), 421-432.

[158] R. Jothiy and B. Raghavachariy, Approximation algorithms for the capacitated minimum spanning tree problem and its variants in network design, *Lecture Notes in Computer Science*, 2004, vol. 3142, pp. 805-818.

[159] R. Jothi, B. Raghavachari, and S. Varadarajan, A 5/4-approximation algorithm for minimum 2-edge-connectivity, *Proceedings of the 14th annual ACM-SIAM symposium on Discrete Algorithms*, 2003, pp. 725-734.

[160] A. Kahng and G. Robins, A new family of Steiner tree heuristics with good performance: the iterated 1-Steiner approach, *Proceedings of IEEE International Conference on Computer-Aided Design*, 1990, pp. 428-431.

[161] K. Kalpakis, K. Dasgupta, and P. Namjoshi, Efficient algorithms for maximum lifetime data gathering and aggregation in wireless sensor networks, *Computer Networks*, 42(6)(2003), 697-716.

[162] J. N. Al-Karaki and A. E. Kamal, Routing techniques in wireless sensor networks: a survey, *IEEE Wireless Communications*, 11(6)(2004), 6-28.

[163] R. M. Karp, Reducibility among combinatorial problems. In *Complexity of Computer Computation*, R. E. Miller and J. W. Tatcher (eds.), Plenum Press, New York, 1972, pp. 85-103.

[164] M. Karpinski, I. I. Măndoiu, A. Olshevsky, and A. Zelikovsky. Improved approximation algorithms for the quality of service Steiner tree problem. *Algorithmica*, 42(2)(2005), 109-120.

[165] M. Karpinski and A. Zelikovsky, New approximation algorithms for the Steiner tree problems, *Journal of Combinatorial Optimization*, 1(1)(1997), 47-65.

[166] A. Kesselman and D. Kowalski, Fast distributed algorithm for convergecast in ad hoc geometric radio networks, *Journal of Parallel and Distributed Computing*, 66(4)(2006), 578-585.

[167] S. Khuller and B. Raghavachari, Improved approximation algorithms for uniform connectivity problems, *Journal of Algorithms*, 21(2)(1996), 434-450.

[168] S. Khuller, B. Raghavachari, and N. Young, Balancing minum spanning

trees and shortest path trees, *Algorithmica*, 14(4)(1995), 305-321.

[169] S. Khuller and U. Vishkin, Biconnectivity approximations and graph carvings, *Journal of the ACM*, 42(2)(1994), 214-235.

[170] J. Kim, M. Cardei, I. Cardei, and X.-H. Jia, A polynomial time approximation scheme for the grade of service Steiner minimum tree problem, *Journal of Global Optimization*, 24(4)(2002), 437-448.

[171] P. Klein and R. Ravi, A nearly best-possible approximation algorithm for node-weighted Steiner trees, *Journal of Algorithms*, 19(1)(1995), 104-115.

[172] J. M. Kleinberg, Approximation algorithms for disjoint paths problems, *Ph.D. Thesis*, MIT, Cambridge, MA, 1996.

[173] J. Kleinberg and E. Tardos, Disjoint paths in densely embedded graphs, *Proceedings of the 36th Annual IEEE Symposium on Foundations of Computer Science* (FOCS), 1995, pp. 52-61.

[174] S. G. Kollipoulos and C. Stein, Approximating disjoint-path problems using greedy algorithms and packing integer programs, *Integer Programming and Combinatorial Optimization*, Houston, TX, 1998.

[175] P. Korthonen, An algorithm for transforming a spanning tree into a Steiner tree, *Survey of Mathematical Proglramming*, North-Holland, 1974, vol. 2, pp. 349-357.

[176] G. Kortsarz and Z. Nutov, Approximating node connectivity problems via set covers, *Algorithmica*, 37(2003), 75-92.

[177] G. Kortsarz and Z. Nutov, Approximating k-node connected subgraphs wia critical graphs, *SIAM Journal on Computing*, 35(1)(2005), 247-257.

[178] L. Kou and K. Makki, An even faster approximation algorithm for the Steiner tree problem in graph, *Congressus Numerantium*, 59(1987), 147-154.

[179] L. Kou, G. Markowsky, and L. Berman, A fast algorithm for Steiner trees, *Acta Informatica*, 15(2)(1981), 141-45.

[180] M. E. Kramer and J. van Leeuwen, The complexity of wire routing and finding the minimum area layouts for arbitrary VLSI circuits, in *Advances in Computing Research; VLSI Theory*, F. P. Preparata (ed.), JAI Press Inc. Greenwich, CT-London, 1984, vol. 2, pp. 129-146.

[181] S. O. Krumke, M. V. Marathe, and S. S. Ravi, Models and approximation for channel assignment in radio networks, *Wireless Networks*, 7(6)(2001), 575-584.

[182] J. B. Kruskal, On the shortest spanning subtree of a graph and the traveling salesman problem, *Proceedings of American Mathematical Society*, 7(1)(1956), 48-50.

[183] H. W. Kuhn, Steiner's problem revisted, In *Studies in Optimization*, G. B. Dantzig and B. C. Eaves (eds.), The Mathematical Association of America, 1975, pp. 53-70.

[184] S. Lee and J. A. Ventura, An algorithm for constructing minimal c-broadcast networks, *Networks*, 38(1)(2001), 6-21.

[185] D.-Y. Li, X.-H. Jia, and H. Liu, Energy efficient broadcast routing in static ad hoc wireless networks, *IEEE Transactions on Mobile Computing*, 3(2)(2004), 144-151.

[186] D. Lichtenstein, Planar formulae and their uses, *SIAM Journal on Computing*, 11(2)(1982), 329-343.

[187] G.-H. Lin, An improved approximation algorithm for multicast k-tree routing, *Journal of Combinatorial Optimization*, 9(4)(2005), 349-356.

[188] G.-H. Lin and G.-L. Xue, Steiner tree problem with minimum number of Steiner points and bounded edge-length, *Information Processing Letters*, 69(2)(1999), 53-57.

[189] G.-H. Lin and G.-L. Xue, On the terminal Steiner tree problem, *Information Processing Letters*, 84(2)(2002), 103-107.

[190] S. Lindsey, C, Raghavendra, and K. M. Sivalingam, Data gathering algorithms in sensor networks using energy metrics, *IEEE Transactions on Parallel and Distributed Systems*, 13(9)(2002), 924-935.

[191] A. Lingas, R. Y. Pinter, R. L. Rivest, and A. Shamir, Minimum edge length partitioning of rectilinear polygons, *Proceedings of the 20th Allerton Conference on Communication, Control, and Computing*, 1982, pp. 53-63.

[192] A. Lingas, Heuristics for minimum edge length rectangular partitions of rectilinear figures, *Lecture Notes in Computer Science*, 1983, vol. 145, pp. 199-210.

[193] Z.-C. Liu and D.-Z. Du, On Steiner minimal trees with L_p distance, *Algorithmica*, 7(1)(1992), 179-191.

[194] E. L. Lloyd and G.-L. Xue, Relay node placement in wireless sensor networks, *IEEE Transactions on Computers*, 56(1)(2007), 134-138.

[195] B. Lu, J. Gu, X.-D. Hu, and E. Shragowitz, Wire segmenting for buffer insertion based on RSTP-MSP, *Theoretical Computer Science*, 262(1-2)(2001), 257-267.

[196] B. Lu and L. Ruan, Polynomial time approximation scheme for the rectilinear Steiner arborescence problem, *Journal of Combinatoriall Optimization*, 4(3)(2000), 357-363.

[197] E. L. Luebke and J. S. Provan, On the structure and complexity of the 2-connected Steiner network problem in the plane. *Operations Research Letters*, 26(3)(2000), 111-116.

[198] D. G. Luenberger, *Linear and Nonlinear Programming*, Second edition, MA: Addison Wesley, 1984.

[199] C. Lund and M. Yannakakis, On the hardness of approximating minimization problems, *Journal of ACM*, 41(5)(1994), 960-981.

[200] W. Mader, A reduction method for edge-connectivity in graphs, *Annual of Discrete Mathematics*, 3(1978), 145-164.

[201] I. I. Măndoiu and A. Z. Zelikovsky, A note on the MST heuristic for bounded edge-length Steiner trees with minimum number of Steiner points, *Information Processing Letters*, 75(4)(2000), 165-167.

[202] M. V. Marathe, R. Ravi, R. Sundaram, S. S. Ravi, D. J. Rosenkrantz, and H. B. Hunt III, Bicriteria network design problems, *Journal of Algorithms*, 28(1)(1998), 142-171.

[203] F. B. Martinez, J. C. Pina, and J. Soares, Algorithms for terminal Steiner trees, *Lecture Notes in Computer Science*, 2005, vol. 3595, pp. 369-379.

[204] F. V. Martinez and B. J. Soares, Steiner trees with a terminal order, *Pro-

ceedings of the 15th International Conference on Computing (CIC), 2006, pp. 254-259.

[205] N. F. Maxemchuk, Video distribution on multicast networks, *IEEE Journal on Selected Areas in Communications.* 15(3)(1997), 357-372.

[206] Z. A. Melzak, On the problem of Steiner, *Canadian Mathematical Bulletin*, 4(1961), 143-148.

[207] A. Meyerson, Online algorithms for network design, *Proceedings of the Sixteenth Annual ACM Symposium on Parallelism in Algorithms and Architectures*, 2004, pp. 275-280.

[208] G. L. Miller, S. H. Teng, W. Thurston, and S. A. Vavasis, Separators for sphere packings and nearest neighbor graphs, *Journal of ACM*, 44(1)(1997), 1-29.

[209] M. Min, H.-W. Du, X.-H. Jia, C. X. Huang, S. C.-H. Huang, and W.-L. Wu, Improving construction for connected dominating set with Steiner tree in wireless sensor networks, *Journal of Global Optimization*, 35(1)(2006), 111-119.

[210] P. Mirchandani, The multi-tier tree problem, *INFORMS Journal on Computing*, 8(1996), 202-218.

[211] J. S. B. Mitchell, Guillotine subdivisions approximate polygonal subdivisions: A simple new method for the geometric k-MST problem, *Proceedings of the 7th ACM-SIAM Symposium on Discrete Algorithms* (SODA), 1996, pp. 402-408.

[212] J. S. B. Mitchell, Guillotine subdivisions approximate polygonal subdivisions: Part III - Faster polynomial-time approximation scheme for geometric network optimization, *Proceedings of the 9th Canadian Conference on Computational Geometry*, 1997, pp. 229-232.

[213] J. S. B. Mitchell, A. Blum, P. Chalasani, and S. Vempala, A constant-factor approximation algorithm for the geometric k-MST problem in the plane, *SIAM Journal on Computing*, 28(3)(1999), 771-781.

[214] J. S. B. Mitchell, Guillotine subdivisions approximate polygonal subdivisions: Part II - A simple polynomial-time approximation scheme for geometric k-MST, TSP, and related problem, *SIAM Journal on Computing*, 29(2)(1999), 515C544.

[215] C. L. Monma, B. S. Munson, and W. R. Pulleyblank, Minimum-weight two-connected spanning networks, *Mathematical Programing*, 46(1-3)(1990), 153-171.

[216] A. Moss and Y. Rabani, Approximation algorithms for constrained node weighted Steiner tree problems, *Proceedings of the Thirty-Third Annual ACM Symposium on Theory of Computing* (STOC), 2001, pp. 373-382.

[217] C. St. J. A. Nash-Williams, Edge disjoint spanning trees of finite graphs, *The Journal of the London Mathematical Society*, 36(1961), 445-450.

[218] T. Nishizeki and K. Kashiwagi, On the 1.1 edge-coloring of multigraphs, *SIAM Journal on Discrete Mathematics*, 3(3)(1990), 391-410.

[219] C. Nomikos, Path coloring in graphs, *Ph.D Thesis*, Department of Electrical and Computer Engineering, National Technical University of Athens, 1997.

[220] C. Nomikos, Routing and path coloring in rings: NP-completeness, *Tech-*

nical Report 15-2000, University of Ioannina, Greece, 2000.

[221] C. Nomikos, A. Pagourtzis, and S. Zachos, Satisfying a maximum number of pre-routed requests in all-optical rings, *Computer Networks*, 42(1)(2003), 55-63.

[222] C. Nomikos and S. Zachos, Coloring a maximum number of paths in graphs, *Workshop on Algorithmic Aspects of Communication*, 1997.

[223] I. Papadimitriou and L. Georgiadis, Minimum-energy broadcasting in multi-hop wireless networks using a single broadcast tree, *Mobile Networks and Applications*, 11(3)(2006), 361-375.

[224] C. H. Papadimitriou and K. Steiglitz, *Combinatorial Optimization: Algorithms and Complexity*, Prentice-Hall, Englewood Cliffs, NJ, 1982.

[225] H. O. Pollak, Some remarks on the Steiner problem. *Journal of Combinatorial Theory* (Ser. A), 24(3)(1978), 278-295.

[226] R. C. Prim, Shortest connection networks and some generalizations, *Bell System Technical Journal*, 36(1957), 1389-1401.

[227] H. J. Prömel and A. Steger, A new approximation algorithm for the Steiner tree problem with performance ratio 5/3, *Journal of Algorithms*, 36(1)(2000), 89-101.

[228] H. J. Prömel and A. Steger, *The Steiner Tree Problem: A Tour Through Graphs, Algorithms and Complexity*, Vieweg Verlag, Wiesbaden, 2002.

[229] J. S. Provan, Two new criteria for finding Steiner hulls in Steiner tree problems, *Algorithmica*, 7(2-3)(1992), 289-302.

[230] Y. Rabani, Path coloring on the mesh, *Proceedings of the 37th Annual Symposium on Foundations of Computer Science* (FOCS), 1996, pp. 400-409.

[231] P. Raghavan and E. Upfal, Efficient routing in all-optical networks, *Proceedings of the 26th Annual ACM Symposium on Theory of Computing* (STOC), 1994, pp. 134-143.

[232] S. Rajagopalan and V. V. Vazirani, On the bidirected cut relaxation for the metric Steiner problem, *Proceedings of the 10th Annual ACM-SIAM Symposium on Discrete Algorithms* (SODA), 1999, pp. 742-751.

[233] B. Ramamurthy, J. Iness, and B. Mukherjee, Minimizing the number of optical amplifiers needed to suppoort a multi-wavelength optical LAN/MAN, *Proceedings of the 16th Annual Joint Conference of the IEEE Computer and Communications Societies* (INFOCOM), 1997, pp. 261-268.

[234] S. B. Rao and W. D. Smith, Approximating geometrical graphs via "spanners" and "banyans", *Proceedings of the 30th Annual ACM Symposium on Theory of Computing* (STOC), (1998), pp. 540-550.

[235] R. Ravi, Rapid rumor ramification: approximating the minimum broadcast time, *Proceedings of the 35th Annual IEEE Symposium on Foundations of Computer Science* (FOCS), 1994, pp. 202-213.

[236] J. Remy and A. Steger, Approximation schemes for node-weighted geometric Steiner tree problems, *Lecture Notes in Computer Science*, 2005, vol. 3624, pp. 221-232.

[237] G. Robins and J. S. Salowe, Low-degree minimum spanning trees, *Discrete Computing Geometry*, 14(1995), 151-165.

[238] G. Robins and A. Zelikovsky, Improved Steiner tree approximation in graphs, *Proceedings of ACM/SIAM Symposium on Discrete Algorithms* (SODA), 2000, pp. 770-779.

[239] G. Robins and A. Zelikovsky, Tighter bounds for graph Steiner tree approximation, *SIAM Journal on Discrete Mathematics*, 19(1)(2005), 122-134.

[240] J. H. Rubinstein and D. A. Thomas, A variational approach to the Steiner network problem, *Annals of Operations Research*, 33(6)(1991), 481-499.

[241] J. H. Rubinstein and D. A. Thomas, The Steiner ratio conjecture for six points, *Journal of Combinatorial Theory* (Ser. A), 58(1)(1991), 54-77.

[242] L. H. Sahasrabuddhe and B. Mukherjee, Multicast routing algorithms and protocols: a tutorial, *IEEE Network*, 14(1)(2000), 90-102.

[243] L. H. Sahasrabuddhe and B. Mukherjee, Light-trees: optical multicasting for improved performance in wavelength-routed networks, *IEEE Communications Magazine*, 37(2)(1999), 67-73.

[244] M. Sarrafzadeh, W.-L. Lin, and C. K. Wong, Floating Steiner trees, *IEEE Transactions on Computers*, 47(2)(1998), 197-211.

[245] M. Sarrafzadeh and C. K. Wong, Bottleneck Steiner trees in the plane, *IEEE Transactions on Computers*, 41(3)(1992), 370-374.

[246] P. J. Slater, E. J. Cockayne, and S. T. Hedetniemi, Information dissemination in trees, *SIAM Journal on Computing*, 10(4)(1981), 692-701.

[247] W. D. Smith, How to find Steiner minimal trees in Euclidean d-space, *Algorithmica*, 7(1)(1992), 137-177.

[248] J. M. Smith, D. T. Lee, and J. S. Liebman, An $O(N \log N)$ heuristic for Steiner minimal tree problems in the Euclidean metric, *Networks*, 11(1)(1981), 23-39.

[249] J. M. Smith and J. S. Liebman, Steiner trees, Steiner circuits and interference problem in building design, *Engineering Optimization*, 4(1)(1979), 15-36.

[250] I. Stewart, Trees, telephones and tiles, *New Scientist*, 16(1991), 26-29.

[251] A. Suzuki and M. Iri, A heuristic method for euclidean Steiner problem as a geographical optimization problem, *Asia-Pacific Journal of Operations Research*, 3(1986), 106-122.

[252] A. S. Tanenbaum, *Computer Networks*, Prentice Hall PTR, Upper Saddle River, NJ, 1996.

[253] H. Takahashi and A. Matsuyama, An approximate solution for the Steiner problem in graphs, *Mathematics Japonica*, 24 (1980), 573-577.

[254] R. Tarjan, Decomposition by clique separators, *Discrete Mathematics*, 55(2)(1985), 221-232.

[255] M. Thimm, On the approximability of the Steiner tree problem, *Lecture Notes in Computer Science*, 2001, vol. 2136, pp. 678C689.

[256] W. T. Tutte, On the problem of decomposing a graph into n connected factors, *The Journal of the London Mathematical Society*, 36(1961), 221-230.

[257] V. V. Vazirani, *Approximation Algorithms*, Springer-Verlag Berlin Heidelberg, 2001.

[258] P.-J. Wan, G. Călinescu, and C.-W. Yi, Minimum-power multicast routing

in static ad hoc wireless networks, *IEEE/ACM Transactions on Networking*, 12(3)(2004), 507-514.

[259] P.-J. Wan, D.-Z. Du, and R. L. Graham, The Steiner ratio on the dual normed plane, *Discrete Mathematics*, 171(1-3)(1997), 261-275.

[260] P.-J. Wan and L. Liu, Maximal throughput in wavelength-routed optical networks, *DIMACS Series in Discrete Mathematics and Theoretical Computer Science*, 46(1998), 15-26.

[261] P.-J. Wan, G. Calinescu, X.-Y. Li, and O. Frieder, Minimum-energy broadcasting in static ad hoc wireless networks, *Wireless Networks*, 8(6)(2002), 607-617.

[262] L.-S. Wang and D.-Z. Du, Approximations for a bottleneck Steiner tree problem, *Algorithmica*, 32(4)(2002), 554-561.

[263] L.-S. Wang and T. Jiang, An approximation scheme for some Steiner tree problems in the plane, *Networks*, 28(4)(1996), 187-193.

[264] L. Wang and Z. Li, An approximation algorithm for a bottleneck k-Steiner tree problem in the Euclidean plane, *Information Processing Letters*, 81(3)(2002), 151-156.

[265] J.-P. Wang, X.-T. Qi, and B. Chen, Wavelength assignment for multicast in all-optical WDM networks with splitting constraints, *IEEE/ACM Transactions on Networking*, 14(1)(2006), 169-182.

[266] W.-Z. Wang, X.-Y. Li, and Z. Sun, Design differentiated service multicast with selfish agents, *IEEE Journal on Selected Areas in Communications*, 24(5)(2006), 1061-1073.

[267] J. F. Weng, Steiner problem in hexagonal metric, *Technical Report* (unpublished).

[268] J. E. Wieselthier, G. D. Nguyen, and A. Ephremides, On the construction of energy-efficient broadcast and multicast trees in wireless networks, *Proceedings of the 19th Annual Joint Conference of the IEEE Computer and Communications Societies* (INFOCOM), 2000, pp. 585-594.

[269] J. E. Wieselthier, G. D. Nguyen, and A. Ephremides, Energy-efficient broadcast and multicast trees in wireless networks, *Mobile Netowrks and Applications*, 7(6)(2002), 481-492.

[270] P. Winter and M. Zachariasen, Two-connected Steiner networks: structural properties, *Operations Research Letters*, 33(44)(2005), 395-402.

[271] Y. F. Wu, P. Widmayer, and C. K. Wong, A faster approximation algorithm for the Steiner problem in graphs, *Acta Informatica*, 23(1986), 223-229.

[272] G.-L. Xue, G.-H. Lin, and D.-Z. Du, Grade of serveice Steiner minimum trees in the Euclidean plane, *Algorithmca*, 31(4)(2001), 479-500.

[273] G.-L. Xue and W. Xiao, A polynomial time approximation scheme for minimum cost delay-constrained multicast tree under a Steiner topology, *Algorithmica*, 41(1)(2004), 53-72.

[274] G.-L. Xue and Y.-Y. Ye, An efficient algorithm for minimizing a sum of Euclidean norms with applications, *SIAM Journal on Optimization*, 7(4)(1997), 1017-1036.

[275] A. Z. Zelikovsky, An 11/8-approximation algorithm for the Steiner problem on networks with rectilinear distance, *Coll. Math. Soc. János Bolyai* ,

60(1992), 733-745.

[276] A. Z. Zelikovsky, The 11/6-approximation algorithm for the Steiner problem on networks, *Algorithmica*, 9(5)(1993), 463-470.

[277] A. Z. Zelikovsky, Better approximation bounds for the network and Euclidean Steiner tree problems, *Technical Report*, CS-96-06, University of Virginia, Charlottesville, VA, USA.

[278] A. Zelikovsky, A series of approximation algorithms for the acyclic directed Steiner tree problem, *Algorithmica*, 18(1)(1997), 99-110.

[279] A. Zelikovsky and I. Mandoiu, Practical approximation algorithms for zero- and bounded-skew trees, *SIAM Journal on Discrete Mathematics*, 15(1)(2002), 97-111.

[280] Z.-M. Zhu, X.-J. Chen, and X.-D. Hu, Minimum multicast time problem in wireless sensor networks, *Lecture Notes in Computer Science*, 2006, vol. 4138, pp. 500-511.

[281] Z.-M. Zhu, W.-P. Shang, and X.-D. Hu, New algorithm for minimum multicast time problem in wireless sensor networks, *Proceedings of IEEE Wireless Communications and Networking Conference* (WCNC), 2007.

Index